Japanese Cybercultures

Japanese Cybercultures is the first book to look at the specific dynamics of Japanese Internet use. Examined from a variety of interdisciplinary perspectives by both established and up-and-coming scholars, this genuinely cutting-edge study analyzes the development of the Internet in Japan, looking at the particularities of Japanese-language use on the Net and the different ways in which users log on. Unlike the English-speaking world where most people access the Internet via computers, in Japan the Internet is overwhelmingly accessed via a variety of portable devices, particularly mobile phones. Japan's ubiquitous "cute culture" has colonized cyberspace, and students are shown to have embraced this technology to the extent that life without mobile Internet access would for many be inconceivable.

Much Internet use in Japan is recreational, and this book considers the role of the Internet in different musical subcultures. But it is equally influential for social and political activism. Women's networks and a growing men's movement, for example, are using this technology in an attempt to highlight problems of harassment and bullying in the workplace, otherwise overlooked by mainstream media.

Moreover, other marginalized groups and subcultures – including gay men, those living with AIDS, members of many new religious movements and Japan's hereditary sub-caste, the Burakumin, have found the Internet a valuable tool. It allows increased networking among disenfranchised individuals who now have access to a powerful technology that enables them to represent themselves in their own voice and challenge public misconceptions. At the same time, mainstream organizations and government bodies also use cyberspace to further advance their own agendas, suggesting that in Japan, as elsewhere, the Internet is being used as a tool to promote both difference and conformity.

Nanette Gottlieb is Reader in Japanese and Head of the School of Languages and Comparative Cultural Studies at the University of Queensland. She co-edited *Language Planning and Language Policy: East Asian Perspectives*, and is the author of *Word-Processing Technology in Japan: Kanji and the Keyboard*; *Kanji Politics: Language Policy and Japanese Script*; and *Language and the Modern State: The Reform of Written Japanese*.

Mark McLelland is Postdoctoral Fellow in the Centre for Critical and Cultural Studies at the University of Queensland. He is the author of *Male Homosexuality in Modern Japan: Cultural Myths and Social Realities* and a contributor to many journals on the topic of Japanese sexual minorities and their use of the Internet.

Asia's Transformations
Edited by Mark Selden
Binghampton University and Cornell University, USA

The books in this series explore the political, social, economic and cultural consequences of Asia's transformations. The series emphasizes the tumultuous interplay of local, national, regional and global forces as Asia bids to become the hub of the world economy. While focusing on the contemporary, it also looks back to analyze the antecedents of Asia's contested rise.

This series comprises several strands:

Asia's Transformations aims to address the needs of students and teachers, and the titles will be published in hardback and paperback. Titles include:

Korean Society
Civil society, democracy and the state
Edited by Charles K. Armstrong

The Making of Modern Korea
Adrian Buzo

Asia's Global Cities is an interdisciplinary series, drawing on the latest thinking in urban studies, cultural geography, anthropology, sociology, history and Asian studies. Each book explores the interaction of the local with the global in the history, development and future of a significant city in Asia. The books are designed to appeal to the general reader seeking a textured picture of a city as well as students and academics.

Bangkok
Place, practice and representation
Marc Askew

Beijing in the Modern World
David Strand and Madeline Yue Dong

Singapore
Carl Trocki

Shanghai
Global city
Jeff Wasserstorm

Hong Kong
Global city
Stephen Chiu and Tai-Lok Lui

Asia.com focuses on the ways in which new technology is changing society in Asia. Titles include:

Japanese Cybercultures
Edited by Nanette Gottlieb and Mark McLelland

RoutledgeCurzon studies in Asia's Transformations is a forum for innovative new research intended for a high-level specialist readership, and the titles will be available in harback only. Titles include:

1 The American Occupation of Japan and Okinawa *
Literature and memory
Michael Molasky

Japanese Cybercultures

Edited by Nanette Gottlieb and Mark McLelland

With a preface by David Gauntlett

Routledge
Taylor & Francis Group

LONDON AND NEW YORK

First published 2003 by Routledge
11 New Fetter Lane, London EC4P 4EE
Simultaneously published in the USA and Canada
by Routledge
29 West 35th Street, New York, NY 10001

Routledge is an imprint of the Taylor & Francis Group

© 2003 Nanette Gottlieb and Mark McLelland for selection and editorial
matter; individual chapters the contributors

Typeset in Times by
HWA Text and Data Management, Tunbridge Wells
Printed and bound in Great Britain by
The Cromwell Press, Trowbridge, Wiltshire

British Library Cataloguing in Publication Data
A catalogue record for this book is available from the British Library

Library of Congress Cataloging in Publication Data
Japanese cybercultures / edited by Nanette Gottlieb and Mark McLelland
        p. cm. — (Asia's transformations)
    Includes bibliographical references and index.
        1. Internet – Social aspects – Japan.   2. Popular culture – Japan.
    I. Gottlieb, Nanette, 1948–   II. McLelland, Mark J.   III. Series.

    HN730.Z9 I564 2003
    303.48´33´0952–dc21   2002068280

ISBN 0–415–27918–6 (hbk)
ISBN 0–415–27919–4 (pbk)

Contents

Contributors

Costa Caspary received a scholarship for the Municipal University of Yokohama in 1995 but spent most of the year wrecking his ears in "live-houses." He returned to Japan in 1999 to investigate the Japanese Noise scene from a more academic point of view. E-mail: *costacaspary@yahoo.com*.

Joanne Cullinane is a doctoral candidate in Anthropology at the University of Chicago where she is working on a dissertation on the medical, social, and political responses to HIV/AIDS in Japan entitled *HIV/Eizu in Japan: Media Event, Medical Crisis, and Lived Experience.* Her research interests include gender, medicine, stigma, and social movements. E-mail: *rujobooboo@ yahoo.co.jp*.

Romit Dasgupta is an Associate Lecturer in the Department of Asian Studies at the University of Western Australia where he is completing his doctoral thesis on the transformation of notions of masculinity in contemporary Japan. E-mail: *romit@arts.uwa.edu.au*.

Isa Ducke is a Research Fellow at the German Institute for Japanese Studies, Tokyo (DIJ). She has also worked in a news agency in Zurich and as a games tester in London. Her research focuses on issues of moral authority in the history of relations between Japan and Korea. E-mail: *ducke@dijtokyo.org*.

David Gauntlett is Professor of Media and Identities at the Media School, University of Bournemouth. His books include *Web.Studies: Rewiring Media Studies in the Digital Age* (Arnold, 2000; second edition, 2003) and *Media, Gender and Identity: An Introduction* (Routledge, 2002). He produces the award-winning Web site www.theory.org.uk.

Nanette Gottlieb is Head of the School of Languages and Comparative Cultural Studies at the University of Queensland where she researches, among other things, the impact of computer technologies on the Japanese language. E-mail: *Nanette.Gottlieb@mailbox.uq.edu.au*

Larissa Hjorth lectures in Art and Design History and Theory and Critical Studies at Swinburne University of Technology and the Victorian College of the Arts, Australia. E-mail: *larissahjorth@hotmail.com.*

Todd Joseph Miles Holden is Professor of Mediated Sociology in the Graduate School of International Cultural Studies at Tohoku University in Sendai, Japan. His research interests embrace social theory, semiology, advertising, gender, political communication and comparative culture. For the curious, his various writings – including academic articles, book chapters, magazine columns, and flights of fiction – can be found on his Web site: *http://www.langc.tohoku.ac.jp/ ~holden/index.htm.* E-mail: *t_sensei@hotmail.com.*

Petra Kienle is a doctoral candidate in the Japanese Department of the University of Tübingen in Germany working on a research project entitled "Self-representation and self-understanding of religious communities on the Japanese Internet." Her dissertation is about Christian new religions and the Japanese Internet. In 2002 she began work at the Centre for Japanese Studies, Marburg University. E-mail: *kienle@mailer.uni-marburg.de.*

Vera Mackie is Foundation Professor of Japanese Studies at Curtin University of Technology in Western Australia and has recently held visiting positions at the Institute of Gender Studies at Ochanomizu University in Tokyo and Victoria University in Melbourne. Her research interests include the history of feminisms in Japan, and issues of citizenship in contemporary Japan. E-mail: *mackiev@ spectrum.curtin.edu.au.*

Mark McLelland is a postdoctoral fellow in the Centre for Critical and Cultural Studies at the University of Queensland where he researches the intersections between media, technology and sexuality in Japan. E-mail: *markmclelland@ graduate.hku.hk.*

David McNeill has taught at universities in Ireland, England, China and Japan. He received his doctorate from Napier University, Edinburgh in 1998 and has been both a doctoral research student and a foreign research fellow at the Institute of Socio-Information and Communication Studies, University of Tokyo. He currently works as a freelance writer in Tokyo. E-mail: *davidaamcneill@ hotmail.com.*

Brian J. McVeigh is chair of the Cultural and Women's Studies Department at Tokyo Jogakkan University. He received his doctorate from Princeton University, and has researched religion, education, gender, material culture, consumerism, nationalism, and the anthropology of deception and simulation in Japan. E-mail: *bmcveigh@gol.com.*

Wolfram Manzenreiter is an avid long distance runner who appreciates fast beats as much as fat bass lines and mind-boggling lyrics to ease his long training sessions. He is also assistant professor in the Institute of East Asian Studies, University of Vienna. E-mail: *wolfram.manzenreiter@univie.ac.at*.

Junko R. Onosaka is a doctoral candidate in the Graduate School of Education at the State University of New York at Buffalo where she specializes in the history of women, information and education. E-mail: *junko604@hotmail.com*.

Gretchen Ferris Schoel is an American Studies doctoral candidate at the College of William and Mary in Williamsburg, Virginia. Her current scholarship grows out of more than ten years as a founder and director of three cultural exchange programs between Japan and the United States. (All now have Internet components.) She taught at the Shonan Fujisawa Campus (SFC) of Keio University, Japan, from 1996 to 2000 and while there launched the global jam project introduced in this collection. Her dissertation is entitled *@america.jp*. E-mail: *gretchen@wm.edu*.

Birgit Staemmler is a doctoral candidate in the Japanese Department of the University of Tübingen in Germany working on a research project entitled "Self-representation and self-understanding of religious communities on the Japanese Internet." Her dissertation is about spirit possession in Japanese new religions. E-mail: *staemmler@japanologie.uni-tuebingen.de*.

Takako Tsuruki is a lecturer at Shokei Women's Junior College in Sendai, Japan. She has a masters degree in sociology from Syracuse University and has conducted a variety of large-scale research projects, including prenatal care in the United States and the diversification of the iron and steel industry in Japan. E-mail: *tsuruki_t@yahoo.com*.

Preface

David Gauntlett

When I was asked to write the preface for a book called *Japanese Cybercultures*, I was not sure that I could do it: as a Western academic with an interest in the Internet and its users, I was already aware of, and embarrassed about, my lack of detailed knowledge about non-Western cybercultures. But the editors assured me that I cannot be alone: this is, in fact, the first English-language book dedicated to life on the non-English-language Internet, even though, on today's Internet, only two-fifths of the content is in English. Happily, when I came to read this fascinating collection, I found it to be an excellent remedy for (part of) my ignorance.

Popular media give us, in the West, a particular image of this topic: the Japanese are a super-technological people, fascinated or even obsessed with the latest computers and gadgets. We imagine, perhaps, that their homes are gleaming temples to the latest cool technologies, and we might expect that they would have embraced and wholly mastered the Internet some time ago. This book shows that this vision is not really true to the everyday lives of most Japanese people (apart, perhaps, from the interest in little gadgets). Indeed, as Nanette Gottlieb and Mark McLelland point out in their introduction, even by 2000, Japanese take-up of the Internet in the home was less than half that of countries such as Canada, Iceland, and Sweden.

This relatively slow adoption of the Internet on PCs in Japanese households perhaps explains why a book such as this has been so long in coming. But it could also reflect the complacency of Western scholars: we assume that people in other countries, using other languages, are probably doing things with Internet technology that are *pretty similar* to those applications we are familiar with. This book shows how wrong that assumption is in many ways. Most striking, for me, was the way in which Japan has embraced the *mobile* Internet. The WAP protocol for mobile phones – which enabled users to access ultra-basic versions of Web sites, which had to be prepared especially, and which were too small and too slow to be much use – has already been pretty much discarded in the West, having enjoyed a brief period as the much-touted "next big thing" in the run-up to the Christmas of 1999. In Japan, however, a variation of this technology has been wholly embraced and is being put to a fascinating range of uses, creating another kind of cyberculture which – by virtue of both its reach and its complexity – is rather unique to Japan.

As Brian McVeigh shows in this book, the Internet-enabled mobile phones which young people use – to interact with friends via e-mail, and to arrange their lives using online friendship sites, timetables, and information services – are not faceless machines for accessing data, but take on considerable amounts of personal meaning. The devices are used to express the self and to connect with others in new ways. This occurs in the virtual space, in written messages and "cute" graphics, and also in the real-world space in which the physical shell of the mobile phone is adorned with stickers, decorative straps, trinkets, dolls and other personalized ornamentation. This "culture of cute" is discussed in Larissa Hjorth's chapter, where we see traditional gender roles changing as both men and women embrace the cute kitties and traditionally "feminine," fluffy and fun forms of electronic greeting.

McVeigh shows that some students even considered the mobile phone to be at the heart of the transformation and detraditionalization of Japanese society. The mobile tends to foster an ever-more busy social life, whilst at the same time making it much easier to rearrange or cancel appointments. The act of writing thoughts and feelings down seems to offer another discrete avenue for self-expression, McVeigh notes, and e-mail communication (via mobile phones) creates a new private virtual space where young people can interact without the scrutiny of their elders. This picture is reinforced in the chapter by Holden and Tsuruki, who discuss Internet sites for meeting people and dating (again, often accessed by mobile phone), where the self is constructed, presented to others, and explored through text. Here, as well as young people, other marginalized groups such as divorced people find a community to be a part of, and a place to express themselves. It is ironic that, since Japanese people seem to have adopted mobiles in such a unique way, McVeigh's study found that many Japanese students subscribed to a common downbeat view that Japan merely copies these pop-culture developments from other countries.

This book's coverage of online activists provides another healthy challenge to the Western view of a conformist Japan where everyone accepts the status quo. At the same time, however, the contributors show that the activists tend to be technically conservative, and are also constrained by internal hierarchies where the high-status activists are older people unfamiliar with the technological possibilities and opportunities, and unwilling to seek help from younger campaigners. Thus in Japan (as elsewhere, but in different ways), the "digital divide" takes on a *generational* dimension as well. Meanwhile, the Internet activists discussed in different chapters of this book seem to have created a few online spectacles and impressive petitions, but not to have had much impact on actual policy. This situation is similar to that in other parts of the world, where we are finding that, although the theoretical possibilities of Internet activists meeting big business and governments on the "level playing field" of cyberspace sound exciting, they are not matched with many clear-cut victories. Nevertheless, there is no doubt that the Internet has been an extremely valuable tool for organization, and sharing information, for many activists around the world.

Japanese Cybercultures considers whether, in the words of the editors, "Internet use really improves communication or just facilitates the flow of information." It is perhaps unfair to expect either of these always to be the case. What we *do* see, in this volume, are many instances of both. There are a range of ways in which the Internet opens up new forms of communication which can be valuable or beneficial in certain contexts – for example, allowing users the opportunity to meet new people; or fostering activist activity; or enabling gay people to access a new world of social and sexual arrangements; or allowing people with certain conditions to share their experiences (such as the HIV+ patients discussed in Cullinane's chapter, for whom the Internet was invaluable for connecting them with this "hidden" community of otherwise isolated and stigmatized individuals). The Internet also, clearly, is of considerable value as an information service alone. Fans of Foucault's power/resistance correlation will be pleased to note that more than one chapter in this book shows that whenever one voice promotes something online, another voice appears to resist it. For example, Kienle and Staemmler show that Japanese Jehovah's Witnesses (the Watch Tower Bible and Tract Society) are cautious about promoting their own Web presence, since interested parties might also stumble across vitriolic Websites set up by opponents of the religion, including the children of Jehovah's Witnesses who feel that the experience was far from being a blessing. Similarly, Gottlieb's chapter on the Burakumin minority shows that, although the Internet has given Burakumin groups the opportunity to organize, and to challenge stereotypes, the positive Burakumin sites are outnumbered two-to-one by hate sites reviling this group.

The Internet cannot always be expected to "improve" communication, then: as an illustration of this, Gottlieb and McLelland point to Ducke's finding that posts on campaign bulletin boards are often too polemical to be of use to the campaign organizations. *However*, the fact the people have taken the opportunity to engage with a particular issue, and express their feelings about it in a public forum, is of some value in itself, and cannot simply be dismissed as "useless" communication. Similarly, in their chapter on the Noise scene – a genre of not-necessarily-musical music – Caspary and Manzenreiter conclude that the Web has not made the community stronger or richer, and may have contributed to its fragmentation. *However*, one could say on the contrary that this fragmentation is perhaps just a consequence of the opportunities offered by the Internet: the chance to create and distribute new works in a purely digital form, without need for offline distribution networks or clubs or fanzines, and also the heightened interest and discussions about the Noise scene which the Internet has fostered, which have led to more diversification.

In the case of women's consciousness-raising – to use a phrase from Western feminism of the 1970s – Junko Onosaka shows that the Internet has clearly played an important role, not just as a provider of information, but as a space where liberating discussions can take place, freed from the gender and status-related conventions which may traditionally constrain communications for Japanese women. Here, the Internet's potential for "self-expression" – which in the sphere of personal homepages or dating sites can sometimes just mean personal vanity –

takes on a more serious and socially important role, allowing women to share intimate experiences and problems which previously were buried in the undiscussable "private sphere." Of course, as Dasgupa notes in his chapter, the Net can offer the same opportunities for anonymous discussions about personal matters to *men* as well, further reinforcing the theme found in various chapters in this book that the Internet is contributing to the general transformation of gender roles in Japan.

One more often-overlooked issue which *Japanese Cybercultures* places on the agenda is that of *language*. The Internet is always spoken of as a "global" phenomenon, but there is not a global language through which all users can communicate (except in the sense that a *Hello Kitty* animation, based on a site in Tokyo, can be enjoyed from around the world). Japanese sites are generally in the Japanese language and are mostly visited by people in Japan, so global accessibility is less important in these studies than the local context and meanings.

Far from being "just another" book about the Internet, this volume offers a valuable opportunity for students of cyberculture – and anyone else interested in media culture, popular culture, and politics – to rethink some of their assumptions. As a valuable guide to the ways in which Internet use has emerged in Japan, it reminds us that human creativity is always diverse, interesting, and able to surprise.

September 2002

Acknowledgments

We are indebted to the many people who have supported the publication of this book. Nanette would like to acknowledge the support of a University of Queensland Research Development Grant that enabled her, as a busy Head of School, to employ as a research assistant the indefatigable Akemi Dobson, whose impressive library and Internet searching skills contributed greatly to the completion of her chapter on time. Mark is also grateful for a University of Queensland New Staff Research Start-Up Grant that enabled the project to first get off the ground. He would also like to thank Graeme Turner, Andrea Mitchell and other fellows and staff of the Centre for Critical and Cultural Studies at the University of Queensland for their interest in and support of the project.

Both of us were amazed and delighted with the speed at which Professor Mark Selden, our Routledge reader, returned his comments on each chapter and with the insight and erudition of his comments, all of which helped shape the final version of the volume. We would also like to thank Professor David Gauntlett for reading through the text and providing us with his thought and insights about the book in his Preface. To everyone else who commented on drafts, listened to seminar and conference presentations of our chapters and contributed useful feedback, we thank you very much.

A note on language

All Japanese names occur in Japanese order: surname first. Macrons designating long vowels have been included in roman transcriptions of Japanese text, including personal names. However, common place names such as Tokyo and Kyoto have been rendered without macrons as is customary.

1   The Internet in Japan

Nanette Gottlieb and Mark McLelland

INTRODUCTION

Working on this book has been a continually exciting and absorbing process. As editors of this first study of the Internet in Japan in English, we have been delighted, impressed and challenged by the scope and depth of the work presented here. We have also become conscious of the vast and constantly shifting terrain which still awaits the researcher. Whereas studies of the Internet in western societies now form a well-established genre of scholarly inquiry, English-language work on the Internet in Japan has until now been hard to locate. This first effort to provide an overview of aspects of the use of the Internet in Japan also addresses the problem raised by Baym: "One of the most troubling shortcomings of the many analyses of online community to date has been their reliance on personal anecdote and hypothetical theorizing in place of close study."[1] All our contributors not only offer analyses based on close readings of actual sites and interactions that provide insight into what is actually happening in the communities under study, but also engage with the wider theoretical issues underpinning the developing field of Web Studies.

Our original plan for the book was to look at subcultural appropriations of the Internet by Japan's various minorities in order to assess the impact this new communications medium has had on groups and individuals whose access to print and broadcast media was limited. Well-aware of Japan's long tradition of grass-roots activism that is often overlooked by the mainstream press and by books on Japanese society that stress its homogeneity, we were interested to see how the Internet was being adopted by diverse communities and to what effect. However, as submissions for the book came in, it became clear that an introductory volume to the Japanese Internet had to be much broader in scope and that a more inclusive focus was necessary. Rather than look at minority subcultures, we decided to broaden the scope to look at *cybercultures*. We understand a cyberculture to come into being when like-minded individuals meet online in order to pursue a common interest or goal irrespective of whether the "community" that develops through this interaction maintains an offline presence. As Arjun Appadurai has pointed out, the Internet has fostered an exponential increase in "communities of imagination and interest" which are otherwise "diasporic."[2] The ability of computer-mediated communication to collapse both time and space has enabled those individuals with access to the technology to maintain contact with, and more importantly, feel a part of networks that are truly transglobal. Onosaka points out how excited one Japanese activist who had participated in various international women's conferences over the years felt about the first-time use of the Internet at the Fourth World Conference on Women in Beijing in 1996. She comments:

> In 1985, I sent some reports via airmail from the Third World Conference on Women in Nairobi, Kenya. It took at least one week; worst-case scenario, two weeks. Now I can read about it in real time on the Internet!

In Japan, as elsewhere, the Internet has helped bring together people from all walks of life who share common interests, ideals or anxieties. For instance, Cullinane mentions an online romance between two HIV-positive individuals, a Japanese woman and a South-American man, which resulted in marriage, and Ducke refers to an alliance between Korean and Japanese activists who united online in an attempt to block official endorsement of a revisionist history textbook by local Japanese school boards.

However, despite the fact that the Internet is often discussed in terms of its "global" reach and its "borderless" frontiers, it is important to remember that individuals who log on are real people in actual locations with specific purposes; the meaning of the Internet, which is in part defined by technology, is thus also partly the product of diverse social contexts. As Hine points out:

> … in thinking of the Internet we should not necessarily expect it to mean the same thing to everyone. It could be said that ideas about what the Internet is are socially shaped, in that they arise in contexts of use in which different ways of viewing the technology are meaningful and acceptable.[3]

As chapters in this volume illustrate, there are many contexts in which Internet use in Japan is, if not unique, then at least influenced by factors strongly pronounced in that society. For instance, one of the most remarkable factors affecting Internet adoption in Japan has been the popularity of hand-held, portable Internet-access devices such as the now ubiquitous mobile (or cell) phone (in Japanese *keitai denwa*). With adoption rates of over 80 per cent among young people, the use of such Internet-friendly devices as NTT's *i-mode* phones can hardly be described as a subcultural phenomenon and yet, as the three chapters dedicated to this important technology show, Internet access via mobile phones has produced specific modes of relating that differ in significant ways from offline life, leading Hjorth to speak of a "*keitai* culture" whose playful appropriation of "cute" characters has given rise to "new multivalent vernaculars." McVeigh, too, notes the important effects that online communication via mobile phones has had in his students' offline lives: for many, life without their *keitai* is now quite unimaginable. Holden and Tsuruki also see mobile phone use not simply as indicative of changes taking place in Japanese modes of communication but as actually producing changes in the way men and women meet and establish relationships – the implications of which are not yet clear.

A focus on purely subcultural uses of the Internet would also fail to take into account the way in which Internet technology has been appropriated by big business and the state. In Japan, the Internet is not yet used so much for e-commerce but itself constitutes a business since users generally pay for the amount of information they download as well as for the amount of time spent online. Students, in particular, run up large monthly bills downloading chimes, melodies and characters to "customize" their phones as well as messaging their ever-expanding circle of "mail friends" (*mēru tomo*).

If the increasing commercialization of the Net in Japan, as elsewhere, is not sufficient to dampen the enthusiasm of cyber-optimists who see the Internet as a revolutionary communications system that will level the playing field between big business, major corporations and the state on one side and the little guy on the other, then the adoption of the Internet by government agencies in order to pursue their particular agendas should cause concern. As Mackie shows, Japan's Prime Minister Koizumi (or rather his spin doctors) are proactive in the use of Internet technology to present a specific image of Koizumi as the "lion-hearted" reformer. Although readers of Koizumi's online magazine are encouraged to e-mail the Prime Minister their concerns and questions, only two or three of the many thousands are answered each month. The e-mail magazine is thus little different from a press release since it simply reduplicates the one-to-many mode of communication characteristic of previous print media. What is presented by the government as an initiative in cyberdemocracy is, in fact, an attempt to portray a very specific (and positive) image of the Prime Minister that is unmediated by potentially critical press comment or, indeed, any dissenting voices.

But before looking further at issues raised by contributors, we turn first to a brief discussion of the development and features of Internet use in Japan.

THE DEVELOPMENT OF THE INTERNET IN JAPAN

Japan's introduction to the Internet came with the launch in 1984 of the Japan University/UnixNETwork (JUNET), a research network linking Keio University, the Tokyo Institute of Technology and the University of Tokyo. Four years later, the Widely Integrated Distributed Environment (WIDE) project, in which the private sector also participated, was established to develop network technology. These networks were maintained by government agencies and research institutes for their sole use; private and commercial use did not begin until 1993,[4] the same year that the development in the USA of the Mosaic WWW browser broke the dependency on text and technical know-how formerly characteristic of personal computers by providing a user-friendly graphical interface available on a variety of platforms. Internet use surged ahead overseas as a result.[5]

In Japan, Internet take-up was relatively slow at first, despite the fact that the technology for electronic character retrieval had been available since the early

1980s. Several factors contributed to this: the strength of the yen, which meant that even expensive American software developers were cheaper to use than Japanese companies; the high cost of land-line phone calls in Japan; conflict between Ministries over policy; and the slower diffusion rate of personal computers, due perhaps in part to the preference in the 1980s and early 1990s for the use of stand-alone word processors.[6] In 1992, however, the opening of the Japanese market to American computer companies saw Compaq clones sell for half the price of the local NEC machines, which led to a rapid increase in the number of PC users. By 1994, PC sales surpassed those of stand-alone word processors for the first time by a large margin, although the PC penetration rate[7] remained lower than in many developed countries: by 1997 only around 20 per cent of Japanese had PCs, compared to over 40 per cent in the United States and about 35 per cent in Australia.[8]

The number of Japanese Internet users grew quickly once PCs were more readily available, nearly doubling between 1996 and 1997 and continuing to grow rapidly. We can see just how rapidly by looking at the growth in the household penetration rate during the late 1990s. By the end of fiscal 1998, it was 11 per cent (up by 4.6 per cent from the previous year) and by March 1999 had risen to 13.4 per cent. It had taken only five years after the Internet became commercially available in 1993 to attain a household penetration rate of 10 per cent, compared to 13 years for the PC, 15 years for the mobile phone, 19 years for the fax machine, and 76 years for the telephone.[9] Of course the social, political and cultural environment of the late 1990s was very different from that of some of those other technologies, in particular the telephone, but nevertheless the speed of Internet spread among private users was remarkable. By the end of 1999 the household penetration rate had reached 19.1 per cent[10] and in February 2000 had reached 21.4 per cent. This placed Japan thirteenth among those countries of the world where the rate exceeded 10 per cent, in a field led by Iceland (45 per cent), Sweden (44.3 per cent), Canada (42.3 per cent), Norway (41.3 per cent) and the USA (39.4 per cent).[11]

Local instances of creative use of the Internet abound. Japanese consumers, on the whole, prefer cash to credit payments and are particularly hesitant about using credit cards online. To develop a mode of Internet shopping that would appeal to Japan's consumers, in 1997 Japan's Lawson Inc. (a convenience store chain) developed a network of over 7,000 in-store touch-screen Loppi electronic shopping kiosks. The terminals allow shoppers to purchase tickets for entertainment events or travel, books, CDs, videos, cosmetics, PCs and a host of other items which are then paid for at the checkout and delivered to their homes or to the store for pick-up. This has now been supplemented with an Internet site which allows customers to order items online and visit any Lawson store to pay for and collect their purchases.[12]

As mentioned earlier, a distinguishing feature of Internet use in Japan, partly due to the cost of land-line calls, has been the high take-up of wireless technology. NTT DoCoMo's *i-mode* service, among others, allows users to access the Internet through a variety of text-based information sites accessible only by mobile phones. By February 2000, around a year after *i-mode*'s launch in 1999, the company had grown to become Japan's leading Internet service provider (ISP) in terms of number of contracts.[13] Other carriers have since also offered this service. In March 2001 there were 36.94 million wireless Internet users in Japan, up from 9.35 million twelve months earlier,[14] and by April 2002 that figure had grown to almost 70 million, not far from the estimated market saturation point of 80 million.[15]

Japan may have lower PC penetration rates than some other developed countries, then, but it leads the world in the mobile Internet and uses this technology in distinctive ways. Rikkyo University in Tokyo, for example, has launched in tandem with DoCoMo a Web site accessible to students from their Internet-capable mobile phones which enables them to check lecture cancellations, communicate with staff and catch up on missed lecture material. A university survey had found that while only 35 per cent of students had personal computers, 92 per cent had mobile phones. Although normally *i-mode* users cannot access ordinary Web sites owing to programming language differences, Rikkyo designed a system to convert standard PC Web sites to *i-mode* reception.[16] The Lawson chain referred to above likewise has targeted the mobile Internet, choosing to launch its Lawson Ticket service in *i-mode* from the beginning because of the perceived link between the service and younger people (the chain's main customers) who are both pop music fans and mobile phone users.[17]

This rapid expansion in Internet use means that the Japanese language has established its own presence on the Net. By September 1998, reported the Ministry of Posts and Telecommunications (MPT), there were more than 18 million Web pages in Japan, up from 10 million only seven months earlier. Working from the assumption that each page contained 2,000 characters, the MPT calculated, that equated to a total of 36 billion characters, exceeding the total number of characters published in Japanese newspapers and magazines in a year[18] – no mean achievement given the size of the printing and publishing industries in Japan. The mobile Internet market, too, has had implications for language use, extending the domain of on-screen Japanese beyond the desktop or laptop computer to the tiny *keitai* screen.

Within a comparatively short period, Japanese rose to become the second most widely used language after English on the Internet, a position from which it was only toppled by Chinese (9 per cent) in March 2001. Three months later, it had regained second place, far ahead of Spanish despite the vastly greater number of Spanish speakers. The latest statistics (March 2002) show Japanese (9.2 per cent) just behind Chinese (9.8 per cent), in third place behind English (40.2 per cent).[19] The early dominance of English on the Internet is clearly in decline as more countries come online and achieve what Crystal calls "a critical mass of Internet penetration" and "a corresponding mass of content … in the local language."[20] Japanese-language pages are, however, rarely accessed by non-Japanese people, other than the small numbers of scholars interested in Japan who are proficient in the language. The vast majority of Japanese-language pages on the Web are used by Japanese.

THE NATURE OF INTERNET RESEARCH

> Internet-mediated communication is too recent a social phenomenon to have
> provided the opportunity for scholarly research to reach firm conclusions on
> its social meaning.
>
> (Manuel Castells)[21]

Despite the relative newness of the phenomenon, Internet research in English is
already voluminous, with a number of important anthologies bringing together a
wide range of material that has appeared in numerous journals. Internet research
is conducted across an enormous range of disciplines including sociology,
anthropology, psychology, politics, economics, linguistics, cultural studies, media
studies and communication studies. Indeed, as other researchers working in the
emerging field of Internet studies have pointed out,[22] the application of a single
methodological approach to a medium as complex as the Internet would be quite
impossible and it is no surprise that there has been no consensus on the implications
or meaning of Internet use.

Studying the Internet in Japan has its own special problems given the almost
total absence of English-language material on the topic. Primary material in
Japanese is, of course, abundant, but there has been little systematic attention paid
to "Web Studies" in the Japanese academy and what exists tends to be empirical.
As can be seen from the contextual information discussed above, research on the
Internet in Japan involves a number of interesting themes. The digital divide and
ways of overcoming it through the use of mobile phones, and language issues
relating to the construction of international and national presences on the Web are
just two of these.

While researchers may either live in Japan or conduct fieldwork there, it is no
longer necessary to be physically present in any wired country in order to conduct
research on its use of the Internet. Web site analysis and e-mail contact, allied
with more traditional modes of research, extend the boundaries of research in a
manner undreamed of fifteen years ago. Cullinane raises the important point of
what this means for a project based on fieldwork: "Since my return from the field,
I have used the Internet to stay in communication with many of the people I met in
the field. Such continuity calls into question the notion of closure in fieldwork,
and muddies the very definitions of 'home' and the 'field.'" So it does, and the
point merits further discussion for anthropologists and others whose projects now
find the boundaries between the two zones of research blurred by the reach of this
technology.

What Internet-based research does not enable, of course, is online contact with
the unwired segment of the population. One might argue that the very fact of their
being unwired puts them outside the scope of this book, except of course that it
doesn't: it is impossible to evaluate the impact of the availability of the technology
without an understanding of the extent to which it is available to those community
groups under discussion. Cybercultures by their very nature are online groups,
but how many others are there of like mind who are unable to access the technology,

and how can the undeniable opportunities offered by the Internet be celebrated when many of those meant to be empowered remain offline? The digital divide is of course just as real in Japan as elsewhere, as Ducke, Gottlieb, Onosaka and Kienle and Staemmler show in their respective discussions of the textbook controversy, the Burakumin[23] community, women's online activism and new religious organizations. It is an important factor in any discussion of the Internet in Japan, and our contributors have taken this into account.

However, access to the Internet alone does not automatically enable an individual to participate fully in online life since exploiting the full potential of the technology requires a high degree of computer literacy – accessing a Web page is far simpler than creating one's own. The studies in this book point to the fact that Internet use is still in its nascent stages in Japan and that neither activists (on either side of the left–right divide) nor government agencies have managed to fully engage with the technology. Internet use by "progressive" activists in Japan has been hampered by the limited access that most non-governmental organizations (NGOs) have to both finance and skilled personnel. Anxieties about online security and the possibilities of virus attacks, too, have encouraged a cautious attitude toward developing a significant online presence for many groups. On the other side, many government-sponsored agencies have yet to develop their own online presence, limiting the scope for online activism. The digital divide is therefore not simply between those with network access and those without but it also separates a wide range of individuals who are to greater or lesser degrees "digitally enabled."

However, as discussed earlier, the fact that the majority of Internet traffic in Japan takes place via mobile phones and not PCs has meant that there exist vibrant and distinct cybercultures built around *i-mode* pages which, although not as numerous as those pages accessible to PCs, are nevertheless prolific. With regard to *i-mode*, the digital divide is less of an issue since among some sections of the population, particularly young people, having access to the Internet has become the norm. Given the important role that the mobile phone plays in Japan's cybercultures, the first three chapters in this collection are accordingly dedicated to a discussion of this phenomenon.

MOBILE PHONES AND IDENTITY

One of the most striking features of Internet use in Japan is the manner in which mobile phones have been used to access *i-mode* pages, thereby overcoming to some extent the digital divide between those with computers and those without (although, as noted above, the number of Internet pages which can be accessed by mobile phones is limited). Mobile phones are today of course a ubiquitous feature of any landscape, urban or non-urban, but what makes them particularly interesting in Japan is the differentiation of this technology for personal expression.

McVeigh describes the first generation of Japanese young people for whom life without their *keitai* (mobile phone) would be inconceivable. The students in his survey express incredulity at how people could ever have organized their social lives before the advent of the mobile phone. Many express their dependency on the technology: from waking them up in the morning, giving them something to do while killing time commuting (or sitting in the lecture hall) and helping them to rendezvous with friends. He also points to the extent to which mobile-phone technology has become a significant factor in modes of "individuation" – that is, how individuals come to structure and experience their interiority.

An important factor in Japan's youth culture that has been the topic of many studies is young people's affection for things which are *kawaii* or "cute."[24] The manner in which "cuteness" has been appropriated as a fluid marker of identity is manifested through mobile phone use. Young (and not so young) people everywhere download ring tones, icons and other interesting options to personalize their mobile phones – in the words of Castells, "customizing the technology"[25] – and Japan is no exception to that. The culture of the "cute" has until now been usually associated with women. Hjorth, however, in her discussion of the mobile-phone culture of cuteness, argues that the boundaries between genders as producers and consumers of "cute" culture are blurring and that the mobile-phone arena provides evidence for a possible shift in gender roles in Japan.

Holden and Tsuruki discuss the way in which mobile phones facilitate new ways of meeting people through *deai* (personal-introduction) sites, some of them interest-based, others more blatantly romance-oriented. As in the case of activist groups on the Internet, the consequences of the use of the technology in this way are not always what the user may have hoped for; the authors show, however, that the widely-held disapprobation of the practice of *deai* is not always warranted, and that there are positive benefits to be gained as well. They view participation in *deai* as a means of asserting individuality and autonomy, discussing the practice as a kind of barometer for changes in Japanese society. Both this chapter and those by Hjorth and McVeigh explore the link between technology and identity. All show how customizing the technology plays a large part in personalizing what began as an impersonal communications device by associating it more closely with the owner through a variety of individual uses and markings. In this way, a sense of individual identity develops, organized around the technology's ownership and use.

THE REACH OF THE INTERNET

An often-heard claim is that because Internet technology allows communication of information over a much wider area and at much greater speed than previously possible, subcultural groups make use of this both to gain wider exposure for their particular causes and to enhance communication among members and with the general (wired) public. But is this really true? An important question for this book is whether Internet use really improves communication or just facilitates the flow of information.

There is no doubt of the importance of the agency of voice being given to players by the possibilities afforded (to those who can access them) by the Internet, but at the same time, who does the voice reach? As Crystal notes, "Most Internet interactions are not global in character; we are not talking to millions when we construct our Web pages, send an e-mail, join a chatgroup, or enter a virtual world."[26] Are we even talking to anyone outside the ambit of our own particular interests and concerns?

While the enhanced flow of information is not in doubt, the assumption that such an enhanced flow somehow enhances notions of community is problematic. Some early work on the Internet adopted an overly celebratory tone regarding the application of the technology, which more recent research has not borne out. More from Crystal, who confirms what studies worldwide have discovered: "Certainly, the mere fact of having engaged in an Internet activity does not produce in a user the sort of sense of identity and belonging which accompanies the term *community*."[27] Some of the studies in this volume of Net presences established by various groups confirm the truth of this in Japan as well, as in the case of the Burakumin groups who have found the Net to be more often a source of unwelcome discrimination from the wider society than an opportunity for debate about their situation with non-Burakumin Japanese. Neither has it provided any enhanced sense of community amongst the wider (e.g. unwired) Burakumin community.

Ducke, too, in her discussion of Internet use by organizations campaigning for textbook reform notes that information made available on the Net can be accessed and used by enemies as well as by sympathizers. Web site owners also related that much of the communication that takes place on their bulletin boards is overly polemical and quite "useless" for campaign purposes, suggesting that although the volume of discussion about controversial issues may be increased by Internet access, the technology has done little to encourage a similar increase in quality.

Other chapters, however, like McLelland's study of an online gay community, focus on the possibilities for enhanced interaction among gay men, a phenomenon that has been recognized the world over. McLelland shows how a gay "cruising" Web site that is ostensibly about facilitating sexual interaction among gay men has been instrumental in generating a sense of communal responsibility organized around following the correct cruising "etiquette" so as not to draw negative attention to gay men's appropriations of public spaces like parks and bath houses. Cullinane's chapter, too, looks at how another stigmatized minority, persons living with AIDS (PWAs), have used the Internet to overcome the sense of isolation and secrecy enforced upon them by NGOs which, in Japan, are not staffed by PWAs themselves, but by "the 'healthy' [who] are charged with helping and protecting the 'sick.'"

What the studies in this book do show beyond doubt is that the Internet allows people in different parts of the country (as well as Japanese living abroad) to link up more easily than before without the necessity for physical interaction. Onosaka's chapter on activism among women's groups makes the point that "this is especially important in Japan, where almost every powerful establishment institution – from governmental institutions to the main offices of large corporations – is located in Tokyo, the capital, and people in outlying areas find it difficult to convey their views to citizens outside their communities." And not just in different parts of the country; sometimes in different parts of the same city, where people are prevented by illness from leaving their homes.

On the international scene, the global reach of the Internet (at least among wired developed countries) enables groups with an ax to grind to reach out to the international community to attract attention to situations of concern in Japan. This of course has implications for language use; unsurprisingly, we find that those organizations wishing to make a case for international support often use English-language Web sites in addition to their regular Japanese-language sites. Burakumin groups, for instance, use English strategically to achieve their aims of heightened international visibility and activism, highlighting the reduced human rights of their members in Japan through multilayered English-language sites which provide detailed information on the state of human rights issues, and in particular Burakumin issues, in Japan. Using English on the Internet in this way allows such groups to put their case out on the international stage, where somebody may see it and make contact. This is not always without problems: Ducke, on the role of language in the recent textbook controversy, notes that Network 21, a Japanese Web site set up as part of an ongoing campaign against the distortion of Japan's wartime history in textbook revisions, maintained only a very limited and seldom updated English version of its Web page. In this instance, she relates, English-language input from other countries sometimes posed difficulties, although some foreign contributors to the site, notably Koreans, were able to communicate in Japanese.

At the same time, the Internet provides a reasonably comprehensive source of information on relevant issues for those Japanese users able to read other languages, particularly English – women activists seeking information about women's issues in other countries, for example, or members of Japan's Men's Movement seeking information about activities and approaches being developed by Men's Movements outside Japan. Indeed, in this respect, the Internet is invaluable in making available what Mathews refers to as "the information and identities available from the global cultural supermarket."[28] Yet, as always, there is another divide here, between those who can and cannot understand information presented in English. Furthermore, to the extent that the vast majority of Net traffic in Japan takes place within Japanese sites, the extent to which the Internet is capable of acting as a global medium in Japan needs to be questioned.

The chapter that demonstrates in the most vivid manner the range of Internet communication is Schoel's discussion of "what appears to have been the first Blues jam ever to connect musicians over 10,000 kilometers apart." Schoel is careful to point out, however, that despite the border-crossing nature of the Internet that enabled Blues musicians in Tokyo and Mississippi to meet online and engage in a virtual jam session, the experience of the "other" is always already mediated by local knowledge. This means that interaction via the Internet does not take place in some disembodied cyberspace, but is still tied up with local understandings, interpretations and prejudices that mean events can have quite different meanings to participants on different sides of the modem.

DISCRIMINATION AND ACTIVISM

Whether in Japanese or English, the Internet has certainly expanded the repertoire of avenues open to activist groups with political, social or cultural agendas. In this volume, those agendas often stem from experiences of discrimination by and exclusion from the wider society. The Internet allows the marginalized a voice which mainstream media do not, as we see in the discussions of persons with AIDS, Burakumin, women's movements, men disaffected with notions of hegemonic masculinity and the gay community. Exclusion is a recurrent theme throughout this volume, whether related to gender, sexuality, social status or illness. It is clear from these chapters that the groups under discussion, lacking full access to the mainstream media, were already making use of alternative means of communication long before the advent of the Internet, usually by producing their own *minikomi* (newsletters and magazines circulated to supporters and other interested parties – the word is a pun on *masukomi*, or mass media). Unlike the mass media, the tone of many *minikomi* was more personal and direct and ideally suited to communicating the frustrations and difficulties faced by members of marginalized groups.[29] Some such groups then transferred their *minikomi* online, but at the same time, mindful always of the digital divide, continued hard-copy production (the *Buraku Liberation News*, for example, is available in both formats). Others discontinued the arguably more expensive production of hard copy, while the volunteers who produce *HIVOICE*, a *minikomi* which has been produced by HIV-positive haemophiliacs since 1993, are divided over whether to move the newsletter online on the grounds that to do so might result in a loss of immediacy and leave a less lasting impression than the printed word. As Manuel Castells points out, "New electronic media do not depart from traditional cultures: they absorb them,"[30] and the chapters in this collection illustrate the many ways in which the Internet is interfacing with and in many cases superceding more established modes of communication.

McNeill gives a bleak picture of the decline of opposition politics in Japan during the last decade, tempered only by his discussion of the increased scope for campaigning that the Internet has provided for many grass-roots networks. Indeed, as he points out, "for some groups and campaigners that only exist in cyberspace, the Net has replaced all other forms of social action." However, he sees the potential of Internet activism in Japan as being held back both by the high cost of Internet access as well as by the timidity of many groups who have failed to fully engage with the potential offered by online communication. This latter point is also reinforced by Ducke who puts it down to a generational digital divide – established activists in Japan tend to be middle-aged or older and lack the familiarity with computers that the younger generation takes for granted.[31] Similarly, younger activists lack familiarity with more traditional modes of campaigning and it will not be until both kinds of experience are brought together, Ducke argues, that the Internet will come into its own as a truly effective campaign tool in Japan.

It must also be remembered that not all use of the Net for activism or proselytization is problem-free for its users, i.e. those who post the material. Just as the

Internet opens up the potential reach of the information, so it opens its users up to the kind of unwanted attention previously delivered through graffiti or hate mail but now more easily aimed right into the heart of the target organization. McLelland notes that certain gay sites run "gay checks" on intending users, asking a series of pertinent questions delivered as a quiz that only individuals with some prior experience of the gay scene would be able to answer. After all, providing unlimited access to one's Web site can be dangerous: one female activist mentioned by McNeill closed her BBS after receiving virtual rape threats for daring to brand Tokyo's controversial governor Ishihara Shintarō a racist. However, despite the fact that there are various means to limit who can read or post messages on a BBS (such as requiring visitors to first obtain a password or having all postings vetted by a moderator), many groups discussed in the collection have chosen not to provide a messaging facility and have chosen to avoid interactive features. This cautious approach to the new technology has severely limited the potential of these groups to develop fully functioning online communities, suggesting that the digital divide is perpetuated as much by psychological as economic factors. Concerns about unwanted postings meant that Network 21 in the textbook affair did not set up a BBS, and such are the levels of discrimination against Burakumin on the Internet that Burakumin groups have set up their own watchdog organizations aimed at combating cyberdiscrimination.

Another factor that needs to be considered is that not all groups and organizations have necessarily welcomed the advent of the Internet since the freedom it gives individuals to access information and communicate with other like-minded people can be seen as a threat to existing power structures. This is particularly the case with some hierarchically-organized religious groups that wish to maintain a strictly centralized hold on the production and dissemination of their doctrines. Kienle and Staemmler argue that new religious movements in Japan have not embraced the Internet as a medium for proselytization to the extent that might be expected. They show that, in the case of the Jehovah's Witnesses in Japan, members have been discouraged from using the Internet and that the only official Web site is a Japanese adjunct of the main US-based site containing little more than Japanese translations of print publications. Interestingly, they show that the Internet is being used more by apostate members, non-believing children and other family members of Jehovah's Witnesses to network and support each other as well as by mainstream Protestant churches that offer counseling services to ex-Jehovah's members. They also point out how other religious organizations with less centralized structures have been more proactive in using the Internet to disseminate information about local and national events.

THE INTERNET AND SOCIAL STRUCTURES

Does using the Internet, something relatively new and in theory empowering, actually free its users from the social structures within which they live? Do people denied a voice in mainstream society somehow miraculously achieve a hearing through publicizing their interests on the Net? The answer, of course, is a resounding no. No matter how widely and how actively different groups promote their causes through Web sites, the matter of uptake is what determines whether they actually achieve any kind of connection with people from groups they were not already in contact with. Burakumin groups may disseminate as much information about themselves as they wish, in either Japanese or English, but unless somebody "out there" takes the trouble to find the site, read its contents, and establish constructive contact, they are arguably no better off than they were before, using printed *minikomi* to achieve the voice denied them by the mainstream media. Even among not-particularly-stigmatized groups such as the Noise alliance, the same is true: Caspary and Manzenreiter question the ability of Noise alliance sites to create alliances, given the lack of communication they afford: "The home page of Tanno, for example, is primarily meant to promote his own activities. Yet, as a secondary function, the integrated bulletin board should help to 'construct an international network.'" The word "should" here is open to question, for the reasons outlined above, although fans of Noise are more likely to go looking for opportunities to communicate than are people wishing to remain blissfully ignorant of other less fortunate sectors of society.

Web sites and associated mailing groups can create a sense of community among those by whom they are run, enabling communication across distances without the need for face-to-face interaction, as has happened in the case of women's groups (although as always, this is constrained by access to the technology). They can educate those willing to find them and read; they can serve as a means of support and information for those interested in/affected by their guiding concern, as in the case of persons with AIDS, gay men interested in cruising for sex, or members of religious sects. They can empower their users, allowing them a greater degree of autonomy in the way they arrange their lives. But they cannot usually create a sense of connection between the group constructing the site and the wider society where that group has been marginalized. The mere fact of raising a voice does not mean that anyone other than the already-converted will listen to what it says. The consumers of the technology have to have the motivation and the interest to go online and find the sites before any connection can be established, and even then, that connection may result from mere curiosity, from the desire to confirm ingrained prejudices, or worse, from the intent to verbally harass others through leaving offensive messages on a site's BBS or by flaming the owner.

For the reasons outlined above, the Internet is likely to have little substantial impact on existing social constructs of discrimination and stigmatization for the foreseeable future. As Holden and Tsuruki note in their conclusion:

> *Deai* may be a new strategy for encounter as a means of solving the problems incident to modernity. We are left to wonder, though, whether it actually delivers its users from the structuration inherent in the society in which they exist.

CONCLUSION

What, then, does this book tell us about the Internet that is specific to Japan? The Net (and in particular the Web), in Japan as elsewhere, has changed the ways in which we approach the gathering and presentation of information, and has offered both individuals and groups new media for the construction of identity. What this has meant in the Japanese case is that voices not usually heard in day-to-day life have found new ways to assert themselves, free in theory from those social restrictions which usually apply but in practice just as constrained by lack of uptake. Structures of social exclusion remain just as powerful whether the Net is available or not. This may stem in part from the fact that Japan has not had as strong a tradition of civil rights activism as the US or neighboring societies such as Korea,[32] although grass-roots activism, particularly in the area of citizens' movements, has been responsible for many changes at the local level.

However, the Internet has not really leveled the playing field in Japan. Despite giving an increased voice to previously silenced minorities, it is still government and big business who, with their access to greater finance, facilities and expertise, are best able to exploit the range and power of the Internet. Far more people access Prime Minister Koizumi's e-zine each month than ever pass through the sites of the activist groups described by McNeill. Lack of time, skills and finance are mentioned repeatedly by volunteer-run NGOs and activist organizations as reasons why they cannot further expand or develop their Internet sites. Lack of foreign-language ability on the part of activists, too, has limited the ability of activist sites to reach a global audience.

Furthermore, as Holden and Tsuruki point out, the sites that are accessed by most people in Japan are those run by the ISPs themselves that offer (for a fee) downloadable chimes, ringer melodies and cute characters with which consumers can customize their Internet-enabled mobile phones. Also, despite the fact that the growth of e-commerce in Japan lags behind that of the US and Europe, the Internet itself is big business given the high cost of logging on and a system which charges for all information downloaded. The current state of the Internet in Japan, then, hardly supports an earlier vision of the potential provided by the Internet, a space in which "information wants to be free."

Given the intractable nature of the social discrimination they face on the home front, some of the groups discussed in this book (and many others as well) have used some English-language on their Web sites as a strategic tool for attracting international attention. But responses from the international community, usually offered in English, often pose a problem for many groups whose ability to function in English is limited by access to skilled personnel and other resources. Although the Internet has done much to remove spatial and temporal barriers to communication, other borders still remain. What of the future then, for activism and the Internet in Japan? As McNeill concludes:

> For those who remain indignant at the failures of the current system and who continue to believe it can be changed for the better, the Internet might better be understood as a tool rather than a panacea.

NOTES

1 N. Baym, *Tune In, Log On: Soaps, Fandom and Online Community*, Thousand Oaks, CA: Sage, 2000, p. 198.

2 A. Appadurai, *Modernity at Large: Cultural Dimensions of Globalization*, Minneapolis: University of Minnesota Press, 1996, p. 195.

3 C. Hine, *Virtual Ethnography*, Thousand Oaks, CA: Sage, 2000, p. 30.

4 Although privately-run BBS systems had been around before this. *GayNet Japan*, for example, started as a BBS in 1989 before going online in 1993; *http://www.gnj.or.jp* (13 May 2002).

5 Ministry of Posts and Telecommunications, *Communications in Japan 2000: Expanding Frontiers – IT in the 21st Century*, 2000. Online. Available HTTP: *http://info2.mpt.go.jp/policyreports/ english/papers/2000-index.html* (in English), and *http://www.yusei.go.jp/policyreports/japanese/ papers/h12/index.html* (in Japanese) (27 September 2001).

6 See N. Gottlieb, *Word-Processing Technology in Japan: Kanji and the Keyboard*, Richmond, Curzon Press, 2000, for an account of this.

7 The penetration rate of a technology is defined as the percentage of the total population who use it.

8 International Telecommunications Union, 1997. Cited in "Emerging market indicators," *The Economist*, 8 August 1998, p. 88.

9 Ministry of Posts and Telecommunications, *Tsūshin Hakusho (White Paper on Communications) 1999*. Online. Available HTTP: *http://www.soumu.go.jp/joho_tsusin/policyreports/japanese/ papers/99wp/99wp-1-index.html* (15 February 2002).

10 The rate for enterprises and small businesses was of course much higher, at 88.6 per cent and 31.8 per cent respectively.

11 Ministry of Posts and Telecommunications, *Communications in Japan 2000*.

12 W. Auckerman, "Japan's Lawson eyes Web sales," *Business News*. Online. Available HTTP: *http://www.internetnews.com/bus-news/article/0,,6_160831,00.html* (24 December 2001).

13 Ministry of Posts and Telecommunications, *Communications in Japan 2000*.

14 NUA Internet Surveys, *Ministry of Posts and Telecommunications (Japan): Growth for Japan in Dial-Up, Broadband, Wireless* (12 June 2001). Online. Available HTTP: *http://www.nua.ie/ surveys/?f=VS&art_id=905356858&rel=true* (27 September 2001).

15 "Cell phones nearing 70M, KDDI regains No 2 spot," (9 May 2002). Online. Available HTTP: *http://sg.biz.yahoo.com/020509/15/2p8wa.html* (11 May 2002) © Dow Jones.

16 See link from *www.rikkyo.ac.jp* (1 November 2001) to *Financial Times* article on this subject.

17 M. Okada, "2001 iForce Heroes," *iForce Initiative*. Online. Available HTTP: *http://www.sun.com/ 2001-0829/feature/profiles/okada.html* (24 December 2001).

18 Japan MarkeTracker: Internet and Multimedia Report, *18 Million Japanese Language Web Pages* (1 December 1998). Online. Available HTTP: *http://www.internetjapan.com* (15 February 2002). Use "language" as a search term to find this article.

19 Global Reach, *Global Internet Statistics by Language: Online Language Populations March 2002*. Online. Available HTTP: *http://www.glreach.com/globstats/index.php3* (4 May 2002).

20 D. Crystal, *Language and the Internet*, Cambridge, Cambridge University Press, 2001, p. 220.

21 M. Castells, *The Rise of Network Society*, Oxford, Blackwells, (second edition), 2001, p. 385.

22 See, for example, N. Wakefield, "New media, new methodologies: studying the Web," in D. Gauntlett (ed.), *Web.Studies: Rewiring Media Studies for the Digital Age*, London, Arnold, 2000, pp. 31–42.

23 The Burakumin are Japan's largest minority group. Physically indistinguishable from other Japanese, they continue to be subject to status discrimination for reasons originating in the hereditary ostracism from mainstream society of their pre-modern ancestors because of their association with occupations involving blood, death and other impurities.

24 See for example S. Kinsella, "Cuties in Japan," in L. Skov and B. Moeran (eds), *Women, Media and Consumption in Japan*, Richmond, Curzon Press, 1995, pp. 220–54.

25 M. Castells, *The Rise of the Network Society*, vol. 1, Oxford, Blackwell, (first edition), 1996, p. 2.

26 Crystal, *Language and the Internet*, p. 59.

27 Crystal, *Language and the Internet*, p. 59.

28 G. Mathews, *Global Culture/Individual Identity: Searching for Home in the Cultural Supermarket*, London, Routledge, 2000, p. 1.

29 Wesley Sasaki-Uemura notes that small size is not necessarily a determining factor of *minikomi* but that they are rather characterized by their anti-authoritarian stance and personal, individualist disposition, factors characteristic of many Web sites, too. *Organizing the Spontaneous: Citizen Protest in Postwar Japan*, Honolulu, University of Hawaii Press, 2001, pp. 146, 245 n. 80.

30 Castells, *The Network Society*, (second edition) p. 401.

31 On the problems caused by middle-management failure to engage with Internet technology, see "A generational divide is holding back Japan," *Asiaweek* (Hong Kong edition) (11 May 2001). Online. Available HTTP: *http://global.umi.com/pqdweb?TS=990586...&Dtp=1&Did+ 00000072869169&Mtd=1&Fmt=3* (23 May 2001). Here, I (McLelland) cannot resist telling a story about a young salaryman friend of mine who was given the task by his boss of finding information about the Japanese movies that were to be premiered at the Cannes Film Festival in 2000. When my friend presented the material one hour later, his boss was incredulous, asking "How could you find all this information so quickly?" My friend replied "On the Internet."

32 The Korean government has been much more proactive in making Internet access available to its citizens than has the Japanese government which left the development of Internet links largely in the hands of private business. Particularly effective has been a government initiative to make Internet access available in all post offices where citizens can access e-mail as well as apply online for passports, tax files, car registration and other government services. See W. Kang, "The engine for the next economic leap: the Internet in Korea," in S. Rao and B. Klopfenstein (eds) *Cyberpath to Development in Asia: Issues and Challenges*, Westport, CT and London, Praeger, 2002, pp. 111–36.

Part I
Popular culture

2 Individualization, individuality, interiority, and the Internet

Japanese university students and e-mail

Brian J. McVeigh

INTRODUCTION

What does the explosion of e-mailing and Web surfing via cellular phones in Japan tell us about the relationship between the individual, social norms, and technology? What does their coming together produce? A cornucopia of choices, an abundance of information, and freedom from time and space? Or, are there certain parameters of Internet use shaped by the very nature of psychology and social expectations?

In this chapter, after briefly introducing the technical aspects of cellular phones, I discuss the practices associated with e-mailing and surfing via mobile phones (e.g. number of messages sent and received, to whom and from whom are messages sent and received, purpose of messages and their content, amount of money spent for services, and the general impact upon daily life). I also examine common themes – convenience, saving time, access to information, privacy, changing social relations – that emerged from semi-structured interviews that I conducted with sixty students about how they view and use mobile phones for e-mail and Internet access. The interviewees, students who I taught during two semesters, were roughly divided between males and females and were all from one university in Tokyo. The aspect of cyberstructure that I focus on concerns "individualization:" the emergence during the last several centuries of practices that isolate the individual from the social mass in order to facilitate sociopolitical analyzing, scrutinizing, and identification. Such practices, motivated by state-formation and economic rationalization, have resulted in increased control and surveillance of persons and populations through systematic record keeping, numbering, listing, cataloguing, and measuring.[1] The IT revolution and digitalization of information can be understood as a continuation of the emergence of individualization. Ironically, despite the intention of control behind individualization, it has become deeply implicated in other grand intellectual projects that have expanded the sphere of personal activity and subjectivity, i.e. individualization has augmented the "interior life" of the individual, constructing autonomous selves of wants and wishes ready-made for emerging capitalist consumerism. The emergence of psychological interiority has afforded a "space" in which a unique self dwells that expresses its own "individuality" by pursuing consumerist desires.

My aim, then, is to show how consumption patterns involving cyberspace – specifically e-mailing/Web surfing on cellular phones by Japanese university students – increase a sense of interiority (e.g. personality, personal traits, distinctiveness), thereby highlighting individuality. In other words, e-mailing is employed as a form of self-expression and impression management. Discussions with students indicate that this accentuation of individuality, as a manifestation of individualization, may be described in positive terms – "personalized individualization" (expansion of social networks, increased access to information, consumer empowerment), or in negative terms – "atomized individualization" (fragmented social structure, trivializing information, simulated social relations). Personalized individualization, being an individual-centric ideology, encourages "individuality."[2] As an example of personalized individualization I discuss how the very object – cellular phones – that connects students to the Internet is also used for purposes of self-presentation and is related to an esthetics of impression management (e.g. being "cool" and "cute"). According to one observer, "No longer will a person's choice of cell phone revolve around issues as mundane as weight, size and battery life – in Japan, at least."[3]

THE TECHNOLOGICAL BASICS OF JAPAN'S CELL PHONES

Cell phones, rather than being mere telephonic devices, are mobile Internet devices, and like many other similar telecommunication tools, they are capable of multi-tasking, combining video camera, radio, TV, camera, personal digital assistance, MP3 player, fax machine, and Web browser. In addition to providing access to the Internet, many cell phones have built-in watches, calculators, and dictionaries.

In Japan, the company most associated with cell phones is DoCoMo (an abbreviation of "Do Communication over the Mobile network"). DoCoMo was established in 1992 by Japan's telecommunications giant, NTT (Nippon Telephone and Telegraph), and has become Japan's principal mobile-phone carrier (another big player is J-Phone, Japan Telecom's wireless carrier).[4] In 1996, DoCoMo launched a mobile Internet service called "*i-mode*," which is considered a "2.5G technology:" something halfway between mobile telephoning and full-fledged mobile Internet. This "service has spawned a whole new style of phone usage, particularly among Japan's youth."[5] DoCoMo opted for a technical standard called Personal Digital Communications (PDC). However, PDC is incompatible with Global System for Mobile Communications (GSM)[6] (as of this writing, the closest GSM has to *i-mode* is Wireless Application Protocol, or WAP, a global standard for developing applications over wireless communication networks). In 2001, "3G" (third-generation) phones were introduced. Such devices have cameras for video conferencing, double as MP3 players, and can conveniently display short video clips of sports highlights or movies.

Japan has the world's biggest Net-linking mobile phone market.[7] By March 2001, 20 million people had subscribed to DoCoMo's *i-mode* (this figure does not include other companies offering similar services).[8] One professor's investigation found that 90 per cent of 915 students surveyed possessed cell phones. Thirty-six per cent said they receive at least one e-mail message during class, while another 30 per cent said they receive two to four messages per class. Only 17 per cent said they refrain from sending messages while in class.[9] A poll conducted in 2000 found that among 3,100 high-school students, half exchange ten or more e-mail messages a day and 48.8 per cent "made new friends" via the Internet.[10] A poll conducted in 2001 found that 25 per cent of elementary and junior high kids in the Tokyo area possessed a cell phone.[11] Why have Internet-capable cell phones become so popular? Much of the explanation has to do with how Japanese spend much of their time when not working: DoCoMo predicted that people would not want to use the Internet as they do at home. Rather, individuals would be happier with "bursts of information during 'in-between' time: while waiting for a train, riding in a taxi, sitting alone in a coffee shop."[12]

One place where cell phone use is rampant is the university classroom. Japanese university students are infamous for chattering while professors lecture. This is why one professor became "so intrigued by the intense silence in his classes that he decided to investigate what had caused the change. He found out by turning off the lights, revealing the glow of myriad *keitai* monitors. All of the students were busy e-mailing."[13] Indeed, "Long gone are the days when students babbling amongst themselves plagued university professors' classes. Instead, students are now more likely to ruin a class by being totally silent."[14] Professor Shimada Hiroshi of Konan Women's University claims that, "University teachers have been bothered all the time by students chattering away, so the quiet has made them relax. Perhaps they haven't realized that mobile phones are taking over the classroom."[15]

The physical characteristics of Internet-capable cell phones probably account for their explosive popularity. These devices are small and featherweight: standard dimensions (height, width and thickness) may range from 123 × 39 × 15 to 130 × 48 × 24 millimeters, and they weigh anywhere from 60 grams to 105 grams. Most are water and shock resistant. A wide range of accessories and peripherals can complement cell phones, including battery packs, single battery chargers, universal serial bus (USB) cables, straps, carrying cases, car holders, quick-charging cigarette lighter adaptors, quick-charging AC adaptors, desktop charging holder for data transmission, earphones, and attachable digital cameras. Because the number of keys is limited (about the same as a phone), a small keyboard can be attached to cell phones, expanding the range of characters and typing capability.

For many, cell phones are attractive because they have the same basic functions as Internet-capable desktop computers: storage capacity for in-box messages and sent messages, mail reply and forwarding, signature, and bookmarks. Typically, a cell phone can receive a maximum of 2,000 characters per message and can send a maximum of 250 characters per message. The in-box message storage capacity might be a maximum of 200 messages. Other capabilities include: *chaku merodī* or *chakumero* (from "incoming melodies," tunes played when mail arrives that can be downloaded from the Internet), *anime* (cartoon characters that can be stored and displayed), and *nabirinku* (links to a car's navigation systems). Cell phones can also connect to other cell phones or be linked to computers (which in turn are linked to the Internet). One model can switch between English and Japanese displays, has a clock (digital, analog, or picture), a calendar display for twenty-nine cities around the world, simple English vocabulary words and thirty-six English conversational phrases, a vibrating function to signal incoming calls, a calculator, dictionary, and dial locks.

A wide range of services is available for Internet-capable cell phone users: voice instruction, automatic answering, call forwarding, call waiting, three-way calling, caller identification, and voice mail services. Currently there are over 23,000 Web sites designed to fit the tiny screens of mobile phones. Significantly, from a business perspective, cell phones are important not so much for e-commerce, but rather for i-commerce, i.e. paying to use the Internet itself rather than ordering things by e-mail.

COMMON THEMES AND USAGES

The three most common words that I heard from students about the usage of cell phones were *ureshii* (happy), *tanoshii* (enjoyable), and *benri* (convenient). Comments included: "Life would be hard without it," "I can't imagine life without it," and "If I can't find public telephone booths, they're quite convenient." Also, "with a cell phone, I don't need a wrist watch." Cell phones are also characterized as *kantan* (simple) and *raku raku* (easy): they are "easier than a regular phone." "Unlike the telephone, I can send the same message to many people" but for long conversations, "the telephone is still better." One individual explained that "Using e-mail is better than talking, because it doesn't take time, doesn't cost much, and is very convenient." Some people are so skilled at using their cell phones that they are referred to as *oyayubi-zoku* (literally, "thumb tribe"), which is meant to describe how they move their thumbs with incredible speed to press the keys.

Making the best use of time was another common theme. Stated simply, cell phones "save time" and in this sense add a type of convenience to one's life. Informants commented that: "Because e-mail allows me to communicate information with my friends effectively, I don't waste time," "I don't like to write letters and cell phones are faster than regular mail, and another advantage is that we don't need stamps," "If I'm late, I can conveniently contact friends," "It saves time when looking for friends." Cell phones provide distraction when passing the time while waiting for a bus, commuting on trains or subways, or lingering at meeting places: "I can kill time (*hima tsubushi*) by sending e-mail messages," "When I have free time, I toy with my cell phone," "When I find it intolerable waiting to meet my friends, I can pass the time surfing." Indeed, one informant commented that, "If there were no cell phones, I wouldn't know what to do when I had free time." However, there are some demerits to "saving time," because cell phones "make people impatient, since they are always waiting for an immediate answer." In the end, cell phones "make everyone busier." People's "sense of time is changing. They don't need to decide on specific times and places to meet with friends if all of them have cell phones," and because individuals "no longer have to worry about deciding in advance when and where to meet, the importance they place on time seems to have diminished." Thus, they no longer become nervous when running late for appointments.[16]

Another key theme, again related to convenience, is access to information. Students explained that "it's fast for getting information anytime, anywhere." Indeed, *jōhō* (information) and *saito* (sites) were also very common words students mentioned when discussing cell phones. Examples of the kinds of information and services that can be accessed include: news, weather, mobile banking, credit cards, securities, insurance, travel, traffic, maps, navigation (*nabi*), shopping, recipes, melodies, images (to be downloaded and used to signal the arrival of mail), games, fortune-telling, entertainment, town information, administration, regional information, and ordering tickets.

One last theme that deserves comment is how cell phones relate to "Japanese culture." According to one student, "Japan is gradually losing its traditions because of cell phones, and I'm worried about the future of Japan's young people." At the same time, another student said that cell phones haven't "penetrated Japanese culture yet," as if there were an essentialist, immutable core to Japanese society. Many students reinforced the stereotypical view that Japanese can only "copy" from foreign countries and that cell phones are imitative, an adopted invention.

What do students use Internet-capable cell phones for? Almost all explained that they use them to stay in touch with their friends and girl/boyfriends. "Family members" were rarely mentioned. What do students actually communicate to each other via cell phones? A surprising number told me "nothing serious," but many said that they inform friends about what has recently happened, what or who has made them angry or happy, about part-time jobs, and, very commonly, when and where to meet.

To the question "What is the content of the messages you send and receive?" many students answered that it consisted of arranging dates and outings and "telling my friends that they're late for an appointment." Also common was "What are you doing now?" (*ima nani shiteru no*), "Where are you now"? (*ima doko*), and "Let's play" (*asobō*). Next in frequency came: "reporting on recent news" (*kinjō hōkoku*), "what I did on a certain day," or "events of the day." More personal matters were also a topic of conversation: "reporting about myself" (*jibun no hōkoku*), "personal things" (*kojin teki no koto*), "things that are troubling me." Also very common was: *aisatsu* (greetings, e.g. "How are you?" "Hello," "Good morning," and "Good night"), "daily events" (*nichijō no dekigoto*), "what happened during the day" (*ichinichi okotta dekigoto*), "daily things" (*nichijō teki koto*), "ordinary things" (*futsū no koto*), "everyday conversation" (*fūdan no kaiwa*), "chitchat" (*zatsudan*), and "gossip" (*seken-banashi*). Many students said their conversations were *kudaranai* (trivial, silly, useless) and that they discussed *karui mono* (unimportant things) and engaged in *karui kaiwa* (unimportant talk) or *chottoshita yōji* (unimportant business). According to one survey of university students, the subject of the mail was "everyday events" 56 per cent of the time, "while 43 per cent of mail was likely to be arranging times to meet up. Worries accounted for only 7 per cent of mail the students sent."[17]

Some students said they engaged in "romantic conversation" (*airen no hanashi*) and exchanged "love messages" (*ai no messēji*). "Unlike on the telephone, messages remain on the screen, like 'I love you,' so I can look at it later, and it makes me happy." A few mentioned how handy their cell phones were during "emergencies." Some students said they search for information about their hobbies and leisure. A few students complained about receiving *chiēnmēru* (chain letters) and *iyagarase* (nasty) messages. Other common complaints were "slow service" and "useless messages." More than one said "I want the prices to come down."

To the question, "About how many times a day do you send or receive e-mail messages?" a very small number said on average 1–2. The most common responses ranged from 5–10 and 10–20, though some said 30–50. When asked "About how much do you spend on your cell phone per month?" students gave varying responses. On the low end, 1,000–5,000 yen was not unusual, though 4,000–5,000 was relatively common.[18] Estimates in the 10,000, 10,000–15,000, and 10,000–20,000 ranges were also common. On the high end, not a small number (20–25 per cent) said 20,000, 30,000, or even 40,000. One student said "depending on the month, I'll pay anywhere from 5,000 to 20,000 yen."[19]

Many students stated that the use of e-mail had improved their lives, though usually they could not give very exact reasons as to why. However, some did say: "I can converse on a train without bothering others," and "I don't come to class late as much as I used to, since incoming messages wake me up." Cell phones have "definitely made my life enjoyable," enthused one female student. In the words of one student, "In times like these, if cell phones and e-mail disappeared, wouldn't the world fall into chaos?"

INDIVIDUALIZATION

Individualization is useful as a general concept for how the forces of modernity – industrialism, capitalism, statism and science – plotted the movements, placements and behaviors of the individual, thereby demanding an increasingly specified, itemized and categorized conception of the person. Such operations produce and control students, laborers and other roles required for modern economic production, citizenship and kinship systems. Turner[20] uses the term "individuation," which Shilling defines as "a set of practices by which individuals are identified and separated by marks, numbers, signs and codes which are derived from knowledge of the population and related to the establishment of norms. Individuation also facilitated the control and surveillance of people."[21]

The views of Turner and Shilling, it should be pointed out, only focus on what may be termed the "external" dimension of individualization, or on how the individual is constructed as a subject from the outside – as it were – by sociopolitical and economic forces. Another way to view individualization is as a type of subjectivity constructed from the "inside out." In other words, the more social pressures demand compliance from an individual, the more psychological "internalness" emerges as a response to individualizing outside forces. External socialization and internal psychologization, then, are two sides of the same coin. The forces of individualization, it can be argued, set the groundwork for the invention of various positionings of the person, such as "individualism" (belief in the inherently autonomous self with inalienable property rights), "citizen" (personalized units for mobilization demanded by national states that emerged in the nineteenth century), the "subject" (the focus of research psychology and the nexus of unconscious desires), and "character" (protagonist of the novel). There is one dimension of individualization that is much more difficult to theorize than its more obvious developments in the political realms. This aspect of individualization concerns how economics, more specifically, capitalist consumerism, has cultivated a definition of the self which possesses an individualized interiority. This personal interior space not only requires expression and exploration but is also filled with demands, desires, and cravings that need satisfaction.

As already mentioned, individualization may be either positive or negative. Individualization may be thought of as positive, or "personalized individualization," which means to affirm one's identity, explore one's individuality, or perhaps even resist perceived social control. Personalized individualization may be a playful response, derisive reaction or frantic backlash to dominant sociopolitical forces. Such counteractions may be spontaneous and semiconscious, studied and intentional, or calculated and conscious. Individualization may also be thought of as negative, – or "atomized individualization." But whether positive or negative, individualization increases "psychological interiorization," encouraging the view of the individual as a walking encapsulated world of complex dynamics.

PERSONALIZED INDIVIDUALIZATION

Personalized individualization as it relates to cell phone use has three aspects that deserve attention: (1) as an endeavor to be fashionable (or one's individuality – *kosei*); (2) as an attempt to more clearly articulate one's sentiments (*kimochi*); and (3) as an effort to acquire one's own space (i.e. privacy). These issues (which are my own categories, i.e. not provided by students), concerning individualization, individuality, and internal psychological processes, all converge on interiority.

Individualization as fashion

According to Otsuki Takahiro, a commentator on Japanese culture, "People are using cellular phones not just as a communication box but as a tool with which to express themselves."[22] As both Hjorth and Holden and Tsuruki (this volume) illustrate, methods of personalizing phones and expressing one's individuality include choosing colors, phone straps, sizes, shapes, functions, payment packages, ringing tones and melodies, and background colors for screens, as well as selecting all sorts of *akusesarii* (accessories). "Character goods" – small, colorful figurines that are hung from phone straps – are also another piece of material culture that allows for personalization (with Disney characters being very popular). Stores stock a multitude of cellular phone straps (some individuals collect the straps themselves); ditto for the tiny character dolls which are attached to straps. One observer describes *chakumero* as "just another example in which [young people] can express themselves. It's a logical extension of the way kids (and older folks, too) personalize their cellular phones by decorating them with *puri-kura* (print club) stickers (small pictures made in photo-booths that are traded and used to decorate surfaces) and trinkets looped into the phones' carrying strap."[23]

As reported in an article entitled "Adding personality to cellular phones," decorative plastic sheets or coating (called *tere-shiiru*, from "tele" and "seal") are put on phones. "The coating peels off easily, so it can be changed at whim."[24] Typical designs include floral patterns, fruits, polka dots, stripes, or animals. The manager of a "nail-art salon" that also paints designs on cell phones and other accessories explains that, "People want a cellular phone that no one else has." One client had his phone painted "because it's cute." Many clients are in their thirties and forties, and male customers, "like businessmen, say they want to show the phones off to their subordinates."[25]

As Hjorth (this volume) also argues, cell phones and similar devices can be characterized as "techno-cute."[26] Techno-cute merges the two esthetics/ethics of cold, complicated contraptions, unfeeling objects, and serious efforts and labor, with warm, understandable social relations, emotional connections, and playful times and leisure. After all, the "Internet scares people. It makes people think they need a PC, a modem, an ISDN line, it costs too much," says Yukiko Takahashi, a manager at Bandai.[27] Techno-cuteness, by coating complex devices with a patina of bright colors, attractive shapes, and non-threatening, even cuddly creatures, firmly positions technology within the social nexus. One student summed up this esthetic by describing her cell phone as "*kyūto de kawaii, petto mitai*" (like a pet, it's cute and adorable). Evidence of the synergy arising from techno-cuteness is apparent in how DoCoMo has teamed up with Bandai (maker of the famous Tamagotchi), which provides content (e.g. cartoons, characters, and other cute images). Individuals are able to download pictures of pop stars, sports figures, animated characters, and cartoons that move when the phone rings. These characters are e-mailed back and forth between friends. "There's even a Hello Kitty for every day of the year."[28] "The Bandai–DoCoMo link has been a match made in marketing heaven, marrying two things Japanese adore above all – cool and cute."[29] One observer has even gone so far as to say that without "Hello Kitty, games, horoscopes and the dating services that have all become *i-mode* fixtures, DoCoMo would have had nothing to sell but a fancy paging service."[30] Indeed, the "biggest money-making sites in Japan are for downloading tinny electronic melodies and cartoon images."[31]

Another example of techno-cuteness involves a popular e-mail software called PostPet, in which "cyberspace pets organize and deliver e-mail for their owners."[32] These "pets" have minds of their own and "write in 'secret diaries' and send you letters while 'on the job.' Sometimes they post unsolicited mail to your buddies." These computer creatures "can betray, rebel, or be sickeningly nice." They can also earn money doing part-time jobs to buy themselves treats or furniture for their own rooms. According to one young man, "I absolutely adore them, even though I'm a man. They're just so cute."[33]

Individualization as expression of one's feelings

According to students, "communication works well" and "it's now a great era because communicating is so convenient" via cell phones. Indeed, not a small number of students said something along the lines of how they saw cell phone usage as "deepening relations with friends" because "I send my feelings quickly to others" (*kimochi o sugu aite ni okuru*), "I can't transmit feelings through the written word anyway [i.e. formally written letters], so I can transmit my feelings (*kimochi*) to others without apprehension," "I can send things that I can't say on the phone," and "[spoken] words are not used, so that the meaning of the message becomes deeper." Some students explained that, "If I can't say it, I can write it," or "I can't use words well to say what I want, but e-mail allows me to say what I want directly." Feelings, as an aspect of psychological interiority, are more easily shared among friends. In the words of one student, "If I didn't have a cell phone to communicate with, my friends would become strangers." One student explained that cell phones are "very convenient for people who have a hard time speaking" while another said that "I can say what I want to because I can't see the other person's face." However, one student cautioned that "there may be some things that can only be conveyed through the spoken word."

When asked if the use of e-mail had expanded their circle of friends, a few said that they saw little expansion. However, the majority stated that their number of friends greatly increased because of cell phone e-mail use (one student believed there was a different relation between cell phone e-mail messaging and his social life: "On the contrary, as my number of friends increases, my e-mail increases"). Some students have more than one phone, each phone apparently corresponding to different roles and personas – e.g. concerning business, friends, family, romantic affairs – that suggest the demands or burdens of increased interiorization.

Individualization as personal space

One last example of personalized individualization concerns the theme of increased privacy or, as if to emphasize its rarity by adopting a foreign-sounding expression, "*puraibēto*" affairs. "Everyone has their own individual life, and we should be allowed to communicate with our friends without interference," said one student. Cell phones "allow us to lead our own lives," they "give us personal space" commented another. Such comments must be appreciated within the concrete contexts of everyday life in Japan: cramped housing, lack of open spaces, and more surveillance at the workplace and in educational sites by authority figures. One male student explained that "my parents don't know who I'm sending messages to so I feel at ease," and many students often discussed the private space afforded by cell phones as if these devices created their own world for them or increased their psychological interiority, or as one young woman expressed it, cell phones are "a personal little telephone booth." One student told me how she wants "to send messages to my professors as an individual," apparently because she wants to ask the professor questions without other students knowing about it.

ATOMIZED INDIVIDUALIZATION

In Japan, some products seem specifically aimed at individuals who spend many hours by themselves and want something portable to be used within tight personal space for leisure purposes. Urban Japan's commuting network has not only produced its own spaces but also its own culture in which individuals, while traveling to offices, factories and schools, occupy themselves with Walkmans, hand-held computer games, Tamagotchi (small "electronic pets"), manga (comic books) and, if one considers calling friends for idle talk, pagers and portable phones (though of course many business people have more "serious" reasons for having such phones). Such objects of material culture have transformed business practices, leisure, subjectivity, interpersonal communication habits and fashion, supposedly adding more convenience, efficiency, and intensity to such endeavors.

Consider cell phones. Though many believe that cell phones connect individuals, widen one's circle of friends, and improve one's social life, others claim that they atomize human relations. Face-to-face networking itself is mimicked and social relations are simulated:

> E-mail reduces people's ability to express themselves and does away with words indicating social status. Mobile telephones, meanwhile, may have increased person-to-person communications quantitatively, but they have had the opposite effect in qualitative terms. Many young people use the devices to jabber away endlessly. In the past, by contrast, conversations on shared land-line phones tended to be more formal and succinct. And possibly more intelligent.[34]

One student said "Some students brag about getting 100 or 200 *mēru-tomo* (mobile phone e-mail friends), whom they communicate with only through e-mail ... but I always wonder if just exchanging short good-morning messages does anything to foster real friendships."[35] According to one young woman, "When many numbers are stored in the memory, I'm glad, feeling as if all of them are interested in me." However, "About 100 of these so-called friends registered in her cell phone are 'cleaned up' each month to be replaced by new ones."[36] Because individuals can change numbers so easily, relations become shallow, and as technology evolves, "people to people communication" (*ningen tai ningen komyūnikēshon*) will disappear. One student described this as "scary." Moreover, because individuals can communicate readily with friends outside the home, family relations may suffer. Students claimed they make friends over the Internet (or cyberpals). However, they also claimed that it is not unusual to never meet *mēru-tomo*. As Holden and Tsuruki (this volume) discuss, there are also many Internet matchmaking Web sites.

There are serious concerns about the heavy dependence on *mēru-tomo* and consequent "addiction" to cell phones. Indeed, some report a panicky feeling if separated from their cell phones. This has been called *keitai-izon* (mobile phone addiction). A survey by DoCoMo of 1,000 men and women found that though 50 per cent liked having *keitai denwa*, the same percentage "also admitted to feeling anxious when they were without their phones."[37] In the opinion of one observer, an obsession with "collecting" *mēru-tomo* "seems to reflect their fear of becoming isolated from others."[38] Professor Tomita Mitsuyasu of Sapporo Gakuin University coined the term *minna-bochi* (persistence in being together) to describe how young people, because they feel neglected, "prefer being linked with as many people as possible. However, they dislike deeper ties with others because such ties are energy-consuming, so they want to keep adequate social distance in their relationships with others for their own convenience." Ironically, though they are connected to many people, they can be at the same time *hitori-bochi* (totally alone).[39] Professor Tomita explains that today's youth "are divided into small groups based on tiny differences. If they can't find a place to stay in that small world, they can't have any place to get away." Therefore, positioning themselves in a world of *minna-bochi* "created by cell phones may be an act of self-protection to prevent being hurt by others or being 'cleaned up'" (removed from a list).[40] Or phrased differently, compulsively connecting one's self to others may be a form of resistance – however oblique – to the pressures of negative individualization (e.g. being incessantly broken down into small units at educational and labor sites).

In addition to encouraging superficial social relations, too much cell phone use leads to the invasion of privacy (cell phones "allow us to peek into other people's lives," claimed one student), causing some to lose their freedom when friends send messages constantly and expect instant replies. "Why do people think they need cell phones in the first place?" asked one student, "after all, there are public phones everywhere."

The constant messaging, especially if it involves trivial information, quickly becomes a nuisance, though many feel they cannot ignore incoming calls and end up always answering. Others describe feelings of being tied down by a small device. In train and subway stations, there is evidence of how new communicative technologies have drawn attention to the need for manners: signs read "Thank you for not using your portable phone." On trains one can now hear announcements such as "Passengers are requested to refrain from using mobile phones in trains" or "Thank you for not using your portable phone." Signs are posted warning how cell phones might affect pacemakers. In the university classroom pagers and sometimes phones go off. "Japanese in particular need them, because they are so busy," a student pleaded after I reprimanded her for talking on her mobile phone during class. One student told me that "I won't answer it if it rings in a crowded train" and others explained that if they do answer their phones on trains, they "speak in a low voice." One student explained that the "tradition of conveying one's feelings through writing letters is being lost" and cell phones "take away the chance to write regular letters."

For some, the cornucopia of information afforded by cell phones is illusory, a type of "miniaturized window shopping." "What good is all this information if you don't have the money to buy anything?" asked one student. A few students believed that cell phones are a "waste of money." Other complaints included how cell phones "take away from my time to sleep," these devices "haven't improved my life much," "I always have to worry about my cell phone," "When messages come they interrupt my concentration while I'm studying," "I've sent the message to the wrong person and ended up in an argument," "Because I'm on my cell phone all the time, I have fewer opportunities to meet and talk with people," "All that surfing causes me to go to bed late." One student simply stated that the explosion of cell phones "isn't a very good trend."

Internet-connected cell phones are part of a continuing technological trend that is increasing individualization and interiorization. In the late 1970s *heddo-hon-zoku* (head-phone addicts) appeared. Then came individuals devoted to video games. "These examples show how media at the time symbolically reflected the characteristics of young people who liked to stay in their own inner worlds and shut out communication with others."[41] Pagers, popular in the mid-1990s, allowed young people to connect with others by exchanging messages with *beru tomo* (bell friends).[42] Now exchanging messages is more direct, immediate, visually based and colorful.

CONCLUSION: CYBERSPACE, CYBERSTRUCTURE, AND PSYCHE

Through an examination of the concrete practices associated with cellular phone e-mailing, I have attempted to explore how the cyberstructural environment configures the use of cellular phone e-mailing and Web browsing by Japanese university students. More generally, I have attempted to delineate some linkages between individualization (disciplinary practices of being numbered, organized, ordered, uniformed, examined, screened, and shunted through the education and employment system), individuality (personal uniqueness), interiority (subjective experience and sentiments), and the Internet (technology of networking and connecting). Linkages between individuality and interiority and commodification are evident in what I have elsewhere termed "consumutopia," the desire-laden consumerist visions fueled by late capitalism, pop-culture industry, and ever-increasing individualization, in which acts of consumption take place in more and more spaces, both external and internal to the person.[43] Indeed, besides being an abbreviation, DoCoMo, is similar to the Japanese word *dokomo*, meaning "every-where." "Even though I'm at a distance from some place, I can still communicate with whomever's there," explained one student.

Despite such almost Utopian imagery, Rimmer and Morris-Suzuki warn about the overly sanguine associations of "cyberspace" that imply a world without the constraints of social structure in which individuals are free from the gravitational pull of wealth, social status, and power. This is why they suggest the existence of a "cyberstructure" in which the accumulated weight of existing technology, knowledge distribution, and social structures limits the use of information networks.[44] More concretely, cyberstructure may be described as constituted by: (1) economics (corporate pursuit of capital accumulation, advertising spin, consumerism, and commercialism); (2) state (investment priorities, infrastructural development, and educational policies); (3) social variables (class, gender differences, access to information and wealth); and (4) personal patterns of use and consumption (preferences, levels of individual affluence, expertise).

Another aspect of cyberstructure concerns the subtle and complex interplay between technology and psychology. Admittedly, this aspect of cyberstructure is difficult to conceptualize, but if technology does not develop in a social vacuum, by the same token, psychology is not unaffected by technological developments: writing (social control and administration), printing (reproduction and dissemination of information), and computerization (rapid and efficient manipulation of data) have all historically increased speculative scientific thinking, hypothetical reasoning, and abstractness of thought (as well as accelerating economic rationalization). Cyberspace – or cyberstructure – having been built on earlier technoscientific paradigms, has vastly increased the speed of information processing, enhanced data dissemination, swelled storage capacities, and expanded the imaginary spaces of the individual psyche as well as consumerist desires and demands.

Cell phones have introduced millions to the possibilities of e-mail and dramatically increased the number of Japanese who are connected to the Internet (at least its *i-mode* version). There are great potentials to position oneself within a global community and to acquire more access to information. However, some wonder about the actual value of the downloaded content, and criticize cell phones as mere gimmicky gadgets, "toys" with little educational or social value. Though the popular media often portrays the youth of Japan as information-seeking, networked into an ever-expanding circle of friends and acquaintances, and connected to a wider world, the actual state of affairs may not be as edifying or exciting. Moreover, another aspect to the criticism of cell phone use concerns the notion – no doubt part truth, part exaggeration – that Japanese youth, reacting to the pressures of too much individualization, become persons who act as if they are psychologically sequestered (called *otaku*, or "nerds"). This turn toward privatism, or an intense interiority, encourages and is encouraged by technology. Moreover, some even see a linkage between the possession of communicative devices and the loss of moral fiber: a rise in bad manners, impoliteness, and crime.[45]

However one decides to evaluate cell phone usage, it cannot be ignored that these communicative devices are part of self-presentation tactics and are regarded as a type of fashion by many. From the perspective of the user/consumers of technology, they may actually counter high-pressure individualization since they articulate one's own individuality. These processes are also visible "in the attire of school children who, though compelled by society to don uniforms, nonetheless demarcate themselves from others via accessories such as buttons, pins, shoelaces and bows. Discreet statements of difference for those wishing to be considered as discrete statements."[46]

Is there anything particularly "Japanese" about Internet-connected cell-phone usage in Japan? Another paper is required to adequately answer this question, but for the sake of argument, a few words of comparison between the US and Japan in regard to Internet-connected cell-phone usage are in order to illustrate how cyber-structures might encourage/discourage Internet-connected cell-phone use. One salient cyberstructural variable in Japan is the mammoth and semi-state-owned Nippon Telephone and Telegraph, which is allowed to charge relatively high rates for telephone bills. Such prices do not encourage the use of modem-linked computers, and indeed, in Japan, only one-third of households possess such computers while in the US about half do (it also takes more time and effort to connect a computer to the Internet than to use a cell phone). At least from a cost perspective, cell phones are more attractive to the average Japanese for connecting to the Internet. Another factor is the infrastructure of commuting in a highly urbanized society: public transportation plays a larger role in Japan than it does in the US. In Japan, individuals are accustomed to long commutes and spending time on crowded trains, subways, and buses, ideal sites for hand-held communi-cative devices.

In spite of any national differences, however, the popularity of Internet-capable phones in Japan does reflect, in its own particular, localized way, a global trend apparent in societies where capitalist consumerism and technological advances compel the individualization and interiorization of the person.

NOTES

1   N. Elias, *The History of Manners: The Civilizing Process*, vol. 1, New York, Pantheon Books, 1978 [1939]; M. Foucault, *Discipline and Punish: The Birth of the Prison*, New York, Vintage Books, 1979.
2   *Kosei*, not "individualism" (*kojin shugi*) as understood in what may conveniently be called the "West."
3   Y. Kageyama, "Cellphone makers focus on fun," *Japan Times*, 5 March 2001, p. 17.
4   Japan's telecommunication industry faces stiff competition from Finland's Nokia, Sweden's Ericsson, and Motorola in the US. (After the handset market, other big telecommunications players are Matsushita, Mitsubishi Electric, NEC, Toshiba, and Sony).
5   A. Ghosh, "Phone Wars: Episode3G," *Time*, 27 November 2000, pp. 43–48.
6   PDC, widely used in Japan, is a second-generation wireless service that uses a packet-switching technology in which messages are split into packets of data for transmission, then reassembled at their destination. GSM, the standard most commonly used in Europe and Asia (but not in the US), is a world standard for digital cellular communications using narrowband TDMA (Time Division Multiple Access) that allows up to eight calls at a time.
7   Y. Kageyama, "Cellphone makers focus on fun."
8   "Subscribers to *i-mode* top 20 million," *Japan Times*, 6 March 2001, p. 12.
9   K. Sawa, "Mobile phones silence chatty students," *Mainichi Daily News*, 22 October 2000, p. 12.
10   "Mobiles big among high schoolers: poll," *Japan Times*, 26 December 2000, p. 3.
11   "One in four Tokyo elementary, junior high kids has a cellphone," *Japan Times*, 7 March 2001, p. 9.
12   T. Larimer, "What makes DoCoMo go," *Time*, 27 November 2000, pp. 50–4.
13   "Online," *Chronicle of Higher Education*, 19 January 2001, p. A29.
14   K. Sawa, "Mobile phones silence chatty students."
15   K. Sawa, "Mobile phones silence chatty students."

16 "Cell-phone users hang up on notions of time," *Daily Yomiuri*, 21 July 2001, p. 8.

17 K. Sawa, "Mobile phones silence chatty students."

18 At the time of writing, $1.00 = 132 yen.

19 According to Y. Kageyama, "Internet-capable phone firms target US," *Japan Times*, 18 January 2001, p. 9, on average, DoCoMo subscribers pay about 10,000 yen per month.

20 B.S. Turner, *Religion and Social Theory*, London, Heinemann Educational Books, 1983.

21 C. Shilling, *The Body and Social Theory*, London, Sage Publications, 1994, p. 78.

22 Cited in K. Takahara, "Art of communication or just cute? Nail salons ringing up cell phone profits," *Japan Times*, 18 September 1999, p. 3.

23 S. McClure, "Newest cellular phone fad rings a bell," *Daily Yomiuri*, 15 January 2001, p. 7.

24 M. Hamasuna, "Adding personality to cellular phones," *Daily Yomiuri*, 3 April 1999, p. 8.

25 K. Takahara, "Art of communication or just cute?"

26 See B.J. McVeigh, "Japan's esthetic of techno-cute: marrying the futuristic and the cuddly" (unpublished manuscript); see also "Creature comfort," *Daily Yomiuri*, 25 March 1999, p. 16; and "Virtual pets II," *Japan Times*, 25 March 1999, p. 11.

27 T. Larimer, "What makes DoCoMo go."

28 T. Larimer, "What makes DoCoMo go."

29 T. Larimer, "What makes DoCoMo go."

30 T. Larimer, "What makes DoCoMo go."

31 Y. Kageyama, "Internet-capable phone firms target US," *Japan Times*, 18 January 2001, p. 9.

32 J. Hanna, "Sequels living up to expectations," *Daily Yomiuri*, 21 January 1999, p. 14.

33 T. Large, "Forget e-mail, get into twee mail," *Daily Yomiuri*, 25 February 1999, p. 7.

34 G. Botting, "IT revolution bastardizing the Japanese language," *Sunday Mainichi*, 19 November 2000, p. 7.

35 R. Kawabe, "Finding the right approach to helping children," *Daily Yomiuri*, 2 December 2000, p. 7.

36 Y. Nagamine, "Isolation fears lead to phone addiction," *Daily Yomiuri*, 13 July 2001, p. 6.

37 T. Itō and N. Chisako, "Attachment to mobile phones reaching point of addiction," *Daily Yomiuri*, 8 July 2001, p. 14.

38 Y. Nagamine, "Isolation fears lead to phone addiction."

39 Y. Nagamine, "Isolation fears lead to phone addiction."

40 Y. Nagamine, "Isolation fears lead to phone addiction."

41 Y. Nagamine, "Isolation fears lead to phone addiction."

42 Y. Nagamine, "Isolation fears lead to phone addiction."

43 B.J. McVeigh, "How Hello Kitty commodifies the cute, cool, and camp: 'consumutopia' versus 'control' in Japan," *Journal of Material Culture*, vol. 5, no. 2, 2000, pp. 225–45.

44 P.J. Rimmer and T. Morris-Suzuki, "The Japanese Internet: visionaries and virtual democracy," *Environment and Planning*, no. 31, 1999, pp. 1189–1206.

45 For instance, "Poll: Japanese youth have more pagers, less respect," *Daily Yomiuri*, 11 April 1997, p. 3.

46 T.J.M. Holden, "Surveillance – Japan's sustaining principle," *Journal of Popular Culture*, vol. 28, no.1, 1994, pp. 193–208.

3 *Deai-kei*

Japan's new culture of encounter

Todd Joseph Miles Holden and
Takako Tsuruki

INTRODUCTION

Deai-kei means "type of encounter." The term is specific to the Internet, reflecting a class of communication, within which two broad types exist: *deai*, where the aim is to meet another person either virtually or in the flesh; and *mēru-tomo*, which literally means "mail friend" – a twenty-first-century pen pal.[1] A cousin of chat, *deai* provides protected encounters which, nonetheless, enable users to drop the veil of anonymity and pair up.[2] The ultimate aim for many is to transform electronic encounters into face-to-face interchange. As a result, *deai* is strongly associated with issues of dating, companionship, sexuality and romance.

In this chapter we'd like to present a glimpse of this mediated world. We do so for a number of reasons. First, as a form of sociation, formal Internet-based encounter sites have only been in Japan since the millennium.[3] Second, as an Internet subculture, *deai* is commonly considered to be among the murkiest. It seems a day doesn't pass that the press and television "Wide Shows"[4] don't report another *deai*-connected crime, another salacious story of unfortunate encounter. What we will show, though, is that *deai* is more than an abject domain. Against the widely held stereotype, Internet-based intercourse is much more than a forum for frivolity, immorality and danger. Our research suggests that there are many constructive uses and more possibilities to *deai* than those reported in the media that prompt the public hand-wringing and pontification about a Japan rotting to the core.

A third, related point we wish to make is that, although *deai* is a tool for sociation, it is also an important instrument in the mediation of identity, the exploration of self, the management of emotions, and the arbitration between the individual and the larger social world. For this reason, *deai* has both practical and pedagogic uses. It is not only a tool employed by Japanese to enhance their everyday lives; it is also a phenomenon that is instructive to analysts who wish to better understand the nature and direction of contemporary Japanese society. *Deai*, in short, has sociological origins, engages sociological operations, prompts sociological outcomes, and stimulates sociological questions.

One reason, perhaps, for *deai*'s sociological bent is that it is highly intertwined with developments in technology – in this case the cell phone that has become a staple of the faddish, mobile, mediated, gadget-centered, youth-oriented, licentious lifestyle of contemporary urbanized Japan. This development, in turn, has articulated well with other changes in Japanese society – in particular the management of romantic relationships. In the past, encounters[5] often transpired within an institutional orbit: through socially-sanctioned *omiai* (arranged marriages presided over by an intermediary); within the context of company outings; via *gurūpu kōsai* (collective get-togethers in which acquaintances or club members would explore potential couplings within the non-binding framework of a group); and via the "club scene" of the 1980s and 1990s, in which singles danced in groups and tended to pair up based on shared activities in a fixed setting.[6] Now, thanks to cell phones, individuals are able to operate in virtual isolation, freer of weighty social structure and claustrophobic external surveillance. At the same time, as we shall consider further on, *deai* affords some of the advantages of the institutional orbit – namely trust and self-defense. In this way, dual benefits are provided: individually-established and managed social connections, as well as a modicum of security.

Before seeing how this is achieved, let's consider a bit more about the technology and its social context.

THE WORLDS OF *DEAI*

Although *deai* is an Internet-based forum, in Japan it is frequently through *keitai* – the cell phone – that *deai* is experienced. It is not uncommon to read analyses attributing this to the nature of Japanese society or the Japanese.[7] Of this ilk are claims that *keitai*'s unprecedented popularity[8] is due to the extensive waiting time and extended commuting in Japan. So, too, are assertions that *keitai* reflects the cultural predilection to miniaturize,[9] the rampant information-orientation of the society, and the commitment to fad and fashion among Japanese.[10]

Keitai's popularity has coincided with the appearance of a second social innovation: Internet-capable *keitai* – what is known as *i-mōdo*.[11] The relationship might be spurious. Nonetheless, consider the following sets of figures. In a 1999 survey of Organization for Economic Co-operation and Development (OECD) nations, Japan's ratio of 44.9 cell phone users per 100 people was good enough for only eleventh place.[12] As of the year 2000, however, *keitai* use among Japanese approached 80 per cent.[13] Patterns for Internet use differed only slightly. While Japan's raw number of 1999 users was second to the United States, scaled to the population, its use rates (of 14 per cent) were closer to fifteenth place.[14] With the advent of *keitai*-enabled Internet, however, the figures grew considerably. By early 2000, the percentage of Japan's Internet users ascended to thirteenth place (one in five, or 21.4 per cent of the population). According to one study, within two years of its debut the total number of subscribers to *keitai*-Internet services had more than doubled the total for dial-up connections on land-based telephone lines (36.9 million to 17.25 million).[15] Well before that, *keitai*-enabled Internet had surpassed the PC as the number-one platform for e-commerce in Japan.[16] In short, *keitai-i* has quickly developed into the preferred means of Internet activity in contemporary Japan.

Certainly, pragmatic factors, such as cost, portability and ease of use have fueled this preference. However, it is also possible that psycho-social factors have spurred the boom. Kogawa[17] forwards a thesis of Japan's current predilection toward "electronic individualism;" in our research we discern in *keitai/deai* not only the opportunity to disengage self from society, but also the desire to create distinct personal enclaves. Certainly as one delves into *deai*, intimations of individuation emerge. At odds with the pervasive theorization that Japan is a homogenous society, the simple act of accessing *deai* suggests specialization and difference. For instance, a keyword search on a popular search engine begets the following message: "31 Yahoo! Categories, 420 Yahoo! Registered sites."[18] From there, users select a thematic area, within which particular sites (and, presumably, users of similar interest) are located. Among Yahoo's categories are: "actors, talent," "comics and *anime*," "event," "illness," "performing arts," "sightseeing," "seniors," "disabled" and "gay, lesbian, bisexual."

Admittedly (and consistent with the widespread media portrayal) a spate of sites are grouped under the moniker "adult." Among the hundreds are "Secret Lover's Café," "LOVE ATTACK," "*rabu sāchi*" (love search) and "BOY MEETs GIRL." Nonetheless, it would be a mistake to read *deai* monolithically. Scores of non-sexual encounter sites can be found, including "*kiizu pāku*" (Kid's Park) – for "children under 15 interested in presenting their creations;" "A World Bridge: *aru sekai no kakehashi*" – aimed at promoting cultural exchange between English-speaking and Japanese mail-friends; and "*jūnanasai no chizu*" (map for 17-year-olds) – for teens beset by problems associated with growing up.

Furthering the notion that *deai* is something other than a simple assignation-finder is the following statistic: of 300 users surveyed in an online poll, those with the goal of actually meeting another person via *deai* amounted to no more than 8 per cent. The other 92 per cent broke down as follows: making a friend of the opposite sex, 9 per cent; making a friend of the same sex, 6 per cent; making a mail friend of the opposite sex, 15 per cent; making a mail friend of the same sex, 15 per cent; talking to people who have the same hobbies, 31 per cent; mere curiosity, 14 per cent; and unspecified aims, 2 per cent.[19] That same poll found that only 34 per cent of *i-mode* users actually accessed *deai*. Of that third, 66 per cent reported employing the service just once; those doing so twice amounted to 22 per cent. And, in terms of meeting fellow users, 39 per cent claimed never to have met anyone, while 37 per cent had met less than three other users; those claiming to have met four to six users stood at 11 per cent; seven to ten users, a scant 6 per cent; and eleven users or more, 7 per cent. In short, curiosity appeared more overwhelming than sustained use.

A second online poll of *i-mode* users provides very similar results.[20] While nearly half (47 per cent) reported engaging in *deai*; 27 per cent had yet to try, but intended to; 26 per cent hadn't and didn't believe they would. Of actual users, only one-third (35 per cent) reported their aim was searching for a "lover," but an equal number (35 per cent) hoped to make "mail friends;" 16 per cent were looking for mere "company" or association; 5 per cent used *deai* for "*gōkon*"[21] and 4 per cent for "chat."[22] Still, for *i-mode* users, *deai-kei* was a popular activity. At 16 per cent, it ranked fifth behind "free news" (47 per cent), "town information" (26 per cent), "traffic information" (18 per cent) and "gourmet information" (17 per cent). Notably, on that list, *deai-kei* was the only sociative activity. In terms of gender, men outnumbered women two-to-one (65 per cent to 35 per cent) in their reported *deai* use.

If one believes the statistics, then, interaction via machine was far more common than in-the-flesh meetings arranged through *deai*. Nonetheless, the actual meeting is what holds a powerful grip over public imagination. In the next section we review two types of stories that occur through *deai* encounters – the negative ones reported in the media, and more positive encounters that often go unpublicized.

COMPETING VISIONS OF *DEAI*

The notion of *deai* as a haven of immorality, subterfuge, criminality and risk has ample visibility. For instance, the National Police Agency (NPA) convened a press conference to announce that the number of crimes associated with *deai* sites hit 302 for the first half of 2001. This compared with 104 incidents for all of 2000. Typical were incidents such as the following: (1) in Kumamoto Prefecture, a man and woman became acquainted through *deai*. After arranging a date he stole her ATM card and withdrew cash from her bank account. (2) In Nagano, three men – aged 29 to 53 – and five girls – aged 16 to 17 – were arrested on charges of prostitution. Their business, an outcall service called "Pastel," was run through a *deai* site. Its genesis stems from the night one of the suspects returned from a *deai* liaison and announced to his companions: "if we were to start a prostitution group (using *deai*), we would strike gold." (3) In Osaka, police uncovered a scheme to distribute obscene images in real time over cell phones. Customers subscribed to the service by accessing a *deai* site. At the time of discovery the service boasted 35,000 registered members, each paying access charges of 80 to 100 yen per minute. Sales were estimated to have reached 300 million yen. (4) In Tochigi Prefecture an 18-year-old boy stabbed a Saitama housewife in her home. The two had met three months before through a *deai* site. After exchanging mail for a time, they met and became intimate. In court the boy claimed that he attacked at his lover's behest. The woman had become suicidal, he alleged, when she began to fear the damage resulting from public exposure of their affair.

Despite all the negative publicity, our research discerns an alternate dimension to *deai*. Encounters can net personally – and socially – beneficial outcomes. Consider the case of Nakashibetsu, a farming community in the eastern tip of Hokkaido, Japan's northernmost island. The nation's 896th largest city, it has a population of 22,300. As far back as 1973, town elders were concerned about the population. Mechanization had reaped great profits, but also generated new opportunities and problems for the community. Women were now freed from the field, they had spare time and their families had disposable income. Increasingly, they were leaving the community to live independently or enter college. With their departure, the chance to replenish the local human stock was also disappearing.

To address this problem, Nakashibetsu created a "third sector organization" aimed at recruiting prospective brides. "Our success was mixed," admits Mr. Kajitani, chief operating officer, "until we created a home page. Thereafter, we really saw dramatic results."[23] Listed in search engines as a *deai* site, Nakashibetsu was suddenly flooded with information requests from all over Japan. The heightened visibility means that Kajitani spends considerable time playing match-maker: screening the many women who apply through the *deai* site, trying to forge a fit with the men who will attend the biannual, three-day pre-marital mixers that Kajitani organizes. From the ninety-six men and women brought together through *deai* in the past four years, thirty-three marriages have resulted; there has yet to be a divorce. Similarly, no problems of stalking or violence have occurred. At a time when rural communities throughout the country are losing population in enormous number, Nakashibetsu has reversed the trend. In Kajitani's opinion it is due to *deai*. This Internet platform has served as a powerful, productive tool for social engineering.

DECODING *DEAI*

Such cases are striking, but how representative are they? Just what is *deai* for the average, everyday user? To gain a better sense, we conducted interviews,[24] then coded the data inductively. Taken together, our findings suggest that *deai* is a mechanism – the situs and instrumentality – for sociological phenomena with wide-ranging, complex implications. Ten, in particular, stand out, each pertaining to negotiations within and between self and society. Collectively, these phenomena shed light not only on *deai* as a medium for social encounter, but also the nature of Japanese society, the character of some of its people, and the status of this research topic vis-à-vis certain central concerns in contemporary social science. The themes we find particularly salient in our data include: stigma, appearance versus reality, identity, emotions, intimacy, personal defence, glocalization, fragmentation/ reintegration, uses and gratifications, and simulation. We consider each, in turn.

Stigma

One difficulty we faced in this research was simply locating people who would admit to using *deai*. Mika, a 24-year-old office worker and avid *i-mode* user, allowed that she engages in chat, but insisted she would never consider *deai*. Like many female Internet users, she believes *deai* smears its users with the mantle of illegitimacy. In this view, one apprehends "stigma" as defined by Goffman: "a term … used to refer to an attribute that is deeply discrediting."[25]

Stigma "constitutes a special discrepancy between virtual and actual social identity," between attributes one could be proven to possess, versus the "character we impute to the individual … a characterization 'in effect'."[26] Important to note, stigma is a relational concept, for, as Goffman observes, "an attribute that stigmatizes one type of possessor can confirm the usualness of another."[27] Thus was it that many of our contacts tended to employ *deai* as a social marker, an indicator of identity, a measure of character. They did this by inversion, in their repudiation and condemnation of it. For instance, Yumiko, a 23-year-old graduate student, loudly exclaimed: "Ehhhh? Me, use *deai*? Gross! No way!" The vehemence of her denial suggested a desire to manage the surrounding social judgment, thereby elevating her own image.

For the stigmatized, there is often nowhere to turn for solace but to others of similar social description. As Goffman explains, the afflicted join to forge a "circle of lament to which (each) can withdraw for moral support and for the comfort of feeling at home, at ease, accepted as a person who really is like any other normal person."[28] Among the hundreds of *deai* communities on the Web, *batsu-ichi* (or "one strike") stands out in this regard. It is an association whose users share a single signal trait: they have all been divorced. To have failed in marriage in Japan is to court stigma – for the mark of divorce provides a window into personal "untidiness," a glimpse of individual imperfection. *Batsu-ichi* reflects a strategic response by those bearing the stigma. What is ironic, of course, is that those stigmatized by one act engage in another which, itself, is stigmatizing, as a means of grappling with the repercussions of the first. This aside, though, one can discern the positive functions of *deai*. Similar to sites that seek to restrict sociation to teenagers, for instance, or work to counsel troubled members of society, defensive spaces such as *batsu-ichi* serve socially productive, wholesome, nurturing aims. A reminder that there is more to *deai* than the widely-held sense of stigma.

Appearance versus reality

In the *Republic*, Plato describes a cave of chained prisoners. Their only access to reality – to the world outside – is images flickering on the wall. These images come from a fire, mediating the figures passing outside. Plato comments that for these captives, "the truth is nothing other than the shadows of artificial things."[29] He then recounts how two prisoners are set free and experience the real world first hand. They find it quite unlike the images on the wall. Plato's implicit query is: "how can they return to the reality they once fervently held as true, when now they understand it to be an imperfect representation?"

Sayaka, a 23-year-old *furiitā*,[30] has occasionally found herself in this position. She has had to reconcile competing images: those obtained through e-messages and those acquired after actually speaking with her partners via telephone. When she has encountered discrepancies between the mail self and the phone self – between oppositional "virtualities"[31] – Sayaka has experienced Plato's prisoner's dilemma: deciding which version of reality to accept.

Such discrepancies are only magnified at the next stage of encounter: between mediated and unmediated selves. Suddenly Sayaka finds herself fully outside the cave. When her companion seeks to jump to the intimacy stage[32] by saying "*kondo, asobō yo*" (Let's go play/Let's go on a date) or "*kondo, nomi ni demo ikō yo*" (Let's go drinking/Let's go out), Sayaka has often found herself in a quandary. At times the embodied reality is so at odds with her *deai* virtuality that she has heard herself say "*ah! kore dame da*" (Oh, this is no good) or "*kyō wa kaeranakuchya*" (Today, I'll have to go home [alone]). Such discrepancies surprise her: "*kitaihazure*" she will exclaim, "this was not what I had hoped."

Appearance versus reality is also at work in the perceptions of the medium itself. Television news, the "Wide Shows," newspapers and magazines distribute morality tales that begin with mail and chat, and culminate in human tragedy following physical encounter. Yet, our research suggests a different reality. Despite the media's fetishistic focus on danger and exploitation, as noted earlier, up to 31 per cent of one survey population reported joining *deai* sites for the purpose of conversing with like-minded hobbyists. Underscoring this dimension, a large number of *deai* sites define themselves in terms of particular themes: bird-watching, hiking, seniors, and the physically disabled. It is against such a shared backdrop that sociation unfolds. Viewed this way, the media's negative visions of *deai* appear akin to the shadows on the cave's wall: they fail to correspond to the socially productive elements of *deai-kei*.

Identity

The fact that *keitai* images often don't correspond to face-to-face realities may mean any number of things. The medium may: (1) be a flawed form of communication – possibly due to excessive amounts of noise leading to signal loss; (2) enable users to engage in impression management or manipulation; or, more positively, (3) provide freedom for users to explore and negotiate potential selves.

The first possibility is unlikely; the second we shall consider further on; the third is a well-substantiated dimension of Internet culture. What makes it worthy of attention here is that Japan is a country where exceptionalist discourse remains quite strong, with one of the core exceptionalist arguments being that individual personality and self-conceptualization are constrained within and determined by the group. It is true that over the past two decades there has been some evisceration in this perspective.[33] To cite but one study pertinent to the present chapter, in the mid-1980s Holden observed numerous instances in an ostensibly "corporate" milieu – denoted by school uniforms, fixed behavioral codes, and widespread social surveillance – in which Japanese youth employed distinctive markers of personal identity. These markers took the form of ribbons, badges, stick-pins, cartoon characters, dolls, decals, and the like, affixed to school bags and clothing.[34] Over the past decade such individuated communications have become manifest in a wider variety of expressive forms: from uniform alteration to body piercing to hair coloration.

Historically, a vestige of Japan's group-orientation has been self-censorship in the face of *mittomonai* (unseemly) behavior. For those cognizant of the social mirror, but nonetheless seeking personal expression, *keitai-i* affords a satisfactory compromise. It is individual, expressive, but also furtive. Users such as Sayaka report spending much of their online time on the kinds of activities McVeigh calls "interiority"[35] – effectively personalizing their electronic world. Sayaka's two most visited sites, for instance, are: (1) a *chakumero* site (which enables her to sample and download various sounds such as voices of famous personalities, animal noises, and popular songs as a means of signaling incoming calls), and (2) *machiuke gamen* (a service offering downloadable pictures and illustrations that greet Sayaka as she opens her *keitai*).[36] The rest of her time is spent online connecting with hand-selected e-mail and *deai* friends.

Compared to simple Web surfing, *deai* affords greater interiority and opportunity for individuation. It is invisible and anonymous. More importantly, in one package it provides a forum to assist the search for self, serves as a tool for users to actively manage identity, and even encourages and empowers the expression of practices that are quite at odds with historical perceptions of self in Japanese society.[37] *Deai* enables a relatively unregulated, unsurveilled stage for people to create and forge new selves. It affords trial, but also enables multiple errors, without retribution and with (often) little personal consequence.[38]

Emotions

Identity is conveyed in part by a tool supplied by the *deai* service provider: comic characters, illustrations, and photos available for pasting into messages, often as a substitute for words.[39] Semiotically, such signs serve as indexes of sentiment; non-verbal signifiers which, nonetheless, communicate a deeper "truth" about the sender's emotional state. More than one informant told us that such depictions are "convenient" and "helpful;" they can convey emotions and thoughts that the sender would never have the courage or ability to express.[40] By inserting an illustration, the sender can depict heartbreak, happiness, loneliness, or love; s/he can modify the impact of words, or even nullify earlier damaging statements – all of this without having to actually write words.

Of course, this surrogate communication is but one small step removed from impression management. Here, then, we stray into the realm of Goffman's self as sign-giver, performer, manipulator.[41] Who, after all, can say that the tears dotting the text are truly being shed by the sender? Are the words of bravado or the *kawaii* (cute) colors or the fonts of aggressiveness or the embedded links to natural scenery a true representation of the inner other? Might these tokens not be skillful, machine-assisted projections of whom the sender wishes the reader to believe him or her to be?

Consider the case of Sayaka. After months of corresponding with a man, and a few face-to-face meetings, she became sufficiently convinced of his sincerity and good character to consent to sexual relations. Her aim, she says, was a long-term relationship – a goal she insists her partner concurred with. After one week of intimacy, however, Sayaka discovered that her new partner was actively engaged in another relationship. Sayaka was devastated. How could she have been so misled? The machine in-between, she realized, had abetted her manipulation.

Intimacy

A central element in Sayaka's story is the search for intimacy. Her encounters aimed at finding a partner who would make her heart pound; a person who would unleash the voice within: "this is the guy, this is the ONE! I *HAVE* to meet him."

This aim was common among all of our informants. Beyond mere sociation, those engaging in Internet encounters seem to want something more than the simple exchange of pleasantries or sharing of daily events (i.e. what chat provides).

Giddens[42] has argued that modern society tends to elicit a greater response in the expression of intimacy. Others[43] have observed that efforts to establish intimacy correlate most highly with the presence of large, impersonal, bureaucratic structures. If both of these claims are true, it should come as little surprise that in Japan (to many, a paradigmatic "corporate society"), one finds extensive intimacy seeking. Importantly, we found this impulse not only in an office worker such as Mika, but also a *furiitā* such as Sayaka, housewives such as the Saitama knifing victim, mentioned above, and Rumiko, who will be introduced further on. In accounting for the *deai* phenomenon, Japan's highly corporatized structure – and the impetus for intimacy which it engenders – may be a key factor.

Personal defense

What Sayaka described as her "betrayal" left her shocked and depressed. It was enough to scare her away from *deai* for months. Ultimately, she returned to online encounters, but was understandably less trusting of the men on the other end. The Sayaka of today states coolly: "in retrospect I see he just wanted to have sex." Consequently, she is now more reluctant to step away from the protective shield afforded by the medium.

This defensive aspect is a central feature of *deai*. Online service providers offer space within which intimacy can be safely practiced. Understanding the character of modernity – and Japan, as well – this should come as no great surprise. For the corporation (here in the form of a service) offers not only a forum for interaction, but a protective structure for participation, as well. The medium constrains all users within its institutional boundaries. It serves as a guarantor of individual safety, thereby stimulating trust.

One distinguishing feature of urban modernity has been the pell-mell increase of outsiders in once-closed social circles. To cope with the dangers of anonymous encounters, humans engage in "civil inattention."[44] Such "polite estrangement"[45] aims at blocking the transmission of "tie signs" – visual cues which risk forging social connection.[46] The danger of connection ranges from physical to emotional, financial to reputational harm. Such danger can be mitigated in only two ways: total avoidance or else (ironically) deeper involvement with the person encountered. The latter demands what Giddens calls "facework commitments:" focused encounters that provide indications of integrity about the person engaged. Such evaluations, of course, are less possible on a telephone and even more problematic on a display printing out words or icons. This is especially true in early stages of encounter, when the potential for (and interest in) image management, identity concealment, and deception runs so high. As we have seen, *deai* is furtive in this way. There is no timbre of voice to gauge, no control over visual and temporal parameters. The medium is designed to intercede – via software that modifies and shapes identity – hence, there is less opportunity for independent verification. With the technology working so well to frustrate external evaluation, sociation can become even more dangerous; the need for personal defense even greater. The case of Sayaka demonstrates how, outside the protective umbrella of the *deai* service, emotional harm can occur. Certainly, as other *deai* stories show, physical, financial and moral damage can result, as well.

There are no perfect solutions for the dangers inherent in human intercourse. Chat-only encounter sites aim at full protection, but they can engender their own problems. As Mika explained: "if you sign into chat at the same time each day, you get used to the people who are there. You get to know them and even count on chatting with them. And then, one day, they're not there any more. And you wonder: 'what happened to them?' You can't find them ... and you start to feel lonely." For people yearning for connection, but seeking institutional protection, they can end up hoist by their own petard.

Nonetheless, the notion of intimate, but protected, connectivity is popular. For instance, there is the extensive *deai* sector known as *gōkon*. Here, in an updating of *gurūpu kōsai*, congregations of one sex will advertise their availability for a night of food, drink, conversation and *karaoke* – an invitation to which a group of like-minded members of the opposite sex may respond. Often, posted pictures fed by camera-enabled *keitai* serve as a pre-screening device, assisting in decision-making, self-defense and, possibly, even deception.[47]

Glocalization

As a medium for sociation, *deai* is uniform in form. Yet, its content is quite diverse. As noted above, heterogeneity is at odds with standard theorization about Japan, leading one to speculate on whether *deai* might serve as fodder for an assault on essentialism. One point that should not be lost, though, is that Japan is a fully developed nation in the throes of high modernity. As such, it *ought to* manifest what Berger[48] calls "the pluralisation of lifeworlds." Such multiplicity means that society's members "naturally" experience a segmented world; their activities carved up into numerous "lifestyle sectors"[49] – one of which appears to be an Internet subculture such as *deai*.

One phenomenon contributing to pluralization, of course, is globalization. We know from Appadurai[50] that despite global flows (like, for instance, cell phone technology) all societies manifest localized responses and applications. In the case of *keitai*, Internet capability is one such localized expression. So too, within this response, services such as *chakumero, machiuke gamen*, and *deai* reflect contextual creations. As for *deai*, even further localization has transpired: in it one finds a variety of hand-tailored environments, none of which are found in these specific incarnations outside the country. In isolation and arrayed combination – and, crucially: subject to the values and will of individual users – these sites provide for greater pluralization of lifestyles in contemporary Japan.

Fragmentation/reintegration

A basic tenet of Durkheimian theory is that social segmentation has the potential to dilute social density, thereby destabilizing social solidarity.[51] The greater autonomy gained by the individual represents a concomitant diminution in the grip of the moral and emotional cocoon containing him. Cut adrift, he can find himself at greater risk, with costs for both the individual and society. It is just such a situation that critics claim has befallen contemporary Japan. One writer has decried, "New Japanese are materially affluent but spiritually destitute."[52] To Murakami Ryū, a highly-acclaimed novelist known for books that portend social problems, Japan no longer possesses clear goals and grounding principles; it is a rudderless nation full of aimless citizens.[53]

Viewing the *deai* phenomenon, one could easily assert that it is all so much personal aggrandizement, social detachment, self-centered and inwardly-directed. Kogawa, for instance, argues: "as far as the present Japanese collectivity is concerned, it is electronic and very temporal, rather than a conventional, continuous collectivity based on language, race, religion, region or taste."[54]

Yet, post-modern though they may be, *deai*-inflected lifestyle sectors may be seen as functioning like traditional clubs or communal associations: tools for steering through the societal soup, for gaining stability and reconnecting. The choice to join a group like *Zenkoku furusato kōryu fōramu* – a site which aims at promoting exchange between "home towns" – may be individual and personal; but at another level it is communal, outward-reaching. Participation in the transitory, electronic community that is *deai* may actually be binding for contemporary Japanese society and, therefore, socially productive.

Uses and gratifications

Throughout this chapter we have argued that *deai* is a mechanism for sociation. To Simmel, who coined the concept, the motive for forging social connections was the satisfaction of human desires.[55] There is, in such intentional behavior, resonance to an early literature in communication studies called "uses and gratifications." This line arose during the period when the model of powerful sender-directed effects began to yield to the paradigm of user-mediated reception. An influential articulation by Katz[56] posited the possibility that humans do not passively receive media messages; rather they actively use media as a means of satisfying certain needs. Although this conception has fallen out of favor, more recently Fiske has advanced a view of the popular that is reminiscent of this earlier position.[57] It seems applicable to the behavior of *deai* users, as well.

Admittedly, some users' behaviors may not reflect purely internal or individually motivated needs. There are the pro-social impulses mentioned above. Additionally, *deai* use may be the result of social factors. One cannot ignore the fact, for instance, that the environment is chock-full of cell phone-aided Internetters. In Japan – fashion-conscious, conformity-aware, surveillance-full – such ubiquity places a certain degree of pressure on fellow users to engage in like-patterns of acquisition and use. Whether externally-imposed or internally-driven, however, it is the uses and reasons for use – the seeking and achievement of gratification by users – which are among the signal aspects of the *deai* phenomenon.

Simulation, or virtu-actualization?

Rumiko is twenty-eight, married, with one child. When she started using *deai* she selected from four potential correspondents: three men and one woman. She chose Shinichi, a 43-year-old man who lived in her area. After exchanging mail for three weeks, they agreed to meet. Shinichi surprised Rumiko. He was better looking than she had expected, witty, loquacious, and fun. Most importantly, unlike her husband, he "treated (her) like a woman." Following three weeks of face-to-face meetings (and at Shinichi's behest), the couple became intimate. Though Rumiko had never really enjoyed sex, she discovered a sudden fondness for it. "A new world opened up to me," she professed. But when her husband discovered the subterfuge he pressured Rumiko to cut it off. With great reluctance, she assented. Today she exclaims: "I love my husband, but I dream about Shinichi."

When commentators on Wide Shows hear such stories they pontificate about the power of the medium to alter human behavior in deleterious ways. "This phone-mail connection reduces our natural defense barriers which usually allow us to block out the dangers of casual face-to-face meetings."[58]

By contrast, some media scholars see such developments as natural. Morse has argued that "as *impersonal* relations with machines and/or physically removed strangers characterize ever-larger areas of work and private life, more and more *personal* and subjective means of expression and ways of virtually interacting with machines and/or distant strangers are elaborated."[59]

In observing *deai*-assisted encounters it is easy to see how these mediated relations differ from other virtualities. Although *deai* does enable the creation of fictional presences, when the machine is cast aside – when *mēru-tomo* or *deai* interchange eventuates in corporeal concourse – then certain fictions cease because actual presences have been conjured. Embodiment has been engineered by the medium. This is Baudrillard's simulation incarnate: artifice that eventuates in the "hyperreal;" people who are different, but "realer" than their originals.[60] Sayaka's sexual betrayal, Rumiko's unrequited love, "Pastel's" prostitution ring, Nakashibetsu's "*omiai* weekends" and thirty-three actual marriages, the Saitama knifing – all of these outcomes represent virtu-actualizations: realizations of virtual scenarios; cyber-sociations which have become embodied, then spill over and play out in physical space. By being created and cultured in the imagination, it is plausible that they exert an even greater pull on human consciousness than corresponding relationships in the mundane, everyday lifeworld.[61]

CONCLUSION

In this chapter we have seen that *deai* is a strategy for embodying the self, satisfying personal interests, expressing emotions, experiencing intimacy, and achieving self-expression. It is a way of asserting individualism and autonomy, while escaping the pervasive surveillance that once dominated and ordered Japanese society. As such, *deai* is a tool of empowerment, affording users the opportunity to forge social connections of their choosing, less fettered by institutional codes and less subject to organizational hierarchies and regulations.

In many ways the impetus for engaging in this behavior lies in problems associated with modernity, most notably fragmentation, alienation, declining trust, and imperiled personal defense. To manage these concerns users have turned to *deai* and, in so doing, have engineered a greater degree of connection than previously possible between strangers passing all-too-quickly on the street.

Viewed this way, we believe that this chapter contributes not only to research on the Internet, but also theorization about society, in general, and Japan, in particular. Through *deai* we are better able to see whence Japanese society has come, where it currently is, and, given the values and behaviors of its people, where it may be going. This said, the recognition should not elude us that sociation is being forged in, sanctioned by, and delivered through an organization – no matter how amorphous, transient or ethereal. Of course, would one expect anything less in a society that, for centuries, has been arranged in accord with corporate understandings and regulated by collective imperatives? *Deai* may be a new strategy for encounter as a means of solving the problems incident to modernity. We are left to wonder, though, whether it actually delivers its users from the structuration inherent in the society in which they exist. One final query: if it did, then what sort of society would Japan become?

NOTES

1 Often the distinction is not sharply drawn and, in such cases, the generic term *deai* is applied. In this chapter, for the sake of parsimony, we will refer to all Internet-assisted encounters as *deai*, unless the distinction is crucial to the analysis.

2 Similar to Internet chat, *deai* requires users to join a community. Unlike chat, *deai* participants scan a list of users, read their self-introductions, and seek to initiate one-to-one message/chat interaction. Once contact has been established, community members are on their own, although their exchanges occur through the *deai* site.

3 The prototype exposition of cyber-sociation is Indra Sinha's *Cybergypsies* (New York, Scribner, 2000), an account of life in multi-user dungeons or domains (MUDs) and human interaction on the "electronic frontier" in the 1980s and early 90s. Such a tale has yet to be written about Japan.

4 "Wide Shows" are "infotainment"-style television programs that run from mid-morning to late afternoon on all major commercial stations. They generally focus on current topics in Japanese society and culture. More often than not scandal or social problems take center stage. A panel comprised of social critics, university professors, journalists and *tarento* (entertainers and personalities) will often weigh in with commentary on the reported matter.

5 This passage presupposes heterosexual relationships. *Deai*, however, has been nothing if not a boon for facilitating homosexual liaisons, due to its focused, "special-interest" character. Certainly, our later claims concerning autonomy, self-defense, self-exploration, emotions and identity attach equally to homosexuals and heterosexuals.

6 The so-called "Juliana's boom" in the late 1980s reflected the values of the bubble economy: conspicuous consumption of designer goods, well-dressed "upwardly mobile" singles, brash public display in glitzy clubs, with women donning short skirts and gyrating to a disco beat on a raised platform with a mirrored floor. "T-backs," "V-backs" and "O-backs" – various styles of undergarments – were part and parcel of this "show me" lifestyle. Widely ballyhooed in the

media, this lingerie served to spur the consumption cycle. The more down-to-earth "hip hop boom" of the late 1990s fit the post-bubble mood: grunge clothing, Rastafarian hairstyles, body piercing, rap music, along with a coarser public demeanor.

7 Historically, the tendency to view Japan and the Japanese as unique has been pervasive. More recently, opposition to claims of so-called *nihonjinron* has been common. For a sampling of the literature on exceptionalism, see H. Befu, *Hegemony of Homogeneity: An Anthropological Analysis of Nihonjinron*, Melbourne, Trans Pacific Press, 2001. For a critique of assertions of exceptionalism, see P.N. Dale, *The Myth of Japanese Uniqueness*, London, Routledge, 1986. Despite the recent "thought style" deriding it, the assertion of Japanese uniqueness persists.

8 "The Internet-enabled mobile phone [is] the fastest-growing consumer technology in Japan's history," *Japan Internet Report No. 45*, January/February 2000. Online. Available HTTP: *http://www.jir.net/jir1-2_00.html* (2 March 2002).

9 See, for instance, O-Young Lee, *The Compact Culture: The Japanese Tradition of "Smaller is Better,"* Tokyo, Kodansha, 1982.

10 As one president of a Japanese Internet consultancy firm says: "cell phones here are fashion accessories and toys above all ... It's unlike anything else in the world." D. Scuka, "Unwired: Japan Has the Future in Its Pocket," *Japan.Inc*, June 2000. Online. Available HTTP: *http://www.japaninc.com/mag/comp/2000/06/jun00_unwired4.html* (2 March 2002).

11 *i-mode* is a particular (read limited) kind of Internet platform – one that is based on satellite access through the parent provider, *DoCoMo*, rather than over land-lines. As a result, *DoCoMo* does not load every URL on the World Wide Web into their satellite – meaning that there cannot be perfect Internet freedom for *keitai* users. A further element mitigating user freedom is cost. Charges for *i-mode* are assessed based on the number of information "packets" downloaded by phone – a fact that can make users reluctant to surf the already truncated sites available.

12 By comparison, Finland ranked first with 65 users per hundred. The United States ranked twenty-second at 31.5 per hundred.

13 The Office of the Prime Minister, *Posts and Telecommunications Policy, 1998–2000*, 2000.

14 The raw numbers: 16,000,000 in Japan to America's 62,000,000. The scaled numbers: 1,323 per 10,000. This compares with, say, Sweden (number one with 3,953 per 10,000), Canada (number four with 2,475 per 10,000), and Singapore (number seven with 1,739 per 10,000). Source: International Telecommunication Union, *World Telecommunication Development Report*, 1999.

15 "Number of Internet connection users," press release, The Ministry of Public Management, Home Affairs, Posts and Telecommunications, April 2001. These figures cover not only *i-mode*, but two rival services, Ezweb and J-Sky, as well.

16 Source: *Japan Internet Report*, no. 45, January/February 2000. Online. Available: HTTP: *http://www.jir.net/jir1-2_00.html* (2 March 2002).

17 T. Kogawa, "Beyond electronic individualism." Online. Available HTTP: *http://anarchy.k2.tku.ac.jp/non-japanese/electro.html* (2 March 2002).

18 Even more astounding, as of February 2002, a Google search yielded 1,420,000 *deai* sites.

19 Statistics from: *Deai-kei mērutomo boshūsaito riyōsha*, online survey conducted by *Japan.internet.com*, 26 April 2001. Online. Available HTTP: *http://japan.internet.com/cgi-bin/archives?c=research* (30 April 2002).

20 Statistics by Soft-Bank V, as reported in *i-mōdo jōhō saito 1000* (*deai and mēru*) (1,000 *i-mode* information sites [meetings and mail]), 12 October 2001, Tokyo, Softbank Publishing Company, pp. 12–15.

21 Explained later in this chapter.

22 The poll lists the remaining 5 percent as "other."

23 Data reported in this section was collected via telephone interview, February 2002.

24 This interview data was collected as part of a reputational/snowball sample of actual *deai* users. Depth interviews lasted on average two hours, and were standardized, but non-scheduled. The population was limited in size to five, and restricted in terms of age (early to mid-twenties), gender (women), and location (Sendai, Japan). Additionally, the data reflects a particular time slice (interviews were conducted in November and December 2001). Given these limitations,

we would term our findings "suggestive" rather than "generalizable." The conclusions drawn from this information are plausible rather than definitive, and merit follow-up with a larger, more diverse population.

25   E. Goffman, *Stigma: Notes on the Management of Spoiled Identity*, Engelwood Cliffs, NJ, Prentice-Hall, Inc. 1963, p. 2.

26   E. Goffman, *Stigma*, p. 2. Note that while the Internet was not yet created at the time of writing, Goffman's distinction fits *deai*-mediated identities and encounters in that he posited "virtual selves" to be unempirical. Instead, they are "unsubstantiated inferences."

27   Goffman, *Stigma*, p. 3.

28   Goffman, *Stigma*, p. 20.

29   Plato, *The Republic of Plato* (ed. Allan Bloom) New York: Basic Books, 1968, Book VII, 515c, p.194.

30   A Japanese neologism meaning a free agent unencumbered by a full-time job and not pursuing any particular career path.

31   "Virtuality" is a term advanced by Morse to denote "fictional presences" that have "come to be more and more supported and maintained by machines, especially television and the computer." M. Morse, *Virtualities: Television, Media Art, and Cyberculture*, Bloomington, IN, Indiana University Press, 1998, p. 11. Nonetheless, Morse pays (pardon the pun) virtually no attention to the greatest medium of virtualization: the Internet. In *deai*, alone, one can see that humans are establishing, exploring and encountering fictional presences willy-nilly. This is even more true when it comes to the design and publishing of Web pages and the ubiquitous use of e-mail.

32   For Sayaka, as with other respondents, we learned that *deai* encounters often proceed through stages. The first is exchange with a faceless correspondent on the other end of a computer or cell phone. Mail is exchanged for a time (in Sayaka's case about 3 months) until one of the two mentions the possibility of exchanging addresses. The rationale is often: "exchanging mails through this (*deai*) site is cumbersome. It would be faster if we could mail one another directly." If both agree, direct interaction (either e-messages or actual phone conversations) will occur. If the voice on the other end of the *keitai* somehow fits the image Sayaka has of her interlocutor, and if the content remains fun or moving, then the time may arrive to step away from the veil of the technology and physically meet.

33   See, for example, T.S. Lebra and W.P. Lebra (eds), *Japanese Culture and Behavior: Selected Readings*, Honolulu, University of Hawaii Press, 1986 [1974]; R. Mouer and Y. Sugimoto, *Images of Japanese Society: A Study of the Social Construction of Reality*, London and New York, Kegan Paul International, 1986; D.P. Martinez, *The Worlds of Japanese Popular Culture: Gender, Shifting Boundaries and Global Cultures*, Cambridge, Cambridge University Press, 1998; and M. Takatori, *Nihonteki-shikō no genkei* (Original types of Japanese thinking), Tokyo, Kodansha, 1975.

34   See T.J.M. Holden, "Surveillance: Japan's Sustaining Principle," *Journal of Popular Culture*, vol. 28, no. 1, 1994, pp. 193–208. Brian J. McVeigh has explored this phenomenon extensively in his book *Wearing Ideology: State, Schooling and Self-Presentation in Japan*, Oxford, Berg Publishers, 2000.

35   See McVeigh's chapter in this volume.

36   At the time of this writing, *chakumero* is a multi-million yen business, with hundreds of sites competing for the same customers, often with different versions of the same re-recorded songs. The cost of a download is about 100 yen. Users can spend up to 20,000 yen per month downloading and changing *keitai* sounds.

37   In this way, studying *deai* enables us to do what Stuart Hall urges: "precisely because identities are constructed within, not outside, discourse, we need to understand them as produced in specific historical and institutional sites within specific discursive formations and practices by specific enunciative strategies." S. Hall and P. du Gay (eds), *Questions of Cultural Identity*, London, Sage Publications, 1996, p. 4.

38   Although a recent case belies this view. Two young men began corresponding, each mistaking the other for a woman. During the course of their *deai* encounters it became clear that neither was what their words had first suggested. Incensed, one boy began sending the other threatening

mail, insisting that bodily harm (as well as public exposure) would result unless money was sent in apology. The other boy, a high schooler, complied with the extortion demand, parting with 20,000 yen, before finally informing the authorities.

39 Larissa Hjorth covers some of this terrain in greater detail in her chapter in this volume.

40 And, of course, well suits what has historically been seen as the Japanese inability or unwillingness to state feelings directly.

41 E. Goffman, *The Presentation of the Self in Everyday Life*, Garden City, NJ, Doubleday Anchor Books, 1959.

42 A. Giddens, *Modernity and Self-Identity: Self and Society in the Late Modern Age*, Stanford, CA, Stanford University Press, 1991.

43 J. Bensfeld and R. Lilienfeld, *Between Public and Private*, New York, The Free Press, 1979.

44 E. Goffman, *Behavior in Public Places*, New York, The Free Press, 1963.

45 A. Giddens, *The Consequences of Modernity*, Stanford, CA, Stanford University Press, 1990, p. 81.

46 Goffman, *Behavior in Public Places*.

47 In fact, a recent ad for J-Phone highlights (and makes light of) this protective function by posing a situation in which four attractive women are heading down the stairs of a pub to meet four poorly-groomed, uncouth men. Just as the women near rendezvous, a competing group of smartly-dressed "junior-executive" types send the women their picture along with an invitation to get together. Uttering quick apologies to the first group, the four women beat a hasty retreat in the direction of the better offer.

48 P.L. Berger, B. Berger and H. Kellner, *The Homeless Mind*, Harmondsworth, Penguin, 1974.

49 A. Giddens, *Modernity and Self-Identity: Self and Society in the Late Modern Age*, Stanford, CA, Stanford University Press, 1991.

50 A. Appadurai, "Disjunction and Difference in the Global Economy," in M. Featherstone (ed.), *Global Culture: Nationalism, Globalization and Modernity*, London, Sage, 1992, pp. 295–310.

51 See E. Durkheim, *The Division of Labor in Society*, (trans. G. Simpson), New York, The Free Press, 1933.

52 Mikio Sumiya, as quoted in E. Hoyt, *The New Japanese: A Complacent People in a Corrupt Society*, London, Robert Hale, 1991, p. 197.

53 J. Sprague and M. Murakami, "Internal exodus: novelist Murakami Ryū sees a dim future," *Asiaweek.com*, 20 October 2000, vol. 26, no. 41. Online. Available HTTP: http://*www.asiaweek. com/asiaweek/magazine/2000/1020/sr.japan_ryu.html* (4 March 2002).

54 T. Kogawa, "Beyond electronic individualism."

55 Simmel's definition was: "Sociation … is the form in which individuals grow together into units that satisfy their interests." He enumerated those interests as "sensuous or ideal, momentary or lasting, conscious or unconscious, casual or teleological." See G. Simmel, *The Sociology of Georg Simmel*, (ed. and trans. K.H. Wolff), New York, The Free Press, 1950, p. 41.

56 E. Katz, "Mass communication research and the study of popular culture: an editorial note on a possible future for this journal," *Studies in Public Communication* 2, 1959, pp.1–6.

57 See, for instance, J. Fiske, *Understanding Popular Culture*, London and New York, Routledge, 1989.

58 Commentary from a Wide Show, *Terebi Asahi*, April, 2001.

59 M. Morse, *Virtualities: Television, Media Art, and Cyberculture*, p. 5.

60 J. Baudrillard, *Simulacra and Simulation* (trans. S.F. Glaser), Ann Arbor, University of Michigan Press, 1994 [1981].

61 And, in fact, recent evidence from a Western context suggests that contrary to widely-held assumptions, relationships that begin online may actually result in strong, long-term unions. See "Chat room chatter may lead to real romance," by P. Hagan, *Yahoo! News*, 15 March 2002. Online. Available HTTP: *http://dailynews.yahoo.com/* (16 March 2002).

4 Cute@keitai.com

Larissa Hjorth

INTRODUCTION

It is the now ubiquitous *keitai* (mobile phone) that has pushed Japan into cyberspace overdrive. This Japanese reworking of the failed European WAP (Wireless Application Protocol) with the ability to perform three functions – cell phone, hand-held computer and wireless e-mail receiver – has managed to dramatically alter the demographics of Internet use in Japan. In 2000, estimates put *keitai* use among the Japanese at 80 per cent, considerably higher among young people.[1] NTT, one of the major telecommunications companies in Japan, leads the Japanese market with its specific *keitai* version called *DoCoMo*. Commanding three-fifths of the market, *DoCoMo*'s domination is even reflected in its name. According to the ads, *DoCoMo* is an acronym for "Do Communication over the Mobile network," but as Rose points out, the pronunciation is similar to the Japanese word *dokomo* meaning "everywhere."[2]

Currently in Japan there are three major companies that offer *keitai* Internet access: NTT, H" and J-phone. NTT and H" produce *keitai* that have the capacity to send e-mail and text messages as well as supporting normal phone operations and H" has recently launched its AU phones that, unlike J-phones and *DoCoMo*s, can be used when traveling throughout the Asia-Pacific region. J-phone, additionally, offers an inbuilt camera that allows users to take, send and receive photos along with their e-mails, text messages and calls. This function has proven particularly attractive to the market of early teens to 30-year-olds as it allows users to "capture" a moment which can then immediately be sent to one's friend or partner, thus rendering the experience more memorable and intimate.[3] Users can send both photos and downloaded cute (*kawaii*) symbols and characters to further personalize and familiarize the space between caller and receiver. As Rose mentions:

> [In 2001] with the advent of animation (made possible by a Java licensing deal with Sun), the range of possibilities shot up dramatically. *Hello Kitty*, the adorable little pussycat that already adorns everything from bank cards to hot dogs, now appears on *i-mode* screens as well, chiming the hour and doing a little dance. *J@pan Inc.* magazine reports that 9 of the top 10 Java downloads

on *i-mode* are games – everything from mah-jong to *Shit Panic*, in which you try to catch the stuff as it falls and flush it down the toilet ... Ring tones and cartoon characters have other uses as well. People set their phones to sound a ring-tone version of the latest pop hit whenever their boyfriend or girlfriend calls.[4]

These additional functions have provided J-phone users with a greater capacity for what Manuel Castells refers to as "customizing" the technology.[5] While the use of *kawaii* features to familiarize new commodities or technologies has been common practice in the material culture of post-war Japan, the ways in which *kawaii* culture is being employed both inside and outside the boundaries of *keitai* cyberspace are creating new multivalent vernaculars. As Sharon Kinsella has noted, the extensive use of *kawaii* culture in Japan is indicative of changes in gender relations.[6] In this chapter I will explore a number of themes arising from a survey of 100 consumers of *kawaii keitai* culture in order to illustrate how this technology and its customization reflect recent changes in Japanese society.

THE CULTURE OF CUTE CONSUMPTION

As Holden and Tsuruki (this volume) demonstrate, the most popular unofficial sites for many *keitai* users are *deiai* or matchmaking services but the dominant official sites are those offering downloadable ringer melodies and screen savers.[7] According to Stocker, toymaker Bandai's site *Doko-demo Kyarappa!* has one million subscribers who receive daily supplies of animated characters for only 100 yen a month.[8] Of course, other more subcultural and subversive sites also exist, such as *Sanriot* – a space dedicated to critiquing *kawaii* characters.[9] On these sites, the characters demonstrate far from the usual passive *kawaii* behavior, the most obvious example being that of *PostPet*, a site where the cute characters are more like *enfants terribles* who fight it out with other cute characters in a Darwinian contest for survival.

The *keitai* landscape is an undulating terrain occupied by *kawaii* characters where more "mainstream" and "established" *kawaii* characters are perpetually being updated, reappropriated and integrated with more recent characters such as *PostPet*. Shifts can be seen in the manner in which both the more traditional and subcultural *kawaii* cultures are being deployed and customized within *keitai* cyberspace. It is this marriage of *kawaii* familiarity with new technologies that can offer insights into shifting gender, sexual and age dynamics within contemporary Japan.

It became clear during my survey of *kawaii keitai* consumers that the usage and appropriation of *kawaii* characters both inside and outside *keitai* cyberspace blurs the lines between mainstream and subcultural, male and female, young and old. Firstly, *kawaii* culture is a paradoxical space where gender and sexuality are reconfigured (as demonstrated by *Hello Kitty*, an obviously female figure that lacks both mouth and genitalia) and often appropriated (men wear *Hello Kitty* to attract "typically" *kawaii* female consumers). As Kinsella points out, *kawaii* culture reflects the specific ways in which different genders and generations interact with consumerism in Japan.[10] Further, the colonization of high-tech spaces such as the Internet by the cute characters usually associated with the female realm in Japan is an important signifier of the power afforded women by this new technology.

In Japan, individual consumers have the power to "customize" the *keitai* to signify their own subcultural and personalized codes. For instance, a 23-year-old female arts administrator states that:

> I use well-known as well as less familiar *kawaii* characters not only as screen savers but also to accompany my e-mails. I think they are important because they give a face and a feeling to the text message. They are capable of accommodating humor. Just as I have special music for when specific people ring, so too do the characters symbolize types of friendships. I use *PostPet* for my two closest friends at the moment, while I use *Hello Kitty* for about ten of my female friends. I have other characters I use, and for me it is important that specific people get particular characters.

Castells argues that the Internet has shifted business models by enabling increased interactivity, flexibility, branding and customization. He states that, "If customization is the key to competitiveness in the new global economy, the Internet is the essential tool to ensure customization in a context of high-volume production and distribution."[11] It could be argued that Japan is years ahead of many other countries in adopting strategies and modes for customization; specifically one dominant mode of customization is *kawaii* culture. The employment of *kawaii keitai* culture clearly demonstrates the elasticity of Castells' discussion of customization, particularly the bipolar flows between global and local, subcultural and mainstream tendencies. In this way, *kawaii* culture operates to tease open stereotypes about gender, age and sexuality. For example, a character such as *Hello Kitty*, a mainstream symbol of *kawaii* culture, can also be appropriated by users to symbolize more subcultural meanings and be deployed as a personal logo shared with other specific users.[12] In turn, *Hello Kitty* can operate not only as a "local" signifier within Japan and the region (notably Hong Kong, Korea and Taiwan) but also "globally" within Western contexts such as Australia.

The relationship between *kawaii* culture and *keitai* cyberspace is far from one way. Just as the *kawaii* can particularize the seemingly universal cyberspace, so too can *keitai* cyberspace reinterpret, recontextualize and reconfigure the meanings and significances attached to *kawaii* characters. Consequently, well-known *kawaii* characters such as *Hello Kitty* can simultaneously operate as universal "cute" signifiers while also being reappropriated by particular users in specific contexts. For instance, a 28-year-old male artist comments that:

> I know other people use the same *kawaii* characters that I use, but they don't use them in the same way. My friends and I have great fun playing with the dominant readings of the characters and subverting them. I don't think that

my usage is that different from many other *keitai* users; I believe everyone has slightly different meanings associated with the same characters.

As the sample survey of *keitai* users discussed below illustrates, the rapid expansion of *keitai* cyberspace in Japan provides a terrain fecund with "glocal" productions that problematize crude dichotomies between local/global, mainstream/ subcultural, particular/universal, masculine/feminine and young/old.

A SURVEY OF *KAWAII KEITAI* USERS

While living in Tokyo in 2000 I noted how many commuters had abandoned *manga* for *keitai* with Internet portals as a means of killing time while traveling to and from the office. In order to discover how the *kawaii* was circulating within *keitai* culture, I conducted a sample survey of consumers. Those surveyed were both women and men living in Tokyo aged between 15 and 50. I first administered a questionnaire to 100 participants and then followed this with a more in-depth survey of a sample group of twenty. Issues such as gender, age and employment were addressed in order to comprehend the complex ways in which individuals customize *kawaii* culture within *keitai* cyberspace. While the survey data is by no means definitive, it does point to a number of themes in the relationships consumers adopt with *kawaii keitai* culture.

It became apparent that the deployment and appropriation of *kawaii* characters within *keitai* cyberspace is paradoxical – despite the obvious popularity of *kawaii* features, out of the surveyed respondents 75 per cent "didn't really care" about the characters they had as ringer-melodies and screen savers. Yet, at the same time, many (including many of those who expressed disinterest) acknowledged that they did spend quite a lot of time reconfiguring and participating in the character culture. Time spent waiting for and riding on public transport was the main period for this preoccupation.

Of the respondents surveyed, 65 per cent claimed they used their *keitai* as their main source for Internet access and sending e-mail. The rest used their *keitai* mainly for phone calls or text messages rather than e-mail or Internet access.[13] One noticeable constant in the survey was that age was a significant factor in how much time was spent participating in *kawaii keitai* culture, with young people more interested in *kawaii* applications than older users. In turn, more women were clearer about their usage and appropriation of *kawaii* characters than men – most men adopted a neutral stance, many claiming they "just kept" the characters they got with the phone. What became apparent was that while both male and female participants used *kawaii* characters to roughly the same extent, it was women who seemed better able to articulate the meanings behind their appropriation of cute characters. As Brian J. McVeigh has noted, the *kawaii* is symbolic of widespread power relations within Japan. He suggests that "cuteness does not merely reflect the social world; rather, via communication, it constructs gendered relations."[14]

A 40-year-old female who works in the arts industry mentioned that:

> I think the stereotypical users of the phone – as with cute characters – are women. But that is changing. The trains are full of older businessmen who have replaced *manga* with mobile phones adorned with cute characters.

A 25-year-old female retail assistant reported:

> My *i-mode* is very important to me. I use it all day and most of the night. Currently I have *Hello Kitty* as my screensaver. The use of characters makes the *i-mode* fun – thus it adds to the communication process. It also, for me, humanizes the technology.

As McVeigh (this volume) demonstrates, for students in Japan the mobile phone has become an important agent of "individuation." It became clear from the survey that k*awaii* characters were used by respondents as much for representing the self as for customizing the *keitai*. A 25-year-old male government worker said that:

> I change my characters a lot. I've had about 20. It helps freshen the object [phone]. The characters reflect my attitudes at the time. And I especially choose ones that are funny.

Kawaii characters are also often used to accompany e-mails and text messages as a form of signature or logo. One female respondent (age 28, administrator) spoke about the increase in dating e-mailing services in which *kawaii* characters were used to stand in for each partner. She spoke about how her friend was very disappointed when she actually met the man she had been messaging who was not nearly as *kawaii* as his symbol! Of the respondents surveyed, 40 per cent said they had up to fifteen *kawaii* characters stored on their phones that they used for the "appropriate" time and person. Twenty per cent had more than forty. Most of the J-phone respondents indicated that they had thirty to forty photos that they would customize – through combining generic *kawaii* characters with their own photos/ images.

A 36-year-old female writer reported:

> A lot of my phone usage is for text messages and e-mails. The usage of cute characters reminds me that there is a receiver ... it personalizes the space and makes the phone mine.

Alternatively, one 30-year-old female respondent claimed:

> Identification with different characters really depends on the individual. That is what is so enjoyable but also frustrating about them. It is a different language system with different rules and meanings.

As noted by the above respondent, *kawaii keitai* culture is able to seemingly reinforce stereotypes while, on the other hand, simultaneously subverting them. Here the respondent is clearly frustrated by the multivalent nature of *kawaii keitai* culture which allows for customization and personalization resulting in diverse interpretations (and possible misinterpretations) of the same characters.

A significant finding from the survey is that gender plays an important role in how the technology is customized. Specifically, the traditional alignment of women with passive characters (epitomized by mouthless characters such as *Hello Kitty* and *Miffy*), is being challenged within *keitai* cyberspace. For instance, one 32-year-old female writer and translator explained that:

> I don't use characters such as *Hello Kitty* as I remember her as I was growing up in the 1970s and she seemed to me to be symbolic of the role that women were made to occupy in Japan: passive and voiceless. I know that some people use her ironically and paradoxically but I can't shift that memory. Although it is funny, it seems to me that more and more men are using *Hello Kitty* in order to attract their imagined ideal of the typical [female] user who doesn't exist anymore.

It became clear from the survey that a number of participants in *kawaii keitai* culture are not just very aware of, but are active in playing with the multivalent meanings attached to cute characters. A 34-year-old male Japanese-American writer stated that:

> It seems that more and more characters are being used on the Internet. For me it is impossible to gauge their real gender or identity; often the cute characters are misleading signs. The rules and etiquette associated with cute figures outside the Internet are very different to those within the Internet.[15]

Despite the stereotypes surrounding *kawaii* in mainstream culture, 80 per cent of the respondents surveyed saw the meanings of *kawaii* culture within *keitai* cyberspace as noticeably different from mainstream readings and uses. The ambiguity associated with these figures is particularly clear on cyber-dating services. As one 25-year-old female administrator stated:

> I have tried Internet (cyber-dating) friendship services which are good fun. But I've heard so many stories about people being sad once they meet the person in the flesh. I'm happy to just have Internet contact. I know that there is no way someone can be as cute as the current character my e-mail friend is using.

Of particular interest are the shifting and complex roles occupied by women as both producers and consumers of *kawaii keitai* culture. As one 36-year-old female curator argued:

I don't see cute characters as being gender specific. Sure, the typical consumer of the cute used to be women but I think that is a myth now. The role of cute symbols on the Internet now is much more complex, you can't make the easy assumption that the user or receiver is a woman.

While women were once the stereotypical consumers of *kawaii* culture, this is no longer the case. In the sample survey, it seemed that equal proportions of men and women used *kawaii* culture to customize their phones in similar ways. However, gender distinctions did operate in relation to the discussion and acknowledgment of *kawaii* culture as a mode of customization. Following this line of logic, it is interesting to note the shifts in gender roles regarding *keitai* production. The symbol of the salary*man* as Japan's archetypal worker is similarly being challenged, especially in Internet-related businesses. One case example of the new prominence of women in this field is *DoCoMo*'s senior manager Mari Matsunaga, who articulates a "new spirit" in Japanese work practices.

As argued above, the specific relationship between *kawaii* culture and *keitai* cyberspace is marked by the undulating flows of "customization." As *keitai* are the main Internet portal for most Japanese users, the manner in which personal phones are adapted for personal use is very revealing. Matsunaga herself is very clear on the definition and representation signified by NTT's *i-mode* devices, gadgets and gizmos. According to Matsunaga, at the heart of this system is a shift from the tradition of "lifetime employment at one company" to a mode of "independence." She says, "For me, i-mode is a declaration of independence. It's 'I' mode, not company mode ... The 'I' in 'i-mode' is about the Internet and information, but it's also about identity."[16]

While *kawaii* culture has been an important tool for customizing and personalizing new technologies in post-war Japan, its deployment within *keitai* cyberspace has seen new roles emerge for women as both consumers and producers. Just as the employment of *kawaii* culture within *keitai* cyberspace is no longer stereotyped as a female preoccupation, men are no longer the prime producers of the technology. In turn, while there may be equal usage of *kawaii* culture among the sexes, arguably it is women who are using *kawaii* modes for customization in a more reflexive – and thus empowering – way. Could this awareness of and overt usage of the *kawaii* herald the shifting position women are occupying in cyber technologies?[17] The fact that the rise of *DoCoMo* is closely associated with one of its senior managers, Mari Matsunaga would seem to suggest so.[19] Stocker points out that:

DoCoMo's i-mode phone has done for the Internet what the Walkman did for hi-fi. The hand-held fashion accessory of the moment is used by 10 million and profits are booming. And the woman [Mari Matsunaga] behind it puts it down to changing the furniture and putting beer in the boardroom.[19]

Matsunaga has attracted attention as a leading example of the new roles being occupied by women in the new economy. Stocker suggests that:

The story of Ms Matsunaga's role in putting the Internet in people's pockets is a parable, of sorts, for a society in the throes of sweeping transformations. It is about the triumph of New Japan over Japan Inc., of concept and creativity over technocratic intransigence. It is about individuality and initiative in a country that has long placed a premium on group identity.[20]

Matsunaga herself argues that:

Japan's traditional system of seniority and hierarchy is breaking down, but the men at its core are still bound up by the system ... There's no room for flexibility. The way things are changing now, flexibility is crucial. It's like a building: in an earthquake-prone country like Japan, you want to have a flexible structure, otherwise it's going to come crashing down.[21]

As McVeigh, and Holden and Tsuruki (this volume) illustrate in their discussions of mobile phone use in Japan, this new technology is very much about individual expression and identity. As Matsunaga so presciently observes, the "i" in *i-mode*, is intriguingly multivalent – signifying both the "i" of information as well as the first-person pronoun.[22] If Castells is correct in arguing that personal power in the "information age" is very much predicated on access to and communication of information, the implications for women in Japan, who have been both key consumers and producers of *keitai* technology, are considerable.[23]

CONCLUSION

Cyberspace in Japan – accessed via *keitai* devices – is being configured and familiarized through very local *kawaii* characters. These characters are often ridiculed and dismissed as meaningless products of a commodity culture. However, the intersections between consumer practices and new technological spaces are far from trivial or mundane but can tell us much about a society. This is particularly the case in Japan where the *kawaii* has become such a ubiquitous presence within *keitai* culture. Yet, despite the current popularity of customized cute images within Japan's cyberspace, it remains to be seen exactly how *kawaii keitai* culture will develop in the cyberspaces of tomorrow. Is the cell phone craze simply a fad, one of Japan's recurring consumer booms? Or, does it signify a more substantial shift in modes of communication, consumption and gender configuration? As Stocker points out:

In this brave new world of wireless Web access, nobody knows whether DoCoMo will reign supreme, or become the Betamax of the cell-phone business. The fun is just beginning.[24]

While the customization of *keitai* cyberspace through the use of the kawaii is a relatively recent phenomenon and thus its role in shifting gender relations is still open to debate, it seems clear to me, at least, that it signifies far more important and lasting changes. As Japanese people increasingly participate in cyberspace, it is the *keitai*, adorned with a plethora of *kawaii* characters, that has become the main portal for Internet access. If customization is, as Castells argues, one key aspect of cyberspace deployment in the "information age," in Japan it is *kawaii* culture that has become the specific mode for particularizing and familiarizing the global space of the Internet.

In this chapter I have attempted to listen to some of the voices of those who participate in *kawaii keitai* culture, not to provide some definitive statement about Japanese Internet usage, but rather to discover some of the meanings people associate with this new technology. In my sample survey, one trend became clear: the link between gender and the cultural capital of *kawaii* is becoming more and more tenuous. Just as the typical consumers of *kawaii* culture are no longer women, so too the typical producers of this culture are no longer just men. Simply put, the intersections between *kawaii* and *keitai* culture provide ways not only for rethinking gender, but also for re-examining a wide range of relationships in contemporary Japanese society.

NOTES

1   The Office of the Prime Minister, *Posts and Telecommunications Policy, 1998–2000*, 2000.
2   F. Rose, "Pocket monster," *Wired*, vol. 9, no, 9, September 2001, pp. 128–35.
3   This relates to the ubiquitous "print club" machines in tourist and entertainment areas where small groups of friends can photograph themselves against a variety of cute backgrounds. The photo is then printed on a sheet of stickers that can be used to adorn letters, cards, memos and other communications.
4   Rose, "Pocket monster," p. 129.
5   M. Castells, *The Rise of the Network Society*, vol. 1, Oxford, Blackwell, 1996, p. 2.
6   S. Kinsella, "Cuties in Japan," in L. Skov and B. Moeran (eds), *Women, Media and Consumption in Japan*, Richmond, Curzon Press, 1995, pp. 220–54.
7   The usage of "official" refers to the sites endorsed by *keitai* companies such as NTT.
8   (As dated October 2000) T. Stocker, "The future at your fingertip." Online. Available HTTP: *http://www.tkai.com/press/001004independent.htm* (20 September 2001).
9   *http://www.comatonse.com/sanriot/hellokarl.html* (15 June 2001).
10 In Japan *kawaii* culture is not a category that appeals only to children. As Kinsella notes, contemporary Japanese *kawaii* culture was informed not only by schoolgirls' development of cute writing but also by male and female student protestors in the 1970s who used it to question traditional Japanese notions of gender and identity. It is important to recognize that *kawaii* culture is not a category just for children as a Eurocentric or Occidental notion of "cute" would assume. See Sharon Kinsella, "Cuties in Japan."
11 M. Castells, *The Internet Galaxy*, Oxford, Oxford University Press, 2001, p 77.
12 One example of *Hello Kitty*'s subcultural reappropriation is "Kitty Flip," a lesbian club night held every third Saturday in Shinjuku, Tokyo.
13 By "mainly" I mean approximately 85 per cent for phone calls, 15 per cent for text messages and e-mails.
14 B.J. McVeigh, "Commodifying affection, authority and gender in the everyday objects of Japan," *Journal of Material Culture*, vol. 1, no. 3, 1996, p. 293. See also McVeigh's "How Hello Kitty commodifies the cute, cool and camp: 'consumutopia' versus 'control' in Japan," *Journal of Material Culture*, vol. 5, no. 2, 2000, pp. 225–45.

15 While the etiquette for public transport is for *keitai* users to turn off their mobile phones, this is not always followed. However, such rules for etiquette mean that many *keitai* users resort to cyberspace preoccupations.

16 Cited in Stocker, "The future at your fingertip."

17 K. Itoi, "Rising daughters," *Newsweek*, 3 April 2000, CXXXV, no. 14, p. 45.

18 Stocker, "The future at your fingertip."

19 Stocker, "The future at your fingertip."

20 Stocker, "The future at your fingertip."

21 Stocker, "The future at your fingertip."

22 The "i" of "*i-mode*" is patterned on the "i" logo of information booths.

23 Rachel Howe points to a number of surveys indicating that the Internet in Japan is becoming "feminized." She notes that advertising firms are beginning to acknowledge that Internet use among women differs in some ways from that of men and "community building" is particularly strong among women: "Cyber feminism in Japan," *WomenAsia.com*. Online. Available HTTP: *http://www.womenasia.com/eng/technology/articles/cyberfeminism_japan.html* (7 June 2001).

24 Stocker, "The future at your fingertip."

5 From subculture to cybersubculture?

The Japanese Noise alliance and the Internet

Costa Caspary and
Wolfram Manzenreiter

INTRODUCTION

The 1990s witnessed an acceleration in the border-crossing nature of Japanese popular culture. At the beginning of the decade, a new kind of avant-garde music called "Japan Noise" appeared on the disc racks of alternative record shops through-out Europe and the US. Soon after, the genre, which is also labelled "Japanoise," or as most Japanese artists prefer it, simply "Noise" (*noizu*), began to spread through the catalogues of independent record mail service companies and enriched the product lines of some major consumer media retail stores. While no narrative account can ever substitute for the experience of listening to a Japan Noise release or a live performance, this definition by American Noise expert Mason Jones is helpful in illustrating the sonic side of what has become a major keyword in the international avant-garde music scene.

> Dominated by the widespread use of distortion and fuzz (effects which add an element of noisiness to any sound), a Noise album washes over you like a tidal wave of sound composed of thundering sheets of hissing, distorted tones. Rapidly shifting shards of sound collide and bounce apart; a heavy clanging may appear and then be washed away beneath rumbling, low-frequency tones; a warbling hum may be cut apart by sizzling, high-frequency feedback.[1]

Unlike manga or J-Pop music, Japan Noise has seemingly transformed itself into "global property" without the distribution channels and support networks of the global media industry. That there are limits, even to the power and control of the major music recording companies, was one of the insights that emerged from the MP3 war of the Napster vs RIAA case at the end of the 1990s.[2] It became clear that technological progress had enabled every Internet participant to become simultaneously a producer, provider, distributor and consumer. The power of the Net to undermine the basic principles of capitalist society has inspired writers on the Internet since the earliest years of computer-mediated communication. Nowadays, hardly anybody would object to the claim that the Internet has potential to impact on power relationships and communication processes. However, Napster's final capitulation to the demands of the major recording companies illustrates that it may be premature to assume the triumph of forces of decentralization, individualism and community over centralization, the market and the state.

Our research into the Japanese Noise network started in 1998 when we sent out questionnaires to various people involved with Japanese Noise in an attempt to get an overview of their activities. While the questionnaires targeted various areas, such as label design, distribution, home page, fanzine editing and, of course, sound production, in most cases all these aspects were controlled by a single person. This initial research generated a whole range of new questions that we put to Noisicians of all generations as well as music journalists and concert organizers in 1999. We carried out interviews and ethnographical fieldwork in Tokyo and Osaka, spending considerable time in live houses and record shops. Only then did we begin to acquire an actual feel for the problems involved in getting a gig, staging shows, and producing and distributing Noise recordings. It became apparent that many of our early assumptions about the Japanese Noise scene, which were often a result of the scene's online presentation, did not live up to "reality." It became clear that the fractures within the subculture that were hardly acknowledged on the Web and avoided in "official" interviews would prove important for resolving many of our original questions.

Our primary impetus was to explore the possibilities for alternative cultural practices and social formations in a world that was dominated by a few major companies.[3] However, the similarities and contrasts that we discovered in alternative and mainstream practices stimulated us to reformulate our research questions for the purpose of this volume. Instead of dealing with the online self-presentation of a loose alliance of people who share a common passion, we chose to focus on the role of the Internet in the production and reproduction of this particular subculture. For this purpose we offer a content analysis of the Noisicians' self-presentation on the Internet dating back to the beginning of our research in 1998, and look at how this has changed in relation to material downloaded from Japanese Noise Web sites in the summer of 2001. Based on this research, this chapter discusses how the new media have been incorporated into the alliance's production modes, distribution chains, communication channels, and meta-narratives. We are especially interested in the social side of this technology-in-use and its influence on creating a distinctive subculture made up of an alliance of artists and fans, producers and consumers.

EMPOWERMENT, RESISTANCE AND THE WEB

Despite the Japanese claim to have coined the concept of the "information society,"[4] and despite the existence in Japan of decades-old bureaucratic blueprints for the information society, academic discourse on the social impact of media and telecommunication technologies has always been heavily influenced by Western traditions (predominantly those in the US). Reasons for this situation are abundant. For example, compared to the US and great parts of the Western world, in Japan, private PC ownership only started to take off in the middle of the 1990s. Further-more, the particular structure of Japan's telecommunication market obstructed private household access to the Internet or other proprietary networks until recently.[5] Because of the comparatively low visibility of network usage, both the social sciences and the humanities turned to Internet research rather late and consequently the theory, concepts, and tools of the Japanese debate are strikingly similar to those pioneered in the West.

No social critic has been more influential in shaping the academic debate about the media in Japan than Jürgen Habermas. His doctoral dissertation from the early 1960s, *The Structural Transformation of the Public Sphere*, revolved around the historic development of the public sphere as a new arena of political participation in a liberal society. The public sphere, as Habermas suggests, may be conceived as above all "the sphere where private people come together as a public ... to engage in a debate over the general rules governing relations in the basically privatized but publicly relevant sphere of commodity exchange and social labor."[6] In the late 1990s, social scientists such as Yoshida Jun, Hamada Junichi, Takeshita Kōshi and numerous other writers on governance and civil society, applied the concept of the public sphere to the analysis of mailing lists, bulletin boards and chat rooms on the Internet.[7] Books like Tachibana Takashi's bestselling *The Internet is the Global Brain* celebrate the empowerment of modern subjects who can gain access to knowledge of the world and articulate their views in front of a world-wide audience.[8] In contrast to what can be seen as sheer techno-optimism, other books such as *Holy Virtual Reality*, by sociologist Nishigaki Tōru, warn against the naïve conception of the Internet as a social space which is free from any political or economic influence.[9]

This oscillation between utopian and dystopian attitudes towards the social impact of the Internet is quite similar to that in the debates undertaken in great parts of the Western world. Japanese society probably also shares with these nations a longing for community and the communal life of the past which might explain the dominance of the idealist approach. Psychologists' research into Internet use among Japanese youth has revealed that they appreciate highly the anonymity of the medium because it frees them from the usual conventions of a status-conscious society,[10] and, as McVeigh (this volume) points out it also frees them from parental surveillance and control. In terms of personal involvement, the Internet offers ample opportunities for direct participation, which small-scale communities could be considered particularly in need of. The fact that participation in Internet forums is less intimidating because membership is more fluid and open, requiring comparatively less commitment, may lead to the development of more widespread ties.[11] Itō Haruki's ethnographic account of "online life" in a virtual fan club suggests that information exchange and even passive participation via bulletin boards may exert the same kind of community-building power as Benedict Anderson in *Imagined Communities* argues was the case with daily newspapers in the late nineteenth century.[12]

Given that the Noise alliance relies heavily on digitalized production work (so-called "industrial music") as well as personal computer usage and computer networks to disseminate its products, the deliberate "cyborgization"[13] of the subculture seems to be inevitable. This is particularly the case when online networks have to compensate for the comparative scarcity of real-life meeting points, normally of vital importance for the survival of a dispersed, small-scale and loose alliance such as the Noise scene. If it is the case that technology changes a subculture, does this transformation amount to the birth of a "cybersubculture?" David Bell's working definition of cybersubcultures as social formations that either signal "an expressive relationship to digital technology ... or make use of it to further their particular project" is too wide because it lacks a clear guideline that distinguishes a subcultural from a mainstream cybercultural formation.[14] For the purpose of our study, this somewhat blurred category will be confined to social formations whose members pursue a non-commercial, subcultural project that is essentially dependent on communication technology for its existence. We recognize a cybersubculture when the relationship between technology, on the one side, and the social structures and communicative processes that constitute the community, on the other, are so intimate that without the technology, this subculture would cease to exist.

MAPPING JAPANESE NOISE

In the 1960s and 1970s, folk and rock music reshaped Japan's popular culture much as it did that of many Western countries. The subsequent punk revolution was spearheaded by the Tokyo Rockers, who were no less provocative than their British or American counterparts, and proved ultimately as disappointing to the fans when they were soon coopted by the music industry. This cooption reached a provisional peak during the so-called *indiizu būmu* (1982–6), which is something of an oxymoron, since "indie" used to be associated with small record companies (and their bands) but was now used to refer to rock bands selling millions. During the subsequent band boom (1986–93) this cooption was completed: when former underground values turned mainstream, the music scene was refashioned into a mainstream of minorities.[15] Driven by a desire to present their individuality through difference from a mainstream that had become harder to define and yet seemed to be more forceful than ever, Japanese musicians again articulated their style of deviance in ways similar to their comrades overseas. Some people, most noticeably in the Kansai area, with its noise-infected bands orbiting the infamous Alchemy record label, searched for sounds that would prove non-commercializable within the confines of conventional rock music. Others turned to club-oriented sounds, beats and samples that were a provocative rejection of the mainstream's obsession with rock. Nowadays, it seems that a strict binary system of underground versus mainstream music cultures (and at the same time groups that can be mapped solely by styles or genres) has vanished completely. The question of whether a musical style, a group of fans or a group of musicians is articulating something different, new or innovative, something that points *out of this place*,[16] needs to be mapped in a much more sensitive way.

The difficulties of mapping and assessing the Noise subculture reflect the same dilemma participants in alternative music themselves face, as can be vividly seen in fan discussions of authenticity and possible sell-outs.[17] To position oneself outside the mainstream and articulate this position in a credible manner has become more difficult than ever before. Our research on Noise has revealed that fans and producers still fundamentally believe that Noise is non-commercializable; it is regarded as something special, which stands outside mainstream popular music. While this belief constitutes an important meta-narrative, other meta-narratives of Noise are clearly visible in its production modes, which are based on a strictly do-it-yourself attitude, and the self-exploiting manner in which Noisicians invest their energy in their art.

"Japanese Noise" is primarily associated with two units: Hijōkaidan, the self-appointed *Kings of Noise* (album title), and Merzbow who toured Europe and the USA successfully in 1989 and 1990. Their violent sounds and challenging performances represent yet another counterpart to the common cliché that Japan is a peaceful and orderly society. This contrast resulted in a rather one-dimensional reception as Western Noise fans searched for the exotic in the experimental sounds. The term "Noise" itself is used in Japan to cover a wide range of experimental genres from industrial music to free jazz. If someone deserves to be credited with having coined the term, then it is Akita Masami. Inspired by the idea of Mail-Art,[18] Akita's artistic alias Merzbow began to produce at home and distribute around the globe his self-recorded tapes as far back as the early 1980s. Since the music scene in Tokyo at that time was preoccupied with the *indiizu* and band booms, there was little space for the development of musical alliances that did not fit in with this "rock band" mentality. Yet, in Osaka, the Alchemy label, as well as the existence of more meeting and performance spaces, supported the development of a Noise scene with very local identification. However, even for this scene neither the label "Noise" nor the prefix "Japanese" is really appropriate since local obligations outweighed musical ones; if anything, diversity was the most important common feature of the local scene. Yamamoto Seiichi, founding member of Japan's well-known alternative act The Boredoms, explained the lack of mutual bonds as due to the widespread sense of individualism in Kansai: "I think people are physiologically not able to come together with others."[19] The Osaka scene gained a reputation for non-conformity, as it was not only opposed to the musical mainstream but also a self-assured opponent to Tokyo, the all-encompassing media center. The avant-garde musicians in Osaka had central hangouts like the Crusade and probably most importantly, the Egg Plant. This club opened in 1983 and combined rehearsal rooms, a live house and a bar in one. In this environment, many experimental bands were formed that released records on Alchemy and regularly appeared live on stage, and personal friendships with visiting artists from Tokyo and other parts of the country were established on the floor or at the counter.

Yet it is important to understand that for the first generation of Merzbow, Hijōkaidan, Incapacitants and other early Noisicians, back in the 1980s Noise did not signify a fixed genre, but rather an "artistic stance."[20] These musicians did not see themselves as producing a distinctive format but as explorers searching for new sounds and ways of expression. The positive feedback from abroad in the late 1980s and early 1990s contributed to the explosive increase of Noise acts in the early 1990s. Eventually, an actual Noise scene emerged, with a second generation comprising bands such as K2, MSBR, Masonna, and CCCC that centered on the sound of their predecessors: "pure" and "harsh" Noise (*pyūa noizu, hāshu noizu*). The pioneers themselves, who had always been motivated by the search for new ways of sonic expression, were full of criticism for the second generation's copycat approach which bypassed the all-important experimental stage. When Hijōkaidan's Hiroshige talks about younger Noise acts, the slightly derogative term *mane* (imitation) is often heard. The generation gap also stretches over the next generation that was formed after the boom period. This generation faces the greatest institutional difficulties. If artists want to get booked by a club in Japan, they are required to buy a certain percentage of tickets in advance. The *noruma* system, which was introduced during the band boom years, makes it very hard and financially risky for young Noisicians, who cannot guarantee a good turn out, like the big names can.

The fact that Noisicians themselves perceive the genre and the scene as being structured by generations may hint at the actual root of the problem. Our classification corresponds to the perspective of MSBR's Tano, while others who put more emphasis on biological age than on effective duration of career, count four generations. The degree of inter-generational cooperation, even communication, is surprisingly low. Hiroshige of Hijōkaidan put it to us most clearly: "Why [do I hardly have contact with younger Noisicians]? Because I belong to the first generation! The founders of the scene!" Tano is equally aware of the gap, and contrary to most young Noisicians, does not hesitate to talk about it:

> The biggest difference [between the Western and Japanese Noise networks] is the big separating ditch between those in their thirties, and those in their twenties. This is a specific characteristic of the Japanese social structure ... There is this unconscious attitude among older people, which stops them from appreciating the merits of people of lower status [i.e. age], no matter what they do. The younger get upset about this, because they dislike this strictness, and it is hard to develop friendly feelings toward them. It is a big gap, especially since the layer of people in their late twenties is rather thin.

Tano tries hard to bridge the generation gap. By organizing shows and publishing the fanzine *Denshizatsuon*, he actively creates space to bring Noisicians from the older and younger generations together. But hierarchy and distance have been powerful forces of social order for centuries in Japan and continue to exert influence even in the setting of alternative lifeworlds. However, different reasons are sometimes given for the divide. Musical reasons, for example, surface when Akita Masami complains about the orthodox Noise fixation among the young "Noise maniacs" or when Mikawa, a former member of Hijōkaidan and founder of the Incapacitants, complains that there is insufficient "powerful Noise" around these days.[21]

Another reason why the younger generations face more difficulties is that the context of alternative musical practice has itself changed. The mainstream is constantly on the lookout for musical novelties, and as provocative acts such as Marylin or The Rich illustrate, extreme musical expressions are no longer "underground" per se. Just as the mainstream once pocketed the Tokyo Rockers, some Noise-related artists who gained popularity in the USA and Europe were invited to release records on major labels.[22] In the current constellation musical style alone cannot serve as a foundation for an alternative alliance like it used to in the subcultures of the 1970s, since the argument that style itself represents deviancy is no longer valid.

If stylistic extremes can no longer even pretend to guarantee artistic individuality, then other substitutes are high in demand. One alternative method is production mode. The young generation of Noise artists therefore sticks to a radical DIY attitude, ignoring all the economic standards of the industry. Producing tapes or CD-ROMs at home by oneself requires more emotional and physical investment. Consequently, this preoccupation with self-production creates a mode of self-presentation in which seclusion and isolation are considered meritorious. Quite separate from the commercialization of the J-Pop scene, this network has its own rules; it is an "uncontrolled and free area" similar to the huge network of manga fanzine producers in Japan.[23] As with the amateur manga fandom, autonomy in production and distribution has been boosted through the spread of the Internet, a medium that seems to be becoming more and more the primary agent in creating both institutional and communicational space for the young Noise scene.

UTILIZATION OF THE NEW MEDIA

Internet usage by the Japanese Noise scene tends to correlate with the generational substructures outlined above. When browsing the Web for Japanese Noise sites, it is particularly notable that none of the original style-defining acts has a personally supervised home page. Instead, one encounters a handful of authorized as well as unauthorized fan pages from around the world that are dedicated to these bands. Only three members of the first generation run their own Web sites: Alchemy Records, Vanilla Records and Noisecapture.[24] However, as all of them have intentionally broken with the orthodox fixation on the release of pure or harsh Noise, they can hardly be considered central to the Noise scene. Deliberately avoiding the term Noise as much as possible, they now deal with a wide range of experimental music. All other Noise Web sites can be assigned to the middle and younger generations of Noise artists.

Web sites that are created for the purpose of communicating alternative musical practices can be divided into the following categories: e-zines, label Web sites and personal Web sites of fans or artists. These categories serve as structuring devices that link the Web site with the world-wide Noise community. It should be noted however, that sites not only *may* transgress from one category to another, they usually do.

Typical label Web sites are run exclusively by the more established labels created by members of the first generation, Alchemy and Vanilla Records. These professionally created Web sites are not produced by the label owners themselves. It is quite obvious that these Web sites serve one simple purpose: to promote and sell products. Hiroshige of Alchemy Records opened a non-virtual record store in Osaka in 1999. He has also installed a typical online store at *www.alchemy.cc* where all kinds of independently produced records are sold. Notably, many of the products released by the autonomous mini-labels of younger Noisicians are not in stock. Instead *alchemy.cc* focuses on indie-labels with established distribution arrangements. Characteristic examples include PSF and Vanilla Records. Both are labels that have hardly anything to do with the young Noise scene.

In contrast to label sites, personal Web sites are created for the purpose of self-presentation. They are not aimed at selling Noise products since they are usually run by fans, not producers. Most feature a large variety of topics that may be Noise-related, but this is not necessarily so. The weaker the focus on Noise is, the fewer the number of links leading from Noise sites towards these peripheral sites. The only non-label/artist site that is frequently linked is *Noisembryo*, which is named after a Merzbow album.[25] The Web master of this site is Noise maniac and collector Itō Gen. Besides offering some information on Gen, this site contains a fairly complete discography of most Noise units, a bibliography on Noise, and listed links of labels, live houses and record shops that feature Noise bands. Gen told us that his home page was originally meant to serve as his personal databank, but it soon developed into a meeting place for Noise fans from all over the world. This is an interesting case of a personal site which may at first glance seem to have been produced out of the fandom-fueled zest for compiling information about one's own favorite music that was later transformed into a communication space for the Japanese as well as the international Noise alliance.

The majority of Noise-related sites, however, are created by the numerous young artists who are producers, label managers and Web masters all in one. Usually these sites focus on releases and live performances of the artists. This emphasis does not imply that the artist's ambition is solely to sell. Rather, the list of self-recorded tapes or CD-ROMs serves to express productivity and creative investment to an audience that rates "output" as the primary purpose, the *raison d'être* of a Noisician: "The production of releases is the most important [motivation], because you get responses and a network starts to develop and to spread. [The production] is a means to communicate. You get to know people and that is what it's all about," said former CCCC front man Hasegawa, who now performs solo under the name Astro.[26] To put it more bluntly, in order to get accepted by the Noise scene, you have to *produce*.

As one Noise observer stated, the distinction between artist and consumer that structures the mainstream is blurred within the underground:

> [I]t's worth noting that avant-garde and Noise scenes everywhere are much more participatory than mainstream music. That is, fans of mainstream bands attend shows to watch the bands, then go home: it's passive entertainment.

That's usually not the case with Noise and avant-garde audiences. These scenes are built around a core of people who are not only audience members, but also participants … Over the years I have found that this sort of participatory scene – at the risk of becoming too insular – helps to inspire members and fosters the energetic exchange of ideas and creativity.[27]

The insider view is based on the classic "bad mainstream" versus "good underground" rhetoric, as differentiation from the mainstream provides a basic source of subcultural identity. The simplified assumption that "fans of mainstream bands" are merely interested in passive entertainment is questionable in itself as it ignores the sometimes close relationships and mutual exchange of talents, ideas and resources that take place between mainstream and underground. The example of the amateur Japanese manga movement clearly demonstrates that followers of highly commercialized mass products have enormous creative potential – something that can be observed among fans of idol singers (*aidoru*) as well. Stigmatized at the beginning of the 1990s for being anti-social, the so-called *otaku* have produced fanzines (in Japan usually called *minikomi* or *dōjinshi* among the manga fans) to such an extent that their output itself has become the focus of another subculture. In contrast to fan alliances of mainstream bands, the Noise scene forcefully demands "output" from its members, based on the punk rock notion: either everybody or nobody is a star.

The step from consumer/fan to producer/Noisician is a small one, since no instrumental skills are needed. Noise is easily generated through the use of distortion and other electronic devices. Lately, usage of PCs and laptops, most famously exercised by Merzbow, has increased. Furthermore, CD-ROM burners and the Internet have facilitated the distribution of low-key releases. The young generation of Noise artists makes excessive use of these technologies, and most create their own tiny labels that cannot even be categorized as indies. Each of these self-produced labels (*jishu-seisaku rēberu*) runs its own Web site where the Noisicians usually provide more than only noise-related information. Private life affairs or activities are featured on pages that are linked via connection buttons such as "profile" or "personal info." How much is revealed about the site owner varies greatly. Most noticeably, hardly any of the owners states their profession, which is usually of primary social importance, but is irrelevant here for signifying the owner's social status. Among Noisicians, only the self-presentation and the output of the artist/label owner count.

Such "personal label" sites do more than just provide a forum for ego-tripping around an individual's musical fetish since they also create communication spaces, where the owners can articulate their own individual stances as well as seek responses, if only in a passive manner. During our first survey in 1999, more than half of the Web sites examined sported bulletin boards, guest books or similar interactive forums but, due to the small size of the scene, most of these forums remained unvisited. In the summer of 2001, less than one third of Noise Web sites continued to provide these services, and only those major forums at central Web-crossings, such as *Msbr* and *Sssm*, were visited regularly.[28] Another example of a central Web site is the *World of Ishigami*, which is supervised by a central Noise-scene figure in Osaka.[29] Large sites like this offer information that is of vital interest to the whole alliance, including regular updates of events, articles on Noisicians and reviews of live gigs. Yet, given the lack of interaction on many of these sites, it is valid to question the ability of this kind of Web site to create alliances. However, writing and posting an article on a Web site is not the only prerequisite for establishing a communication space. Web sites that are linked among members of the Noise alliance are perhaps better viewed as personal rooms with doors leading to the rooms of other artists/label owners. When linked to one another in this manner, they can create a communication space that has multiple possibilities.

Currently, Web sites are either a substitute for or supplemental to what was previously the most important communication device for subcultural groups: fanzines, that is, small, self-produced, and more or less non-profit publications circulated among members. Fenster notes that the "formation of a dispersed community of fanzines has enabled the circulation of emergent identities and helped to establish a network of publications, individuals and groups with which people could identify and get involved."[30] Of particular importance is the aspect of getting people involved. For small-scale scenes such as the Noise alliance, every new recruit is of vital interest for the whole scene. "You need a great scene to get great artists. To get a great scene, you need great listeners. To get great listeners, you need to educate," stated Tano, when he explained why his fanzine *Denshizatsuon* features articles on Noisicians of all generations.

Tano also runs what is currently the most important and central Web site: *msbr.com*, which is the name Tano uses for all his Noise activities. Early in 1999, *msbr.com* used to be divided into two parts: "information on Noise," where new releases or live dates were announced, and the "Noise forum," where Tano invited visitors to respond to questions he posed for discussion. In the meantime this forum has been cancelled. Yet the attempt shows that the medium of the Internet potentially enables Web sites to surpass the possibilities of printed fanzines as a means for getting other members actively involved. Msbr.com still features categories like diary entries or large forums that fulfil the same function as fanzines. Yet Tano would not consider *msbr.com* to be a substitute for the print magazine *Denshizatsuon*, which is the only fanzine dealing exclusively with Noise in Japan today, and has a fairly large circulation of 1,000 copies. "I don't intend to publish *Denshizatsuon* solely on the Internet. [On *msbr.com*] I try to add personal views and comments." According to Tano himself, the home page is primarily meant to promote his own activities. Yet, as a secondary function, the integrated bulletin board should help to "construct an international network."

E-zines are more appropriately seen as the digital equivalent to fanzines. Among the Noise media, however, only *Noisecapture* comes close to the idea of an e-zine.[31] Run by Okazaki, another first generation Noisician and central figure of the infamous improvisation unit Dislocation, *Noisecapture* covers a wide range of improvisational and experimental music. Like many others of his generation, Okazaki dislikes a strict Noise-only attitude. But the e-zine is a very important medium for Noisicians in the Kansai area where Okazaki occupies a similarly central position to that of Tano in Tokyo. He particularly encourages the young generation to write and publish online with *Noisecapture* where Noise releases and concert reviews are regularly featured. Previously, Okazaki also published the print media *Cos* and *Noisecapture*, but came to the conclusion that "the financial risk and the difficult organisation are real problems one cannot escape from." He finally decided to concentrate on the digital version.[32]

CONCLUSION: SOCIAL ORDER IN JAPAN AND THE INTERNET PROMISE

As our brief discussion of the Japanese Noise alliance's utilization of the Web has shown, the relationships between technology, its use and specific cultural projects are very intimate. The Internet in Japan has certainly not fulfilled the dream of emancipating cultural producers from the all-encompassing power and control of the media industries. But it has opened new windows of opportunity. The Web has made it possible to reach for the world without selling out. It has enabled artists to maintain a certain degree of "isolation" without necessarily being isolated. Most Noisicians agree upon the impossibility of making a profit from selling their products and so the Net is used for exchanging audio files or self-made releases rather than for selling the sound. To this extent, the Web has helped Noisicians to maintain their sense of identity and authenticity and to avoid objectionable compromises with the mainstream. It facilitates the maintenance of the community's sense of identity which is largely based on its sense of independence. The central concern of this chapter, however, has been to look at the way the Internet has contributed to the construction and reconstruction of a subculture.

Noisicians' Web sites serve as communication spaces, forums for self-presentation and, to a certain degree, commerce, too. However, we are reluctant to categorize the Noise alliance as a "cybersubculture." In order to understand our reluctance to use this term, it is helpful to inquire how much reliance on the Web is necessary in order to speak of a cybersubculture? It could be argued that the Internet has facilitated the Mail-Art-like production and communication modes exercised by Merzbow in the 1980s. Yet the artist could not represent himself as part of an alliance (and probably did not want to) because no coherent Noise alliance existed at that time. In contrast, the new generation's sometimes orthodox attitude towards Noise certainly helped to strengthen an alliance that represents itself as being much stronger in the virtual world than in the physical one. Many Noisicians have withdrawn into an intimate world where they exercise total autonomy over their production and distribution of Noise, because they want to be part of a world-wide online Noise alliance that no longer needs to rely on physical artefacts such as fanzines, independent record labels or live houses. Therefore they themselves reject the notion of a Japanese Noise cybersubculture and even "Japanoise" itself because they see themselves as participants in a deterritorialized, border-crossing project.

As we have seen, it is no easier to keep track of the Noise scene on the Internet than it is in real life since, due to the generation gap and the absence of any center, this small scene continually fragments into ever smaller units. Therefore, the many possibilities the Web offers for forming alliances are insufficiently utilized. We were surprised to observe how it was in fact mainstream commercial practices that triggered the younger generation's Internet usage. Yet, due to their lack of status, power and centrality within the scene, their Web sites do not generate substantial traffic and therefore fail to fulfil the seminal functions of the marketplace in traditional village life. The fact that offline power relations continue to shape the online community is a severe blow to any idealistic hopes that the Internet is inherently an equalizing medium.

Due to the internal structure of the subculture, it has turned out that the major Web axes are supervised by those members who are the least interested in the notion of a Noise scene. Also, due to the generation gap which limits newcomers' opportunities for status enhancement, the major objectives of Web participation have become self-display and self-promotion. Hence, Internet usage within the community is uni- rather than bi- or multi-directional. Personalized accounts, such as those offered on Web sites or in e-zine articles, only represent the claims of a few specific individuals, while the majority of visitors remain anonymous. The Web sites of the Noise alliance, which are potentially interactive as well as informative, have not lived up to their interactive potential. Thus the drawbacks of computer-mediated communication seem to have a debilitating effect on subcultural users as well.

The influence of computer-mediated communication on social relationships and community development has been analyzed in numerous studies in such diverse fields as sociolinguistics, psychology and communication studies where numerous observations on the negative impact of the technology have been made. It has been argued that conventional notions of community are undermined in situations in which language is reduced to script and loses its non-verbal signs, and where communication takes place in anonymous or pseudonymous situations lacking explicit knowledge of who actually is talking or listening and where the rules of conduct are not clearly understood. Cybercommunities are easily joined, and even more easily left behind as commitment and obligation to participate are comparatively weak requirements of membership. This does not necessarily imply that all cybercommunities are characterized by greater social distance or that all members of a given cybersubculture are uncommitted or just passive observers. Communication in a semi-public, semi-private communication space like the Internet may be for purposes other than the exchange of ideas and opinions. Morioka Masahiro and Kawaura Yasuyuki have pointed out that self-promotion and self-presentation are characteristic traits of computer-mediated "conscious communication." [33] As we have observed in the case of the Noisicians, the Web neither strengthens the community, nor does it necessarily enrich interaction. However, if understood as a supplement to real-life interaction, it opens up new spaces for self-presentation. Yet, if understood as a substitute to the "real marketplace," it can be argued that the Web has actually contributed to the fragmentation of the Noise scene.

NOTES

1 M. Jones, "Noise," in A. Roman (ed.), *Japan Edge*, San Francisco, Cadence Books, 1999, pp. 75–100; p. 76.
2 Napster was developed as an indexing and virtual community service that simplified and expanded opportunities to share and download music files for free on the Internet. At this virtual meeting place individuals could look for compressed MP3 files stored on the hard disk of any other participating computer of the Napster community. Soon, more than 20 million users joined in the biggest music swap in history. In December 1999, the Recording Industry Association of America (RIAA) charged Napster with music piracy and copyright infringement. Napster denied the action because it neither stored the files nor did it know which files were actually traded within the community. A preliminary injunction against Napster was issued in July 2000 which was finally confirmed in March 2001, requiring the company to stop the trading of copyrighted songs. In the meantime, Napster and BGM, one of the five major labels that have legally challenged Napster, surprised the other plaintiffs by announcing the creation of a joint fee-based secure membership service. The controversial deal was consolidated by a loan of $50 million that guaranteed Bertelsmann controlling rights in the new Napster scheme. The new company soon found itself fighting a two-front war: against Sony, Warner Music and the other major labels refusing to team up, and against the customers who switched over to alternative peer-to-peer Internet services such as Gnutella, Swapoo, Mojo Nation, or Freenet.
3 See C. Caspary, "Das japanische Noise-Netzwerk. Eine Diskussion über die heutigen Möglich-keiten Alternativen zu Mainstreamkulturpraktiken zu schaffen" (The Japanese Noise Alliance), Vienna, 2000, unpublished MA thesis accepted by Vienna University (in German).
4 See Castells, M. *The Information Age: Volume III, End of Millenium*, Malden, MA, Blackwell Press, 1998, pp. 236–7; T. Morris-Suzuki, *Beyond Computopia. Information, Automation and Democracy in Japan*, London, Kegan Paul, 1988, pp. 3, 7–8. Morris-Suzuki refers to the former Economic Planning Agency (EPA) advisor and Tokyo Institute of Technology (TIT) professor Hayashi Yūjirō whose main work on the information society was published simultaneously with the first government reports on the issue in 1969. However, Itō Yōichi identified anthropologist and biologist Umesao Tadao, who later was appointed director at the National Museum of Ethnology (Minpaku), as the inventor of the concept as he speculated as early as 1963 about an evolutionary model of society based on the spread of information technologies and services (see Itō's "The birth of Joho Shakai and Johoka concepts in Japan and their diffusion outside Japan," *Keiō Communication Review*, 13, pp. 3–12). Itō's explanation seems to have turned into the stock repertoire of the global discourse, even though Umesao's essay did not contain the term information society as such.
5 For a detailed account, see W. Manzenreiter, "Japan's Digital Unite: Grundlagen und Grenzen des M-Commerce in internationalen Vergleich" (Japan's digital unite: conditions and limitations of mobile commerce in international comparison), in C. Erten and R. Pirker (eds), *Wirtschafts-macht Süd-Ost-Asien: Länderspezifische Erfolgsfaktoren für wirtschaftliches Handeln*, Vienna, Wirtschaftsverlag, 2002, pp. 187–204.
6 J. Habermas, *The Structural Transformation of the Public Sphere. An Inquiry into a Category of Bourgeois Society*, Cambridge, MA, MIT Press, 1991, p. 27.
7 Yoshida J., *Intānetto kūkan no shakaigaku. Jōhō nettowāku to kōkyōken* (Sociology of Internet space. Information networks and the public sphere), Kyoto: Sekai Shisō Sha, 2000; J. Hamada, "The right to information: the core of the network society," *Review of Media, Information and Society*, 4, 1999, pp. 69–78; K, Takeshita, "*Shakai kagaku no tenkan to kindai seiō bunmei. Dejitaru shakai to gendai Ajia*" (The transformation of social science and Western civilization. Digital society and contemporary Asia), in *Keizai Ronshū* vol. 50. no. 3, December 2000, pp. 35–62.
8 T. Tachibana, *Intānetto wa gurōbaru burein* (The Internet is the global brain), Tokyo, Kōdansha, 1997.
9 T. Nishigaki, *Sei naru bācharu riaritii-jōhō shisutemu shakai ron* (Holy virtual reality: discourses on the information system society), Tokyo, Iwanami Shoten, 1995.

10 K. Miyata, "Netto shakai no miraizu: nettowāku komyuniti kara mita seikatsu sekai no henyō" (Future plans of the network society. Change of everyday life, seen from virtual communities), in *Shin Chōsa Geppō* 19, 1999, pp. 54–7; M. Matsuda, "*Keitai ni yoru denshi mēru kyūzō to sono eikyō*" (Rapid increase of mobile e-mails and its impact), *Nihongogaku*, 17 October 2000. Online. Available HTTP: *http://www3.justnet.ne.jp/~misam/nihongogaku.html* (13 May 2001).

11 Some researchers, however, forecast the rising isolation of individuals who fail to cope with real life. Interestingly, the current phenomenon of *hikikomori* is discussed in similar terms as the *otaku ron* a decade before. See Y. Shiokura, *Hikikomori* (Confining oneself indoors), Tokyo, Birēji Sentā Shuppan Kyoku, 2000.

12 H. Itō, *Tsūshin kaisen no on to ofu–aru ongaku fan no baai* (On and off in data transmission networks), 1998, unpublished manuscript. Online. Available HTTP: *http://member.nifty.ne.jo/ haruki/works/bz/onoff.htm* (1 February 2002). B. Anderson, *Imagined Communities: Reflections on the Origin and Spread of Nationalism*, London, Verso, 1983.

13 T. Terranova, "Post-human unbounded. Artificial evolution and high-tech subcultures," in D. Bell and B. Kennedy (eds), *The Cybercultures Reader*, London, Routledge, 2000, pp. 268–79.

14 D. Bell, "Cybersubcultures: introduction," in D. Bell and B. Kennedy (eds), *The Cybercultures Reader*, pp. 205–9.

15 Holert *et al.,* who had observed a simultaneous process happening in the West, later coined the term "mainstream of minorities" (*Mainstream der Minderheiten*). See T. Holert and M. Terkessidis (eds), *Mainstream der Minderheiten. Pop in der Konsumgesellschaft*, Berlin, Edition ID-Archiv, 1996.

16 This refers less to The Animals' song "We gotta get out of this place" and more to Lawrence Grossberg's fundamental discussion of rock and postmodernity: L. Grossberg, *We Gotta Get Out of this Place: Popular Conservatism and Postmodern Culture*, New York, Routledge, 1992.

17 With reference to Japanese Rap, Ian Condry has argued that the reproach of being simply an imitation of an American phenomenon led to the circulation of ideas of authenticity among commercial party-rappers as well as the anti-establishment underground hip hop community, albeit with different results. I. Condry, *The Social Production of Difference: Imitation and Authenticity in Japanese Rap Music*. Online. Available HTTP: *http://www.yale.edu/~condryi/ rap/* (15 June 1999).

18 Mail Art developed primarily out of the 1960s Fluxus Movement and attempted to bypass the established art market by encouraging everybody to exchange and trade their art, ideas and concepts via postal services. Akita traces the appearance of industrial music and Noise back to this conceptual approach in his collection of essays. See M. Akita, *Noise War: Noizu myūjikku to sono tenkai* (Noise war: Noise music and its outlook), Tokyo, Seikyūsha, 1992.

19 Yamamoto Seiichi, interviewed by Matsuyama Shinya, in D. Onojima, *Nyū senseishonzu. Nihon no arutanatibu rokku 1978–1998* (New sensations. Alternative rock in Japan 1978–1989), Tokyo, Myūjikku Magajin, 1998, p. 93.

20 Akita Masami, quoted in *Eureka* March 1998, p.161. Only after the boom of the early 1990s did "Noise" became an established term among fans of avant-garde music when even the Japanese *Music Magazine* featured a Noise special. See "Zatsuon-gunron. Noizu ni torikumu ātistotachi no manifesto" (The group model of Noise: the Noise artists' manifesto), *Myūjikku Magajin*, October 1995, pp. 90–7.

21 Quoted in *Eater*, 1999, pp.7–8, 27.

22 The Boredoms signed to a major company in 1992, and Violent Onsen Geisha in 1995.

23 S. Kinsella, *Adult Manga: Culture and Power in Contemporary Japanese Society*, London, Curzon Press, 2000.

24 Alchemy Records: *http://www.alchemy.cc*; Vanilla Records: *http://www.dance.ne.jp/~vanilla*; Noisecapture: *http://www.web-rain.com/noisecapture/* (1 February 2002).

25 *Noisembryo*: *http://www.asahi-net.or.jp/~er6g-itu* (1 February 2002).

26 Personal conversation, 21 April 1999.

27 Jones, "Noise," p. 91.

28 *Msbr*: *www.msbr.com*, *Sssm/Contagious Orgasm*: *http://plaza26.mbn.or.jp/~sssm/* (1 February 2002).

29 Ishigami Kazuya is a member of the Noise four-piece Billy?, plays solo as Daruin and has organized many live shows in Osaka. His Web site is *World of Ishigami*: *http://www.osk. 3web.ne.jp/~ishigam/* (1 February 2002).
30 M.A. Fenster, *The Articulation of Difference and Identity in Alternative Popular Music Practice*, Ann Arbor, UMI Dissertation Service, 1992, pp. 138, 135.
31 *Noisecapture*: *http://web-rain.com/noisecapture* (1 February 2002).
32 *Noisecapture* 1, p. 120.
33 M. Morioka, *Ishiki tsūshin: doriimu nabigeita no tanjō* (Conscious communication: the birth of dream navigator), Tokyo, Chikuma Shobō, 1993; Y. Kawaura, "Intānetto ni okeru nettowāku no tokuchō" (Particularities of networks on the Web), in *Nihon Shinri Gakkai dai 63 kai taikai happyō ronbun shū*, Nagoya, Nihon Shinri Gakkai, 1999.

6 Filling in the blanks

Lessons from an Internet Blues jam

Gretchen Ferris Schoel

INTRODUCTION

It had seemed like such a good idea. Having grown up in Alabama and Mississippi, I had always loved the Blues. Since taking a job in 1996 at the Shonan Fujisawa Campus of Keio University, Japan's foremost institution for Internet research, I had been delighted to discover so many Blues artists in my adopted homeland. Tokyo was filled with Blues bars. So it was natural enough, I thought, to propose an experiment: an Internet jam session in real time, linking Blues musicians in the two countries. Because Clarksdale, Mississippi, is the legendary birthplace of the Blues, I decided that this global Blues jam would work best during Clarksdale's annual Sunflower River Blues and Gospel Festival, an event known to Blues aficionados world-wide.

The experiment worked, but not at all in the way I planned. On 14 and 15 August 1999, Japanese and American Blues artists participated in what appears to have been the first Blues jam ever to connect musicians over 10,000 kilometers apart. Playing through the Internet, about twenty musicians performed the legendary songs of Robert Johnson, Son House, Muddy Waters, and Johnny Lee Hooker. Artists on both sides of the link also shared some of their own compositions and had the pleasure of hearing their themes picked up by their counterparts across the world. The artists jammed across an ocean, two languages, two cultures, and several time zones, and through a fiber optic cable, a phone line, cameras, mics, computers, a projector, conferencing software, and all the digital bits and their pathways that comprise what we call the Internet. Jamming through the Net turned out to be everything jamming is supposed to be – improvisational, unpredictable, impassioned, exhilarating, exhausting. As at any successful jam, the musicians in Tokyo and Mississippi created new sound through inventive play upon a set of simple patterns.

But there were also problems – big problems. One was that while one side was drinking coffee, the other was getting drunk. After all, the jam happened on Saturday and Sunday mornings in Clarksdale, already late at night the same days in Tokyo. But the greater challenges were technical. Ironically, given Tokyo's technological advantages over rural and impoverished Clarksdale, Tokyo was the weaker link. While my students and I had managed to lease a room and the high-speed lines of Clarksdale's one and only Internet Service Provider, the best we could do in Tokyo's popular Bright Brown Blues Bar was a phone line. The result was that we often lost our audio connection, the video feed, or both – sometimes in one direction, sometimes in both. Systems crashed and audiovisuals fell out of sync. Songs, jokes, and greetings mixed together.

The disorder was painful, but also exciting. If the connection had not worked at all, we could have given up and gone home. What we got instead was something probably familiar to almost anyone first experiencing real time audiovisuals across the Net: the frustrating lure of intermittent clarity. Those sitting in Clarksdale's ISP office and standing in the Bright Brown Bar heard from one another not silence or continuous sound but sporadic bursts of music. At the same time they could see one another, but only in the form of alternately clear and distorted faces. Recognizing that at least they were linked, no one wanted to quit. One incident, characteristic of our exchange throughout both days, elaborates key aspects of our global jam and sets the stage for a discussion of the textured layers of cultural meaning that moved rapidly back and forth across the globe.

WE HAD OUR MOJO WORKIN' …

Thirty minutes into the second day of jamming, still we could not get a smooth connection. Amidst this now routine periodic confusion, an unexpected thing happened. In Tokyo's Bright Brown Bar, the drummer, bass player, pianist, and guitarist were following the lead vocals of a dynamic young man. In Mississippi, where I was, we could see them all well enough on our 40 cm television monitor, but only fragments of the audio were coming through. Most of us kept chatting and ignored the sounds. But Bluesman Hairy Larry, sitting at the center of our Clarksdale crowd, began trying to piece together the Bright Brown Bluesman's song. Hairy Larry stared attentively at the screen, eyes widening, posture straightening as he noticed familiar moves. Without removing his gaze from the monitor, Hairy Larry reached down into his bag for a harmonica. Tapping his feet and nodding his head to the stops and starts of the audio, he began to play along. Hesitantly at first, he matched his chords with what he heard from Japan and jumped in when bits of their music made it through. The key and rhythm grew familiar, so he began, as he said, "filling in the blanks." Several seconds of music came from Japan and then suddenly stopped, at which point Hairy Larry took up the progression and moved it forward with the unique inflections of his own harmonica style. "They're taking a rest to listen to my solo!" he bragged as the gaps continued but a song became clearer.

By this point excitement was building inside the room, and the rest of us had dropped our conversations and gathered closer to the screen. Two minutes into the exchange, Hairy Larry was playing non-stop, following Japan's lead and rendering the gaps in their presence almost unperceived. Just at the moment when he topped a musical bar with three strong E's and exclaimed, as if finishing, "I kept in time with them pretty good!" Japan came across with the celebrated refrain, "I got my

Mojo workin'!" We could hear an excited echo in the Japanese audience's response. The call was repeated. Japan's vocalist sang out, "I got my Mojo workin'," and we in Mississippi, in rough unison with the audience in Japan, mimicked in thrilled reply, "I got my Mojo workin'." From that point on we were all engaged in the jam. Hairy Larry changed roles and became our conductor, turning toward us, moving to the edge of his seat, and waving his arms to lead our response. Back and forth, call and response four times in a row, building until the song wound down with the barely audible lament from Japan, "but it just don't work on you."[1] Hairy Larry finished off the final notes with his harmonica, smiled widely, and remarked, after a pregnant pause during which we could hear only the frantic typing of our technicians, "Well that was fun. Jammin' with Japan! We got that J thing goin' on."

In Mississippi we broke into a lively discussion about how great the exchange had been. Stuttered applause came through from Tokyo. We clapped in return. "Good job! Let's do it again!" one person screamed into the mic. Hairy Larry laughed, so delighted with this success. Gradually the energy wound down and the conversation turned to making sense of what had happened. There were comments about the technology, the time difference, and the skill of Japanese Blues men. I mentioned the parallel with the call and response form of traditional African-American music. Hairy Larry jumped in. "It's the whole basis of the Blues," he said emphatically, and then reflected, "Yea, that's right. The call and response work song ... They were coming through and we were singing along."[2]

Indeed Japan was coming through with Mississippi singing along and this three-and-a-half minute moment became one of our most memorable. For us in Mississippi, the jam was like a global explosion of spontaneously synchronized ideas, feelings, and knowledge stretched halfway around the world. For a brief moment, distance had disappeared and though 9:30 on a Sunday morning we were just as surely inside that Tokyo bar at 11:30 that Sunday night. But none of this happened due to the ease of a fluid connection. It was the uncertainty that made this happen and the endeavor to make things clear. It was the creative mental effort of first Hairy Larry and then the entire group as they worked to grasp fully what was coming across the screen. Jamming across the Net meant thinking through a novel mix of disembodied faces and discontinuous sounds. It meant guessing the rhythms and words and reacting confidently to play along when these were present and fill in when they disappeared. And it meant an inventive application of local knowledge to make whole the broken messages coming through. Knowledge of the key, musical progression, and the participatory nature of *Mojo Workin'*, as well as a sense of when and how to coordinate the local group, became crucial resources for the success of this event. By inserting themselves into the exchange, Hairy Larry and crew closed the ambiguity, gave meaning to the sounds, and turned the jam into a true collaboration.

In this experience of interaction through the Internet, a distinctive style of communication comes into view. It is a style vitalized by absences in the expected flow of information, by tricks played upon the eyes and ears, and by disruption to the conditioned feel for the timing, pattern, and nuance in a human exchange. Its fundamental energy goes toward negotiating this ambiguity and collaborating to find common points of interest, focus and understanding. Fueled by attributes of the technologies themselves as well as the cultural differences and geographic distances that these technologies bridge, this quality of ambiguity encourages quick, imaginative thinking and makes producers out of participants as they work to sustain a connection with one another. As Internet technology improves, gaps and delays will disappear, diminishing the need for the ingenuity witnessed in Clarksdale. But interaction through the Internet may always generate something new – new versions of old songs, unprecedented global camaraderie, remakes of shared traditions, new kinds of jokes or words, or even the creation of a global space. For even when audio and video signals flowed perfectly during the jams, the Clarksdale–Tokyo connection remained full of room to play. So this improvisational tendency is not short-lived, nor is it unique to a global jam. Rather it is something that warrants attention with the increasing use of the Internet as a medium of communication, interaction and collaboration across cultures. The ambiguity pervasive throughout this kind of exchange heightens the role of the imagination not only in the single mind, I contend, but within social and cultural systems as well. The implications, therefore, are broad.

Recall Benedict Anderson's seminal work on the emergence of the concepts of the nation, nationality, and nationalism and the concomitant creation towards the end of the eighteenth century of the nation-state geopolitical system. Anderson attributes these developments partly to the predominant information technology of the time – printing. The significance of printing technology and with it the printed word, Anderson writes, was that it fostered a new style of imagining the world among even the most ordinary people. This style of imagining helped nurture into being political, economic, social and cultural communities as yet unseen and perhaps undreamt of: the sovereign, geographically distinct nation.[3] Considering Anderson's work while at the same time examining the Mojo jam prompts the questions: what kind of imagining does the Network foster and what might we create as a result?

... BUT IT JUST DON'T WORK ON YOU

Certainly the global Blues jam has its limits as an explanatory model; it cannot be used to identify Network behavior that is common to all. There is likely no such thing. In fact, for the participants in the Tokyo bar, nothing close to a global-scale jam took place during the Mojo song. Far from engaging the Mississippians with the same animated enthusiasm, the Blues men of Tokyo stared impassively at a wall where, via live life-size images, the Delta's Blues players once had joined them in the bar.[4] Contrary to expectations and experiences up to that point, however, no one showed up on the wall for this song and Hairy Larry's harmonica solos sounded more like, in the words of one patron, "a radio breaking down." The lead guitarist even looked a little bored. While he kept his body turned hopefully toward the wall, his mouth hung open lazily and his eyes were half shut. The main vocalist, too, finally gave up singing at a vacant wall and turned back to his local audience, rousing from them the collective comebacks that *Mojo Workin'* needs. Some part of the technology had failed the Tokyo side. The only strong evidence of global contact was Microsoft's NetMeeting logo and the panicked dialogue in the chat room that linked my two research teams.

Nothing of Mississippi was clearly there for the Japanese. At the same time Hairy Larry reveled in the "J thing" and those with him boasted of "being" in the Bright Brown Bar, Tokyo's musicians were not aware of, much less engrossed in, a cross-the-globe exchange. It follows, then, that despite appearances on our screen, there were not really any Japanese jamming with the Clarksdale crowd. They did jam, but they did so locally. Such a stark contrast in the two experiences of the same moments in time, the same connection, and the same activity complicates Mississippi's sense of the collaboration that occurred. The discrepancies demonstrate quite dramatically that the international intimacy felt by the Mississippi group depended not on the presence of the Japanese, but on their absence. Or rather, the tie derived from digital bits alone, which gave only representations of activity in Japan. In hindsight we know that those representations hid most of Bright Brown from view. And because of this, we also know that the vibrant connection for Mississippi was only in the mind. For Mississippi the Mojo jam was an imagined event.

THE ABSENT POTENTIAL

Nonetheless, what an event it was! Forget the rush of excitement in Clarksdale's ISP office. The opportunities raised by the dissonance of this jam are in themselves a cause for celebration. The Mojo ironies confront head-on the persistent challenges of culturally divergent perspectives. They inspire serious reflection on the meanings wrapped up in words such as "global" and "real" or even "here" and "there." And they muddle any attempt to say positively which country had the more genuine global jam. If indeed the absences and tricks of the Network heighten the use of the imagination and increase the need to apply local knowledge to make whole the meaning in the messages we receive, does not this suggest a multiplicity of wholes, each shaped within the histories and cultures of the individuals and communities involved? If so, it underscores the limitations of the claim that homogenization is the inevitable and invariable twin of globalization.[5] More critically, it illustrates the need for a conceptual approach to global culture that takes heterogeneity as fundamental and organizes its questions around the complexities therein.[6] The Mojo jam makes it clear that improvisation around heterogeneous elements is a genuine possibility for activity on the Net. What is more, improvisation on a sound, image, typescript or movement can come from multiple directions simultaneously. And finally, agents of improvisation are intertwined. Thus they invent around differences that are constantly in flux. To understand how these shifting positions intersect and the significance of what happens as they do is the primary challenge driving this chapter.

The post-colonial theorist Homi Bhabha reminds us that "cultures are never unitary in themselves." As a basic condition of their existence, they are always intertwined with one another.[7] Cultural theorist Judith Butler agrees, finding in her analysis of gender identity that "the disruption of the Other at the heart of the self is the very condition of that self's possibility."[8] In other words, cultures, like people, define and produce themselves in relation to what they are not. This view on the formation of identity has gained widespread acceptance throughout scholarly communities.[9] Fundamental to Net technology is an interactive capacity and a global expanse. Together, these attributes provide fertile ground for the cultural interconnections that, according to Bhabha, Butler and other theorists, bring identities into being. Considering this, too few scholars of the Internet have explored in sufficient detail its contours as a medium that links cultures. Anthropologist Arjun Appadurai discusses the Internet in terms of the "communities of interest" that it enables among otherwise "diasporic" people and he introduces ways in which diasporas may change in light of new forms of electronic mediation.[10] As well, throughout his three-volume treatise on *The Information Age*, sociologist Manuel Castells highlights the capacity of Internet technologies to connect economies, people, and ideas around the world in novel fashions and with fresh implications. His particular interest is in how these multidimensional, cross-cultural connections "specifically [refer] to the emergence of a new social structure." Castells declares that all societies at the turn of the century are living through a "rare interval" in history that is "characterized by the transformation of [their] 'material culture[s]' [via] the works of a new technological paradigm organized around information technologies."[11]

Castells' bold endeavor to explicate the appearance of a new social order that he ties, in part, to the Internet's faculties as a cross-cultural channel deserves concerted attention. In his keynote speech at the first annual conference of the Association of Internet Researchers (AOIR) in September 2000, Castells requested that more time be given to research in this area.[12] He also urged that whatever facet of the Internet we choose to study, "we need specificity in our research."[13] The steady outpouring of case studies of Internet experiences attests to a general effort to achieve the specificity Castells advised. As for his first appeal, though, attempts to explore the ways that the Internet brings together disparate linguistic, ethnic, racial, and national groups remain few and far between.[14] The subject does appear peripherally, as a suggestion for future scholarship or as a subdued ingredient of general discussions on information technology and global change. In addition, an increasing body of work compares dissimilar geographic, linguistic, cultural and national groups of Internet users with one another. This work offers invaluable insight into what makes each people distinct in its relationships with this new technology, but it does little to investigate the effects of cross-cultural electronic interaction.[15]

Perhaps the reason for a general absence of an explicitly cross-cultural focus in Net research is that there are so few opportunities to actually do this kind of work. After all, linguistic barriers are high and translation systems immature – if available at all. Digital divides, governing controls, and other cultural obstacles also get in the way. Theoretically, though, the Internet is a medium that collapses time and space. Its technologies not only enable but encourage visual, aural, textual, and even kinetic interconnections among people scattered all over the world. In this sense, it is a border space in which diverse cultures can introduce, exchange, negotiate and collaborate around the wide-ranging values, habits, and ideas they each have. While this abstract and glowing perspective on the Internet may seem naïve or a decade old, like at the jam in Mississippi, could it also be that if we engage possibility, reality will unfold? We can only know by assuming the *entire* challenge posed by Manuel Castells: "seek specificity" in the research that you do, he reiterated numerous times, but at the same time recognize that "there *are* genius intuitions" and "you should not lose those," but rather follow them in balance with what you find.[16]

Jumping back to Mojo, then, but considering it through the experience of the Japanese in the Bright Brown Bar, the "real" absence of Mississippians from the Mojo exchange meant that a "fictional" presence of Mississippians was maintained. That is, the inability to confirm whom they were jamming with required the Japanese to jam on the faith that their expectations would unfold. Of those expectations, the most basic was that Mississippians were present on the other side. Unable to see or hear them, though, an imagined "other" had to serve in lieu of the Clarksdale people. As Hairy Larry was filling in blanks to construct the Japanese, so too the Japanese filled blanks to produce him. A vast literature in Japan is devoted to the study of Japanese uniqueness. Called *Nihonjinron*, these "discourses on the Japanese" help disseminate throughout Japan the idea that "cultures possess definable, pure, and uniform essences."[17] No different from conventional thinking almost anywhere in the world, the premise of this literature is that "there is a genetically determined and therefore immutable, proprietary relationship between race and culture."[18] Skin and hair color match habits of everyday life. The habit at hand in our global jam was a talent for the Blues. Tokyo's Bluesmen may have been yellow, but their Blues ideal was the color black. So strong was this ideal that, if "anticipation conjures its object," as Judith Butler maintains in her studies of gender and culture, then when the Network's malfunction was corrected and visual data began streaming into the bar, the color of the Mississippians projected on Bright Brown's wall should have been nothing other than black.[19]

A BLACK FACE

But alas, absent as well on this day were any "real" Blues players, meaning African-Americans. At least, black was the imagined color of Delta musicians in the minds of the Japanese in the Bright Brown Bar. As "bluesy" as Hairy Larry was with his bare feet, simple attire, and Southern inflections, he is not African-American. In fact, Hairy Larry is on the lighter side of white. "*Ara! Hakujin da na*" – "Hey! That's a white guy!" – someone exclaimed when Hairy Larry first appeared on the wall in Bright Brown.

"Eeh?" "Hontō?"	("What?" "Really?")
"Deruta deshō?"	("They're in the Delta, right?")
"Okashii na."	("That's strange.")

The surprise reverberated throughout the bar as heads turned for confirmation. *"Ah, hontō da. ... Jya!"* "Oh yea, you're right ... Well, whatever! Let's get on with the jam!" is the implication of these words. In his discerning study of jazz in Japan, Atkins stresses that "for any Japanese performing or identifying with a musical genre typically characterized as 'black' ... ethnic authenticity and credibility are major issues."[20] Anyone can play the Blues, notes Mie Seno, founder and editor of Japan's *Blues Market Magazine*, but the Japanese presumption is that Blues is, fundamentally, "a black thing." "Not surprisingly, the best Blues is typically thought to be black Blues."[21]

Hairy Larry did not cut it with the Tokyo crowd. But the surprise at his color underscores the gratification Bright Browners felt when actual black bodies came into view. This happened on the first day of the jam, when five of the six musicians in Mississippi were African-American. The most famous of those musicians was Super Chikan (aka, the Chikan,[22] pronounced "chicken"), whom everyone in Bright Brown knew. "Super Chikan is actually there!" a woman screamed from the back of the bar. When Super Chikan leaned into the camera lens and cackled loudly like a chicken, the Japanese roared with laughter and gave a standing ovation. Local Bluesman Razor Blade was also a part of the visual data that streamed into Bright Brown Bar. Grinning mischievously, he sang, "Uuh Huh, Ooh wee wee y'all, Aaw!" as his self-introduction. This electrified the audience in Bright Brown and had nearly every patron yelling for more. *"Hatarake, hatarake! Nanika yatte kudasai yo nanka!"* ("Get to work, get to work! Come on and play something, please!") *"Nanka yare!"* ("Do something!") *"Buruzuman, [buruzu o] yaro!"* ("Hey Bluesman, play some blues!")

As luck would have it, at just about this moment the audio stream to Mississippi broke apart. So we in Clarksdale decided to disconnect the live exchange and send the Tokyo viewers a video recording of the Chikan's festival performance the previous day. Thus they got what they wanted – blues performed by a black Bluesman. Their energy sustained, it was primed to explode when the live connection resumed and an unforeseen player came on the scene. This was 14-year-old Kameisa,[23] daughter of a Delta Bluesman known and admired by the Bright Brown crowd. She is black. The Japanese saw her first in the festival recording. She was the leader of a small troupe of young girls dancing beside the Chikan on the stage. Bright Brown was hugely impressed. Two prominent Bluesmen in Japan, Ishikawa-san and the Harp Man, led the chorus of praise for young Kameisa. *"Yeah, yeah!" "Sugoi ne." "Sugoi na."* ("Great!" "Wow!") People in the back of the bar waved their arms and some mimicked the choreography of Kameisa's dance.

Bright Brown loved Kameisa. So when the two venues reconnected in real-time, the Japanese were thrilled to discover that Kameisa's face was the one projected on the wall. She introduced herself. "Hello my name is Kameisa Carter.

And I play drums, bass, and trumpet and I also dance. I am 14 years old." *"Sugoi ne,"* came the reaction. The Harp Man joked, *"Suiito fōtiinu iiazu ando."* "And never been kissed" was implied, an expression which everyone seemed to know, as almost everyone laughed. When the Harp Man took the stage, he flirted still more blatantly. He was the lead singer in the second song performed in the Bright Brown Bar that day.

> Do me a favor?
> Keep your business to yourself.
> Please, darlin', do me a favor
> Keep your business to yourself.

The Harp Man had chosen a tune in which the singer addresses his lover, urging her to keep their affair undisclosed. Blues musicians always put something of themselves into their renditions of songs, but the Harp Man had taken a step further, assuming the role of choreographer as well as vocalist. Looking straight at the image of Kameisa, the Harp Man belted out these words and added gestures to match. He wagged his finger sternly and shook his head to suggest the gravity of his plea.

> Please baby.
> Woman, you just keep it to yourself.
> Don't you tell nobody
> Don't you mention it to nobody else.

The Harp Man's plea became a demand.

> Don't tell your mother.
> Don't tell your father.
> Don't tell your sister.
> Don't mention to your brother.
> Please babe.
> Just keep it to yourself.

The singing lecture continued but grew more severe. The Harp Man raised both hands in mock exasperation, singing,

> You have a husband.
> I have a wife.

His hands came down and one of them waved across his throat like a straightedge razor. The Harp Man warned his secret lover,

> You start talkin' something
> Go messin' up our lives.

Still looking at Kameisa, he concluded the song.

> Please baby.
> Woman, you just keep it to yourself.
> Don't you tell nobody.
> Yea, don't mention it to nobody else.[24]

Bright Brown's emcee shouted into the microphone, "How was it?" Super Chikan replied, "Great!" The Harp Man wasn't sure. *"Boku no saiten wa?"* ("How about me?" "How did I measure up?") Super Chikan assured him, "Sound[ed] like a true Bluesman!"

A BLACK FACE, ACCUSED

A true Bluesman indeed. By the end of the performance, the Harp Man had rendered the young woman a gossip, a lover, a liar and a whore. Yet his choice of song and his play with words simply showed that he knew the Blues and its conventions. Since its beginnings, the male Blues tradition has objectified black women as mere sexual beings, be they breeders or whores.[25] African-American novelist Gayle Jones indicates how strong this ideology is through the voice of her male protagonist in the Blues-based *Corregidora*: "As long as a woman got a hole," he explains, "[she] ain't got nothing to worry about."[26]

Male musicians at Bright Brown would not say it in the same way, but essentially they would agree. At least, in their personae as self-conscious "Bluesmen" they would agree and as far as they conform to Japan's deep-rooted associations of dark races with sex.[27] "I want a sweet little girl that will do anything I say" was the refrain of the first song played in Tokyo. It came just before the Harp Man's performance of the clandestine affair. The third song was even more blunt. A shy, giggly Japanese man played it by sliding a bottle up and down the neck of an acoustic guitar. From its earliest practice in the Mississippi Delta, "bottleneck guitar" has signified the prowess and ingenuity of the Bluesman as well as the poverty and peril endemic to his world. Fueling these connotations were the words of the song itself.

> Yea, I feel like stuffin' a pistol in your face.
> Yea, feel like stuffin' a pistol in your face.
> I'm gonna leave my drink here, woman,
> Be at a girl friend's place.[28]

Participants in both countries laughed at the bold incongruity of this timid Japanese man belting out the meanest, and most misogynous, message yet. But his performance also shocked them. The owner of Bright Brown was even embarrassed. He looked into the camera and apologized, reasoning that his patron had played left-handed on a right-handed guitar and in addition was completely drunk. *"Sono mama hiitan de, hontō wa mo chotto umai desu."* ("He had to play under these [rough] conditions, so really he's a little better [than what you heard].")

Besides Super Chikan's pre-recorded performance, Razor Blade's "Ooh wee wee y'all," and Kameisa's trumpet solo toward the end of the jam, the only song initiated in Mississippi on this day was entitled "My Black Mama."[29] A white male New Yorker chose the song, played it, and sang the refrain.

> You take a brownskin woman'll make a rabbit move to town.
> Say, but a jet-black woman'll make a mule kick his stable down.

His words addressed blatantly the relation between female sexuality and the shades of black skin. Super Chikan and Razor Blade exchanged glances and grinned. Bluesman Ishikawa-san sat spellbound throughout the song and repeated three times *"Shibui na,"* which suggests that something is cool in a mature way. At the same time, the Harp Man playfully worried that if Bright Browners only stared the Americans would think that Japanese could not swing to this kind of song.

All of these songs, and the actions that accompanied them, tap into a deep reservoir of cultural meaning associated with the Blues and the physicality of black women. The Harp Man's song was only one of many in a similar vein. But it stands out among the others because its performance exposed the mechanics behind the objectification of black women endemic to male Blues and pervasive throughout the cultures of both countries. Kameisa's body was not just discussed, but seen. The visibility of her body enabled the Harp Man to focus the attention of his audience squarely upon it. He pointed at her, eyed her, threatened her, and claimed her, not as a partner in his jam, but as a sexualized hazard to his own life. With his audience as witness, the Harp Man captured Kameisa in the prison of his own ideas and reaffirmed the degraded status of the black female. Feminist historian Patricia Collins calls this process of looking, accusing, and colonizing "highly visible sexualized racism." Often gazed upon, the black woman is stopped short of gazing back.[30] The "gaze" to which this refers is not just a look or stare, but a viewing relationship privileging the perspective of the entity most empowered. As a young black woman raised in a land with echoes of slavery, Kameisa knew well what her actions were supposed to be.

JAMMING OUT OBSTRUCTIONS

But Kameisa looked back, not away, flouting the cultural clout that the Harp Man assumed. Critically, the Harp Man saw Kameisa looking back. What is more, in her eyes, facial expressions, and body movements as she looked back, the Harp Man saw himself and his censure of a black female body. No longer only an object looked upon, Kameisa had become a subject looking back. The Harp Man's own gaze, expressions, and gestures came to mirror those of Kameisa's on the wall. What the two of them independently saw gradually intertwined. They got tangled in a loop of looking relations. Taking cues from one another, Kameisa and the

Harp Man effectively jammed out obstructions to the circulation of cultural conventions that make the black female body a sex object.

At first, the signals between them moved linearly. The Harp Man performed within the parameters of a deeply entrenched sex–race paradigm. Kameisa looked back into his gaze, thereby breaking a cardinal rule. This caught the Harp Man's attention and he performed more vigorously, adding gestures with sexual innuendo. Kameisa kept looking, but also flashed an uncomfortable smile and squirmed in her seat. With this the Harp Man knew that Kameisa knew something was up. His audience also started catching on, adding new pathways to the information flow. They laughed at his acting, which encouraged him. He performed more forcefully, including gestures that gave commands. Still looking, Kameisa shifted her weight, smiled shyly, and darted her eyes from screen to camera. The Harp Man raised both hands in mock exasperation. This time Kameisa turned away, not out of shame or fear, but as if to feign nonchalance or perhaps to check if anyone had noticed her intimate link. The audience laughed some more and followed the lines of the gazes back and forth from wall to stage.

Still looking, Kameisa began to nod her head steadily in time with the music's beat. Both the Harp Man and his audience immediately recognized that Kameisa had joined the game. The Harp Man was performing upon her, but she was performing back. So he puffed up the drama and mimicked the movement of a knife across his throat. Unmoved, Kameisa kept looking and nodding her head. "Ooooh." The audience let out this slow, embellished response acknowledging, in effect, that Kameisa had not only survived the latest challenge, but increased the stakes. Tension rose inside the bar. Kameisa had defied all of the main parameters of the Harp Man's sex–race frame. What the Harp Man was saying and doing appeared more and more absurd. One patron hissed in playful annoyance. Another one booed. The terms of this jam exchange had altered entirely. The Harp Man's reputation as a leading Bluesman was on the line.

As the song wound down, Kameisa applauded the most generously of anyone in either place. Metaphorically, this delivered the final strike to the ideological model the Harp Man had used. In her influential work on African-American Blues women, cultural historian Hazel Carby concludes, "the physical presence and visual display [of female Blues performers] was a crucial aspect of their power."[31] Not only were women aware of their objectification on the part of their male peers, but as audience members or performers themselves, they actively engaged in an ongoing dialogue with the Blues community about these gender relationships and their shared African-American lives. Like the experiences of Carby's Blues women, Kameisa's own "dialogue" with the Harp Man occurred through her physical presence and visual display.

LINES OF LIGHT, LINES OF KNOWLEDGE

Comparisons between Kameisa and Blues women and the Harp Man and Blues men suggest that as Blues people, Kameisa and the Harp Man understood the ideas behind the music and were practiced in the behaviors that matched. But more than this, such comparisons point to the intersections between the Blues and the Internet as two very different, but equally valid information media. This, in turn, is not to imply that there is an absolute correspondence between Blues music and the Internet, but rather it is to open for analysis the significance of social, cultural, ideological, and technological mixtures in the work of the Internet. The central premises of *Nihonjinron*, call-and-response patterns in African-American music, the revelry of Bright Brown late at night, behavioral conventions of "highly visible sexualized racism," linguistic barriers, faulty software, intelligent hardware – all of these elements and more intertwined with the inherent technological capacity of the Internet to interconnect simultaneously a theoretically infinite number of points around the world. This ecology of factors, not simply technological features alone, determined the process, outcome, and meaning of our global Blues jam.[32]

Hairy Larry's exposure as a white man highlights a dominant ingredient in the particular Network ecology of the global jam. That is vision. To understand the obstruction of old knowledge that took place between Tokyo and Clarksdale, one must explore the new kind of looking relationships that the Internet affords and the cultural ideologies that are likely to change as a part of this visual structure. Recall that the "gaze" that the Harp Man first employed is "an historical deployment." It consists of "lines of light and lines of knowledge that are entangled … with lines of power and subjectivity." In other words, "vision and knowledge are not simply causally related, but are entangled with each other in the schemas and complexities of specific histories and specific events."[33] The viewing practice that the Harp Man applied in his interactions with Kameisa situates an object to be looked at and acted upon. It is unidirectional and tied to particular patterns of thinking about hierarchies of race and sex. Yet, the visual structure that the Network presented and that he actually experienced was reciprocal. Instead of the panoptical power and authority that the Harp Man expected, the reciprocal viewing structure interrupted and destabilized his vantage point, putting him into the picture and, via the eyes of Kameisa, making him an object of his own gaze. What became apparent to the Harp Man via Kameisa's look were the specificities of his own position and the inevitable restrictions to his point of view. The visual mutuality of the jam forced him to stop, think, and look again.

Manuel Castells states that "what characterizes the current technological revolution is not the centrality of knowledge and information, but the application of such knowledge and information to knowledge generation and information processing. [There is] a cumulative feedback loop between innovation and the uses of innovation." The "logic" of the Network, Castells explains, is "the immediate application to its own development of technologies it generates."[34] The Kameisa–Harp Man jam – as well as the Mojo jam and most every other jam moment over the two days – demonstrated this loop of innovation in process. Kameisa and the Harp Man did not generate *material* technologies, but they did generate *cultural* technologies, and with these new uses of material information technologies.[35] Kameisa had pushed the boundaries of black female marginality in the Japanese context. The Harp Man's only choices were to continue with his viewing habits in isolation from his local audience or to apply his new awareness and work with Kameisa and his audience to negotiate a new understanding of her relationship vis-à-vis the Japanese in the Bright Brown Bar.

By the end of the jam, Kameisa had become Bright Brown's welcome authority on contemporary African-American music and film. She had also become a translator into slow, clear English of the many comments and questions by Bluesman Razor Blade. In addition, the owner of Bright Brown had extended to Kameisa a sincere invitation to come to Japan as a Blues dance instructor. Along with this invitation came cheers of encouragement from everyone in the bar, including the Harp Man, and a group shout "We love you!" as the cameras were closing down. The Harp Man's choice, then, is obvious. He had reconsidered the ideological lens through which he viewed Kameisa. By the end of the jam, the Bright Brown audience and its musicians had not only accepted Kameisa as an active participant in the jam, they had gone so far as to redraw their cultural boundaries and include Kameisa – and her black skin – in the "in-group" of Bright Brown.[36]

CONCLUSION

The Tokyo-to-Clarksdale global Blues jams of August 1999 bring into view a distinctive style of communication. It is a style vitalized by absences in the expected flow of information, by tricks played upon the eyes and ears, and by interference with the familiar patterns in cross-cultural and interpersonal exchange. Its fundamental energy goes toward negotiating this ambiguity and collaborating to find common points of interest and understanding. Fueled by attributes of the technologies themselves, as well as the cultural differences and geographic distances that these technologies bridge, this quality of ambiguity encourages quick, imaginative thinking and makes producers out of users as they work to sustain their connection with one another.

As Internet technology improves, gaps, delays, static, breakdowns and all sorts of other limitations may disappear. Technological improvement will diminish the need for the ingenuity witnessed in Clarksdale and Tokyo. But as long as the Internet continues to be used as a medium of communication, interaction and collaboration across cultures, interaction through it will always generate something new – be it a highly original version of Muddy Water's *Mojo Workin'* or a positive adjustment to worn racial thinking.

NOTES

1   Legendary Bluesman, Muddy Waters, sang the first recorded version of *Mojo Workin'* in December 1956 under the aegis of Chess Records. Courtesy of the Blues Archive, University of Mississippi, Oxford, Mississippi.
2   In total there are thirteen 60-minute digital videotapes that document the global jamming on 14 and 15 August 1999. The recordings document both sides of the exchange, Tokyo and Mississippi,

and capture a minimum of two and as many as four different perspectives in both locations at all times during the jam. Unless otherwise noted, all quotations and details in this and later descriptions of the global jams come from these recordings. The wide range, top quality, and thorough detail of these recordings explain how I am able to know most of what was done and said in Japan even though I was in Mississippi at the time. I have the superb students of Keio University (SFC) to thank for that good work.

3 B. Anderson, *Imagined Communities: Reflections on the Origin and Spread of Nationalism*, London, Verso, 1983. Other prominent scholars corroborate Anderson's thesis and move it forward. Anthropologist Arjun Appadurai's numerous studies of late twentieth-century globalization and global culture point to the "historically situated imaginations" of once geographically cohesive but now globally dispersed persons and groups as an important source for understanding contemporary developments between and within economies, cultures and politics world-wide. Echoing both Anderson and Appadurai, sociologist Manuel Castells locates the ingenuity behind our "information age" in the millions of people using the era's novel technologies, as much as in those who formally make them. At the first annual conference for the Association of Internet Researchers in September 2000, Castells stressed repeatedly that being alert to the resourcefulness of those millions is an astute way of gaining insight into the meanings a technology might assume within whole cultural groups.

4 My students who worked on the Tokyo side of our global jam projected the video streamed from Mississippi onto a long wall to the front and left of the band. So, for everyone in the Bright Brown Bar, the interface with Mississippi was at eye-level on a wall they commonly faced and with images at least life-size. In addition, the students fed Clarksdale's audio stream through the bar's stereo system. This made the sounds from Mississippi much more robust than those which would have come through the computer alone. Had we had the space and equipment in Mississippi, my students and I could have done the same there too, channeling both sights and sounds beyond the limited scale of the computer. As it was, while Clarksdale's ISP office had the faster Network connection, it had much more meager material surroundings.

5 For an excellent survey of leading theoretical discussions on the cultural consequences of globalization and the ways in which the Internet is affecting and being affected by globalizing processes, see J. Slevin, *The Internet and Society*, Oxford, Blackwell, 2000, pp. 198–213.

6 In *Modernity At Large*, anthropologist Arjun Appadurai states, "the globalization of culture is not the same as its homogenization." Rather, he explains, "globalization involves the use of a variety of instruments of homogenization (armaments, advertising techniques, language[s] ... and clothing styles) that are absorbed into local political and cultural economies, only to be repatriated as heterogeneous dialogues of national sovereignty, free enterprise, and fundamentalism." A basic premise of Appadurai's work in *Modernity At Large* is that theories of global culture and development have been too simplistic to wrestle successfully with the "complex, overlapping, disjunctive order" of the late twentieth-century global cultural economy. Appadurai poses as a replacement to the dualistic models of center–periphery, push–pull, surplus–deficit, consumer–producer or homo–hetero a multidimensional theory of global cultural flows that emphasizes increasingly non-isomorphic paths and deep disjunctures between economy, culture, and politics. A. Appadurai, *Modernity At Large. Cultural Dimensions of Globalization*, Minneapolis, University of Minnesota Press, 1996, pp. 42, 32.

7 H. Bhabha, *The Location of Culture*, London, Routledge, 1994, pp. 35–6.

8 J. Butler, "Imitation and gender insubordination," in L. Nicholson (ed.), *The Second Wave: A Reader in Feminist Theory*, London, Routledge, 1997, p. 311.

9 See the following works to get a sense of the multifaceted debates that frame this assertion: K.A. Appiah and H.L. Gates Jr (eds), *Identities*, Chicago, University of Chicago Press, 1995; S. Hall and P. du Gay (eds), *Questions of Cultural Identity*, London, Sage, 1996; and P. Moya and M. Hames-Garcia (eds), *Reclaiming Identity: Realist Theory and the Predicament of Postmodernism*, Berkeley, University of California Press, 2000.

10 Appadurai, *Modernity at Large*, p. 195.

11 M. Castells, *The Information Age: Economy, Society and Culture*, vol. 1, *The Rise of the Network Society*, Oxford, Blackwell, 1996, pp. 27–9. Also see *The Information Age* volumes two and three, *The Power of Identity* (1997) and *The End of Millennium* (1998).

12 In his ten-point list of the "lessons of Internet history," Castells lamented the dearth of academic projects aimed at unveiling the multicultural influences on the creation of Net technologies. Castells stated, "It is false that [the Internet] is an American creation." M. Castells, keynote speech, Association of Internet Researchers, Lawrence, Kansas, 15 September 2000.

13 M. Castells, Roundtable Discussion, Association of Internet Researchers, Lawrence, Kansas, 16 September 2000.

14 Note that in D. Bell and B. Kennedy (eds), *The Cybercultures Reader*, London, Routledge, 2000, the articles that address links across cultures are themed under the term "cybercolonization" which suggests not interaction or heterogeneity but domination and absorption by one (e.g. west/white/center) of another (e.g. east/black/periphery).

15 Some of the most cutting-edge conceptual work that examines the Internet as a connecting medium across and within its abundant and diverse global contexts comes out of the biannual international conference on Cultural Attitudes Towards Technology and Communication (CATaC), 1998, 2000, and 2002. The expanded versions of 1998 papers can be found in C. Ess and F. Sudweeks, *Culture, Technology, Communication: Towards an Intercultural Global Village*, Albany, NY, SUNY, 2001. Also see G. Hawisher and C. Selfe, *Global Literacies and the World-Wide Web*, London, Routledge, 2000 and *New Media & Society*, 2001, vol. 3, no. 3.

16 Castells, Roundtable Discussion.

17 E.T. Atkins, *Blue Nippon: Authenticating Jazz in Japan*, Durham, Duke University Press, 2001, p. 29.

18 Atkins, *Blue Nippon*, p. 29.

19 J. Butler, *Gender Trouble: Feminism and the Subversion of Identity*, London, Routledge, 1990, p. xiv.

20 Atkins, *Blue Nippon*, p. 27.

21 M. Seno, Interview, 12 December 1998.

22 The meaning of "*chikan*" in the Japanese language is "pervert." It is possible that the association of this meaning with Super Chikan's name fed the license and sexual innuendo of not only the Kameisa–Harp Man exchange (explored next), but also the entire jam session that day. The name did raise eyebrows, and a confused or embarrassed silence ensued for most of the Japanese when they saw the name in print. But beyond this – despite the loaded meaning of the word when seen in writing – the Japanese seemed to ignore "chikan" and both see it and read it as "chicken."

23 To protect the identity of this individual, I have changed her name to "Kameisa Carter."

24 The title of the song from which these lyrics derive is *Keep It To Yourself*. The probable first "author" of the song is Sonny Boy Williamson (aka, Rice Miller). However, according to curators at the University of Mississippi's Blues Archive, there is no guarantee that this song was sung by Sonny Boy Williamson. No date or place of an original recording could be found. The Blues Archive's inability to find complete information about the particular Blues lyrics that the Harp Man chose to sing underscores the African-American oral traditions within which the Blues developed. Most Blues have no original written version and thus no definitive first author. Most Blues share themes, refrains, and rhythms with other Blues. This is not considered copying, but rather an expression of identification with a wider Blues community and culture. For more information on the subject of Blues as an oral tradition, see the works of Alan Lomax, Angela Davis, Hazel Carby, Houston Baker, LeRoi Jones (Amiri Baraka), and William Ferris, among others.

25 See A. Davis, *Blues Legacies and Black Feminism*, New York, Random House, 1998; H. Carby, "It jus be's dat way sometimes," *Radical America*, vol. 20, 1988, pp. 9–24; H. Carby, "In body and spirit: representing black women musicians," *Black Music Research Journal*, vol. 11, no. 2, 1991, pp. 177–92; J. Morgan, "'Some could suckle over their shoulders': male travelers, female bodies, and the gendering of racial ideology, 1500–1770," *The William and Mary Quarterly*, vol. 54, no. 1, 1997, pp. 167–92; and W. Jordan, *White Over Black: American Attitudes toward the Negro, 1550–1812*, Chapel Hill, University of North Carolina Press, 1968.

26 G. Jones, *Corregidora*, Boston, Beacon Press, 1986, p. 100.

27 Japan scholar John G. Russell argues that the position blacks have come to occupy in the Japanese hierarchy of races "not only echoes Western racist paradigms but is borrowed from them." He asserts, "Japan's image of the Black Other is deeply ensconced in the cozy deliriums of Euro-American supremacy." J. Russell, "Race and reflexivity: the black other in contemporary Japanese mass culture," in J. Treat (ed.), *Contemporary Japan and Popular Culture*, Honolulu, University of Hawaii Press, 1996, pp. 17–40. Also see S. Morikawa, "The significance of Afrocentricity for non-Africans: examination of the relationship between African Americans and the Japanese," *Journal of Black Studies*, vol. 31, no. 4, 2001, pp. 423–36 and J. Russell, "Narratives of denial: racial chauvinism and the Black Other in Japan," *Japan Quarterly*, vol. 38, no. 4, 1991, pp. 416–28.

28 Curators at the Blues Archive, University of Mississippi, could find no recorded song with these exact lyrics. However, they guessed that it derived from *Pistol Slapper Blues*, first recorded by Blind Boy Fuller in 1938 under the aegis of Document Records.

29 *My Black Mama* was recorded first by Son House in May 1930 under contract with Document Records. Courtesy of the Blues Archive, University of Mississippi, Oxford, Mississippi.

30 P. Collins, *Black Feminist Thought: Knowledge, Consciousness, and the Politics of Empowerment*, New York, Routledge, 2000. Here Collins explains the historic relationship between race and sexuality as it has affected black women and argues convincingly that the sexuality of African-American women is inextricably tied to their dark color. Also see J. Jones, "The accusatory space," in M. Wallace (ed.), *Black Popular Culture*, Seattle, Bay Press, 1992, pp. 95–8; and B. Hooks, *Black Looks: Race and Representation*, Boston, South End Press, 1992.

31 Carby, "It jus be's dat way sometimes," p. 18.

32 This mixture of social, cultural, and technological Network attributes is an on-going theme in Janet Abbate's work, *Inventing the Internet*. The invention of the Internet, she writes, was "a tale of collaboration and conflict among a remarkable variety of players." The Internet's "identity as a communication medium was not inherent in the technology; it was constructed through a series of social choices." The meaning of the Internet, then, "had to be invented – and constantly reinvented – at the same time as the technology itself." J. Abbate, *Inventing the Internet*, Cambridge, MIT Press, 1999, pp. 3–6.

33 V. Bell, "Performativity and belonging: an introduction," in V. Bell (ed.), *Performativity and Belonging*, London, Sage Publications, 1999, p. 7.

34 Castells, *The Information Age*, vol. 1, pp. 32–4.

35 M. de Certeau, *The Practice of Everyday Life*, Berkeley, University of California Press, 1984; and M. Foucault, *The History of Sexuality: Volume 1, An Introduction*, New York, Random House, 1978.

36 The significance of this change in Kameisa's status vis-à-vis the Japanese group can hardly be exaggerated. Japan historian John G. Russell argues, "the tendency to dehumanize and belittle blacks disguises another tendency ... to employ the Black Other as a reflexive symbol through which Japanese attempt [to] deal with their own ambiguous racio-cultural status in a Eurocentric world." Japanese images of blacks are "inextricably linked to Japan's unequal relationship with the West." The Black Other gives the Japanese a way to "reappraise their status vis-à-vis whites and the symbolic power ... they are seen to represent." Blackness, then, "is employed as a category mediating White Otherness and Japanese Selfhood." Russell, "Race and reflexivity," pp. 20–4. Also see J. Valentine, "On the borderlines: the significance of marginality in Japanese society," in E. Ben-Ari, B. Moeran and J. Valentine (eds), *Unwrapping Japan: Society and Culture in Anthropological Perspective*, Honolulu, University of Hawaii Press, 1990.

Part II
Gender and sexuality

7 Challenging society through the information grid

Japanese women's activism on the Net

Junko R. Onosaka

INTRODUCTION

The Internet is becoming more and more a part of daily life in Japan: currently 37.1 per cent of all Japanese, that is 47.1 million individuals, use it.[1] Moreover, the number of female users has increased drastically: from 1997 to 2001, the percentage of women online increased from 16.5 to 44.5 per cent of all users.[2] Women are rapidly discovering the effectiveness of the Internet for accomplishing a variety of goals.

Japan has long since established itself as an economic superpower, and women in Japan are now becoming a vital part of its Internet community. Yet, the presence of Japanese women is still noticeably lacking when important decisions are made in some social sectors. While Japanese women commonly make decisions about children, their education, family spending, and consumption, they constitute less than 8 per cent of Diet members and less than 10 per cent of all administrative workers in Japan. Accordingly, the Gender Empowerment Measure of the UN places Japan forty-first among 174 countries world-wide which makes it the least gender-balanced among advanced nations.[3] My analysis of Japanese women's Web sites, however, revealed a rich panoply: numerous women and women's groups are using the Internet in order to improve their status both individually and collectively and to expose and challenge myriad forms of social discrimination.

Gender and age differences are implicit in virtually all direct conversation in Japan; to disregard such distinctions, which are indicative of social status, is commonly seen as rude. Communication on the Internet, however, enables women to take part in public discussion while preserving their privacy and anonymity; that is, they have no need to engage in face-to-face debate or to use language "appropriate" to hierarchically-ordered direct discourse.[4] Thus the anonymous character of the Internet has provided space for a variety of new interactive forums for women. For example, one housewife relates that:

> As a housewife, I was always treated as subordinate to my husband ... My opinion was often treated as not mine, but "my husband's wife's." The only place I was treated as an individual was on the Net, probably because I didn't reveal that I am someone's wife.[5]

This woman believed that if she had revealed her status as a housewife, she would not have been treated as an individual even on the Net. Yet, since participants on the Internet only have the information that is made available to them in written text, writers can preserve their anonymity with respect to age, occupation, marital status and much more, or even invent an identity.

As early as 1996, Matsuura Satoko found a diverse range of women's voices on Web sites for Japanese women:

> I was astonished to see that various women's voices are heard and visible on the Internet. These are very different from ordinary conversations among women, which usually have to be modest, self-effacing, and settled privately ... Their voices are lively. They are trying to build real communication and relationships on the Internet, leaving behind "secure" and "soothing" conversations ... Women who were once confined at home to "private" issues such as child-rearing, care of the elderly, domestic violence, sexual harassment and other discrimination, are now stepping out. It's as if a Pandora's box has been opened in Japanese society.[6]

Many women have embraced Internet use as a means for building and maintaining relationships with other people, for protesting against various forms of discrimination such as sexual harassment and domestic violence, and for campaigning against nuclear arms and environmental pollution. On their Web sites, women can subvert traditionally rigorous gender expectations by taking their own decisions as to what personal information to divulge or hide. This enables them to communicate more easily with others in efforts to overcome various forms of social discrimination.

Many Japanese women's groups have actively begun to use the Net for the advancement of women – the number of mailing news groups or Web sites established for this purpose has increased exponentially in recent years. Given its novel capacity for expression unhindered by social strictures, the Internet may prove a powerful tool for minority groups and women's groups in Japan which seek to challenge social prejudice.

For the purposes of this chapter, I examined Japanese-language Web sites that: (1) were created for (and most often by) Japanese women; (2) offered accurate information about women's current social conditions; (3) were relatively enduring (at least one year); and (4) were dedicated to ending inequality, particularly sexism.[7] These sites included research institutions such as governmental women's centers and bureaus at both national and local levels, and women's groups organized around different topics of concern. They included governmental or semi-governmental as well as non-governmental Web sites.

I analyze how these sites facilitate activities such as social protests or campaigns for the advancement of women, networking beyond geographic boundaries, sharing information and resources, and strengthening women's capacity for institution building. The results of my analysis highlight the powerful role the Internet plays in expressing the diversity of women's voices and, in doing so, demonstrate the wide range of Japanese women's activism in cyberspace. As will become clear, the Internet enables women to share information with each other in a free exchange of ideas that promotes their activist efforts; moreover, it provides a stage for female activists from all regions of Japan to make their voices heard and their political actions visible – achievements previously made difficult given the male bias of Japan's mass media.

GOVERNMENTAL SITES: WOMEN'S CENTERS AND BUREAUS

Governmental or semi-governmental sites include those of women's bureaus and women's community centers, some national and others municipal. Most such organizations emerged after 1975, when the United Nations Decade for Women began. In addition to serving as conduits for policy response to women's affairs, they have also worked out their own "programs of action for local women." One such site is the *Gender Information Site* (GIS), which provides coverage of national government activities aimed at realizing a gender-equal society. It offers information on related policies, plans and procedures, workshops and material on women's issues, and, since November 2001, a mailing list service to provide updates of interest to site users.[8]

Today there are over 100 national and local women's centers in thirty-five prefectures and ten cities that maintain their own home pages.[9] The Web sites of national centers, for example, include statistical information on women and families, bibliographic databases, periodical and newspaper indexes, directories of women's centers and other organizations and educational programs for women.[10] Local centers offer more community-based information – in particular, statistics and reports on surveys (usually conducted by the local government) concerning gender roles, aging, family and work.[11] Both center and bureau sites routinely offer chronological surveys of national and local historical and political events relating to women as well as statistics from surveys concerning their living and working conditions.

However, despite the fact that governmental sites provide valuable information not otherwise easily obtained, they rarely offer opportunities for interactive communication. Also, they tend to reflect official views on gender issues, views which, while purporting to promote a gender-equal society, implicitly accept gender difference as essentially genetic.[12] Furthermore, as Matsumoto Yukika points out, government sites often buckle under the pressure of groups opposed to feminist activism.[13] In this chapter, therefore, I focus more on Web sites of non-governmental organizations, many of which use mailing lists and Bulletin Board Systems (BBS) to share information and to generate dialogue among members.[14]

The two kinds of site are, however, often linked since some non-governmental women's groups grew out of local workshops run by official government bodies. Local women have often built on the foundation offered by courses sponsored by women's centers, inaugurating activities that empower women such as the study of gender issues, including local women's history. Eventually, some of these groups began to deploy the Internet in order to expand their networks and promote their activities more effectively.[15] Many women have utilized opportunities originally offered by the government to train and educate themselves to effect social change.

CYBERACTIVISM BY JAPANESE WOMEN

Multiple-issue organizations

Non-governmental/non-profit organizations (NGO/NPO) promoting the advancement of women may be categorized into multiple-issue and single-issue sites. I focus in this section on sites which have the broad aim of mobilizing women through drawing their attention to a variety of issues. Some organizations existed prior to the appearance of the Internet, others began as online networks for female activism.[16] I examined examples of the latter kind, particularly the *Women's Online Media* (WOM) and the *Japan Women's Movement Network* (*fem-net*) which operate mainly via online meetings.[17] They also collect information, advise users on how to help set up women's networks, inform participants of events sponsored by women's groups across the nation, gather group information, and provide a means for exchanging information of mutual interest.

WOM, established in 1995, is a non-profit organization made up of women from diverse backgrounds. It aims to empower women through the use of electronic media, and holds Internet seminars to help newcomers learn about the Net. Furthermore, the information dispersed by WOM has not typically been made available in official publications and government announcements. WOM's first major activity was to post information on the Internet in 1995 about the Fourth World Conference on Women in Beijing, providing day-by-day coverage of the sessions. Inoue Haneko, one of many Japanese women who received this information in Japan via the Internet and who had attended an earlier World Conference on Women, said:

> In 1985, I sent some reports via airmail from the Third World Conference on Women in Nairobi, Kenya. It took at least one week; worst-case scenario, two weeks. Now I can read about it in real time on the Internet![18]

Fem-net, an online network of activist women that has been in operation since 1997, also used the Internet to provide alternative views to those available in Japan's 1999 Official Governmental Report. Like WOM, the aim of *fem-net* is to enable women's groups to understand and utilize information technology more effectively.[19] In 1999 *fem-net* supported the efforts of the NGO Report Preparation Group organized to prepare Japan's National Alternative NGO Report, a report that reviewed the implementation of the Beijing Platform for Action five years on and was prepared by a coalition of members of women's organizations on behalf of the Women 2000 Conference in New York.

Since an online version was available, the members of the committee worked together online. The site further provided information on the committee, with plans, coordinators, participants, volunteers, links and resources in both Japanese and English. Given the indispensable role of the online network in its production, Matsumoto has claimed that "the Report could not have been realized without *fem-net*."[20] Later it also provided information on the Women 2000 Conference in New York. One woman in Japan said, "This site gave me a real sense of the atmosphere at the Conference, more so than newspapers, the television, or faxes could have!"[21]

Not only has *fem-net* thus been instrumental in providing alternative views to official reports, it has also developed an electronic network that can be shared with many grass-roots women – an effective tool for networking to realize a common goal. Recently on *fem-net*, eight women started the online campaign "Give Us Our Half! Women in the Elections." The site was created after the exchange of almost 400 e-mail messages in anticipation of the Lower House general election held in June 2000. It carried profiles of female candidates, results of surveys of candidates and political parties, Lower House election data, press releases on election results written from the perspective of gender, explanations of political subjects, links and a Bulletin Board System (BBS).

The BBS, in particular, was filled with information, opinions, ideas, and also self-introductions from female candidates. For example, one woman made this exhortation to women via the BBS:

> The mass media often say the fact that people don't vote means they don't trust politics, but this idea merely justifies their absence at the polls. We MUST take action and vote! Don't just read this; write your ideas on this site and go vote![22]

Another woman said, "I was simply very much excited at finding a self-introduction from the candidate herself – the Internet is a great tool for us!"[23] Women around the nation offered information on female candidates in their respective local areas (Tokushima, Nara, Ehime and so on), most of whom had been overlooked by the mass media. The site brought together women from disparate regions who had suffered from a variety of obstacles in their respective rural, usually conservative, areas.

Thus, the Internet increased the capacity to form bonds with other women, even those living in different regions. Eventually, the site received over 10,000 hits during the half-year period leading up to the election, many more than the organizers had expected.[24] Eventually, the number of female representatives was more than doubled, although women still represent only 7.3 per cent of Lower House members.[25]

Fem-net's activism has extended to opposition to terrorism and war such as the attack on the World Trade Center and the US military's resulting assault on Afghanistan. *Fem-net*'s anti-war site began with reports on the war in Afghanistan based on information gathered from international NGOs, including over forty related links.[26] The production of the *No Terrorism – No War – No Violence* site was a joint undertaking with VAWW-NET Japan, to which many people contributed. One woman said, "Japan is the only country to have had an atomic bomb dropped on it. Bush suggests he'll use atomic bombs, and Koizumi supports Bush – we have to strongly oppose them." A man wrote that "I totally support *fem-net*! Let me link this site to my home page." Another woman said, "It is time for us to use information technology for building peace. Let's raise our voices against war, and share information on the Web!"[27] In this way, many on the Web shared their wish for peace and took actions proposed on the *No Terrorism – No War – No Violence* site. Eventually almost 3,000 signatures with 297 messages, including seven English messages, opposing Bush and Koizumi's decisions were collected via online messages, fax, and postal mail by the end of October 2001. Meanwhile, the organizers said on the Web site:

> We have been very much encouraged by receiving supportive messages and signatures from around the world. We sent all of them to the House of Representatives and to Prime Minister Koizumi on 29 October 2001 … We understand that the American and Japanese governments will probably not immediately understand our outcry against war, but the most important thing is that each of us has raised and will continue to raise our voices. We firmly believe that this encourages others and, at the same time, empowers us. We will continue raising our voice for peace. Again, thank you for your support.[28]

Web sites such as those discussed above clearly enable women's organizations to improve their networking efforts and strengthen women's bonds with each other by providing online platforms for the exchange of ideas.

Single-issue organizations

Certain women's organizations in both metropolitan and rural areas grapple with one specific issue for an extended period of time. Their Web sites typically appeal to women to understand the urgent social issues they raise and help to resolve them. They provide, in energetic fashion, information that is often overlooked by society at large, given that the general media often overlook groups who are marginalized because of gender, social class and sexuality.

These single issues include, most prominently, civil law reform regarding the right of married women to keep their surnames and the rights of illegitimate children, violence and harassment against women, problems faced by working women, the political empowerment of women and sexuality. Most sites of this kind use the Internet very effectively and aggressively to reach women and bond with them beyond spatial and temporal boundaries, making the most of a technology that enables people to communicate without the necessity of physical interaction. This is especially important in Japan, where almost every powerful establishment institution – from governmental institutions to the main offices of large corporations – is located in Tokyo, the capital, and people in outlying areas find it difficult to convey their views to citizens outside their communities. The Internet allows people to make connections that would otherwise have been much more difficult.

To take an example: a recent workplace harassment lawsuit in Nara, 300 miles from Tokyo, garnered support from around Japan through a Web site called *Nara Medical University Lawsuit* (*NML, Nara Kenritsu Ikadaigaku Akahara Soshō*).[29] The site is a good example of those which deal with sexual harassment, abuse and violence against women and was created by supporters of Dr Ōgoshi Kumiko, in order to finance the 5.5 million yen lawsuit she launched in 1998 against a professor in her department. Ōgoshi said:

> After I took the action, a support group was formed. Especially after the home page for this case went online, those who had been in similar situations to mine started to raise their voices, and the types of harassment they had been suffering, to my surprise, were very much alike.

As a result, she placed examples of the harassment of other women on the Web site and encouraged women both to send messages to her and also to voice their concerns online to targeted governmental offices. The Web site received more than 40,000 visits.[30] Ōgoshi was able to build a community of supporters in part by making the progress of the lawsuit public on the Web as well as publicizing the issues through the media and through other women's networks. Thanks to her efforts, others who had been in similar situations spoke out also, asking for help and explaining their cases. However, in the meantime, the professor filed a lawsuit against her and both the creator and the provider of the Web site, charging defamation of character. At the end of October 2000, Ōgoshi won her suit in the Osaka District Court. Later, on 28 September 2001, Ōgoshi, her supporters, and other victims established *Akademikku harasumento o nakusu nettowaaku*, an NGO for the victims of harassment in academia.[31] On 29 January 2002, she won her second suit in the Osaka District Court.

In a similar vein, the *Campus Sexual Harassment National Network* (CSHNN) would not have been established and would not have been able to expand so rapidly without the Internet within such a short period of time.[32] This network began by listing e-mail addresses on the several universities' home pages, including, for example, national universities such as Nagoya University and Kochi University. One of the founders recalled, "We made guidelines for sexual harassment cases before other universities did, and placed my e-mail address on our university's home page as a window for sufferers. But I received so many serious messages from outside our university."[33] As a result, a national network, CSHNN, and the Web site were established. The site, in tandem with a mailing list, informs visitors of an appeal for a national conference and for the publication of a newsletter as well as carrying reports on various activities including lobbying.

Other single-issue sites dealing with women's careers and their political involvement include *The Society for the Study of Working Women* (*SSWW/ Josei Rōdō Mondai Kenkyūkai*), *Working Women's Network* (*WWN/ Wākingu Uimenzu Nettowāku*), and *WINWIN* (*Women in the New World International Network*).[34]

That such sites thrive is presumably a reflection of the fact that Japan's discriminatory corporate culture includes women in only 8.2 per cent of administrative positions, and that women account for only 10.7 per cent of the national and 6.2 per cent of the local assemblies.[35] One of the founding members of *WINWIN* says:

> During this short period of time, we have been successful in helping to elect several women to Diet seats and two women Governors. This is a small but important step. I think that the new era in Japan will be influenced by two important factors: women and information technology. And as a matter of fact, they are very much intertwined.[36]

Some sites offer information on efforts to amend discriminatory civil laws that prohibit wives from retaining their original family names and that discriminate against children born outside marriage. *The Network for Amending the Civil Law* (*Emu Netto/Minpō kaisei jōhō nettowāku*), the *Organization for Considering Separate Married Surnames* (*Bessei o kangaeru kai*), and the *Single Mothers' Forum* (*Shinguru Mazāzu Fōramu*) are examples.[37] Akaishi Chieko of the *Single Mothers' Forum* reports that the Web site facilitates discussions and dialogues among members, mainly living in the Tokyo area, although face-to-face communication or phone conversations remain crucial to maintaining their relationships. The group extends a helping hand to single mothers around the nation via e-mail. E-mails received are immediately forwarded to the member of the group who knows most about the issue raised by the sender.[38]

The Japanese government is currently trying to cut financial assistance to single mothers. However, the members of *Single Mothers' Forum* realized that many single mothers are not aware of this plan, due to the fact that many of them do not subscribe to newspapers, or do not have time to read them or watch the TV news. Furthermore, the mainstream media are not eager to cover such gender issues in general. For example, the "Women's International War Crimes Tribunal" on Japan's Military Sexual Slavery was sponsored by the Violence Against Women in War-Network Japan (VAWW-NET Japan) in December 2000, but was virtually ignored by many in the mainstream media.[39] Although almost 5,000 women and men gathered for the five-day Tribunal, and over 300 journalists and reporters from around the world attended,[40] according to Matsui Yayori, "The overseas media depicted our tribunal as a historical event, but the Japanese media totally ignored us. One major newspaper didn't even write so much as a single line about us."[41]

The group therefore began to establish a national mailing list and a BBS, providing information related to this issue, and posting various survey results and data showing the present difficult financial situation of single mothers and their households on the site. A woman who started this section says, "I started this because I want to let women know about this urgent problem beyond our area of activity, Tokyo; to know what they think of that; and to send opinions to legislators and the Ministry of Welfare."[42] Simultaneously the group started to collect signatures opposing the governmental plan via the Internet.

On the BBS, fifty messages opposing the plan sent from around the nation were received within a month.[43] One woman said, "Originally, I wanted to keep my distance from 'feminism,' but this time I cannot stand what the government is doing, so I want to sign the petition form and link this site to my home page."[44] On 25 December 2001, several members went to hand the petitions and signatures collected through e-mail, fax and postal mail from around the nation to the Ministry of Welfare. In this way, the BBS enabled many single women to emphasize their need for financial assistance and to encourage each other.

On the other hand, messages criticizing single mothers also appeared on the BBS. The messages said, for example, that "many single mothers receive financial assistance illegally," "You shouldn't divorce for any reason other than domestic violence – especially for your children's sake," or "Our taxes should be used for us, not for you single mothers who hardly pay taxes," and so on.[45] Presumably, messages of this kind were meant to discourage single mothers, but even they were sometimes put to good use. For example, single mothers on the BBS responded to them one by one: explaining why they became single mothers, how they reared or rear their children every day in a society marked by gender inequality, how they support society as tax payers, and so on. Such antagonistic messages on the BBS clearly suggest that there is prejudice and ignorance about single mothers in the wider community. However, in their responses, single mothers were able to speak eloquently of their pressing situations and circumstances beyond their control, and reveal the disadvantaged conditions of the weak in Japanese society. Furthermore, the messages actually ended up encouraging many of the women. As one woman said:

> I actually appreciate all the messages – even those from people against our proposal. This is because I learned a lot through reading them on our BBS. Now I know how to respond to guys who say things like "We don't need financial assistance for single mothers because it simply promotes divorce!" They might be posted as FAQ (Frequently Asked Questions) on our site. Who can live on only 40,000 yen a month with children? Who would want to divorce for such a pittance? Anyway, thank you very much everyone. Don't be sad – be happy!! Please visit our BBS again![46]

The messages numbered over 220 within four months. These accumulated messages, as well as various data including details about many single mothers' households shown on the Web site, can help improve people's understanding of the pressures faced by single mothers. Furthermore, the site became a point of contact between single mothers and others, and between single mothers in Tokyo – who founded the site – and those in other areas.

It is clear that the Internet provides many Japanese women with increased opportunities to improve their working and living conditions. Moreover, it has provided new outlets for communication among women, breaking the isolation that some faced, providing an interactive forum for women's issues and, for some, a women's space. For example, as was mentioned earlier, some female victims of discrimination or harassment found supporters more quickly through creating their Web sites than would otherwise have been possible; moreover they often discovered that the kinds of discrimination from which they suffered were very similar to other women, and went on to create organizations to support their causes. Furthermore, the Internet has helped many organizations that were originally limited in scope to specific geographical communities to establish links with women from other regions.

The ease with which the Internet can be used by many makes it possible for women who might otherwise not participate in activism to take action – for example, by signing online petitions, expressing their opinions on electronic message boards and sending protest e-mails. The Internet, in many instances, has also provided a springboard for face-to-face communications, meetings and organizational activity.

Internet-based groups have clearly built on the experience of previous women's grass-root networking, particularly the role that *minikomi* (mini- as opposed to mass-communication media) played in facilitating such networking.[47] Women's groups originally disseminated information about their activities through such *minikomi* as small pamphlets and newsletters. However, the production and distribution of even small-scale *minikomi* required a great deal of time, effort and resources and these publications were difficult to obtain by those outside the network.[48] An example of earlier women's activism occurred in 1996 when eight women protested against an organization requesting the removal of references to sexual slavery by the Japanese military from a textbook.[49] The women distributed *Onna tachi no kinkyū apiiru* (Emergency appeal from women) via fax, and in just two weeks collected 1,700 signatures against the organization through this method.[50] Women in Japan have a long history of utilizing the latest tools available to them at any given time to take action. It is no surprise that the Internet has emerged as the latest medium for information dissemination, communication and mobilization among women in Japan, given the lack of financial resources of most women's organizations. Without the existence of already strong grass-roots networks, however, deployment of the Internet would have been far less effective. The success of women's online organizations relies on strengthening the relationship between Internet users and the development of face-to-face activities.

CONCLUSION

This chapter has demonstrated ways in which the Internet has become a powerful tool for communication among women in Japan. The sites examined provide information about the current condition of Japanese women that is not typically made available in official publications and government announcements. Furthermore, the sites have assisted in creating networks among different women's groups scattered throughout Japan, many such networks initially being established through links to personal home pages. Such links enable women to discover quickly and easily information about organizations they are interested in regardless of their location.

The Internet can therefore accomplish two broad aims. First, it can facilitate the exchange of information among women and thereby give them an increased voice. Second, it can make women's activism undertaken outside the environs of the so-called "center" of Japan, Tokyo, much more visible. It can facilitate nation-wide awareness of issues, and nationwide campaigns that address issues whose origin may be local. It can also, for some, be an avenue for information about women's issues and activism globally – particularly for women able to read other languages. Since most sites display links to other women's groups in Japan and elsewhere, the Internet has increased women's capacity to form bonds with individuals and groups who would otherwise have been virtually inaccessible.

On the other hand, the Internet has also widened the gap between the haves and have-nots. Only 25.9 per cent of housewives and 15 per cent of people over sixty in Japan have Internet access.[51] Women as a group lag behind male users in their level of information literacy.[52] Moreover, the Internet can be used to exploit women, as evident in the many pornographic and mail-order bride sites. Japan is, in fact, known as the world's greatest importer of trafficked Asian women, and also as the world's greatest creator of child-pornography Web sites.[53]

Yet, it must be remembered that insofar as the mainstream Japanese media rarely report on female activism, the Internet offers a primary and irreplaceable means for Japanese women to make their voices heard and to communicate with one another. The development of and changes in women's online organizations demonstrate the subtle yet persistent efforts of Japanese feminists to challenge the status quo, and suggest that they have the capacity to contribute to the construction of a gender-equal society. In this sense, for Japanese feminists, these Web sites constitute the stage for a dynamic cyberculture that not only reflects but also enables change in Japanese society. The Internet, like no other medium, makes visible the diversity of Japanese women and their activism.

NOTES

1   The Ministry of Public Management, Home Affairs, Posts and Telecommunication, *Information and Communication in Japan, the 2001 White Paper*, Tokyo, Gyōsei, 2001, pp. 3, 100.

2   The Ministry of Public Management, Home Affairs, Posts and Telecommunication, *Information and Communication in Japan*, p. 100.

3   N. Hashimoto, *Women and Men in Japan*, Tokyo, Zenkoku-kyōiku bunka kaikan, 2001, pp. 2–3.

4   D. Kondo, *Crafting Selves: Power, Gender and Discourses of Identity in a Japanese Workplace*, Chicago, University of Chicago Press, 1990, p. 31.

5   A. Orita, S. Miyagawa, and M. Niimi, "Network communication brings opportunity for minority: identity of woman at home." Online. Available HTTP: *http://www.isoc.org/isoc/conferences/inet/99/proceedings/3c/3c_4.htm* (27 April 2002).

6   S. Matsuura, "Josei tachi ga intānetto de hajimete iru koto" (Women are beginning to use the Internet), *Onna tachi no 21 Seiki* (The twenty-first century for women), no. 20, 1999, p. 20. All translations, unless otherwise noted, are mine.

7   In some cases there are English-language pages, but these are never identical in content with the Japanese.

8   Gender Information Site of the Gender Equality Bureau, Cabinet Office. Online. Available HTTP: *http://www.sorifu.go.jp/danjo/index.html* (6 June 2000).

9   Gender Equality Bureau, Cabinet Office, *Chihō kōkyō dantai ni okeru danjo kyōdō sankaku shakai no keisei matawa josei ni kansuru seisaku no suishin jōkyō* (The present state of the promotion of policies relating to women, or the formation of a gender-equal society in local public entities). Online. Available HTTP: *http://www.gender.go.jp/index.html* (7 January 2002). S. Uno, "What are women's centers in Japan?" *Dawn: Newsletter of the Dawn Center*, January 1997, pp. 6–8.

10 Kokuritsu Fujin Kyōiku Kaikan (NWEC [National Women's Education Center], Japan).Online. Available HTTP: *http://www.nwec.go.jp* (27 April 2002); Josei to Shigoto no Miraikan (The Center for the Advancement of Working Women). Online. Available HTTP: *http://www.miraikan. go.jp* (27 April 2002).

11 Kyoto Wings (Kyoto). Online. Available HTTP: *http://web.kyoto-inet.or.jp*; *Will Aichi* (Aichi). Online. Available HTTP: *http://www.will.pref.aichi.jp*; Dawn Center (Osaka). Online. Available HTTP: *http://www.dawncenter.or.jp*; Tokyo Women's Plaza (Tokyo). Online. Available HTTP: *http://www.tokyo-womens-plaza.metro.tokyo.jp/* (27 April 2002).

12 As Nishiyama Chieko stresses, "Civil laws, legislation and policies of the government still encourage Japanese women to be a good wife and a wise mother rather than to be an independent citizen free from the traditional gender roles." C. Nishiyama, "Josei seisaku to jendā" (Women's policies and gender), *Kanagawa Josei Jyānaru* (Kanagawa Women's Journal), vol. 13, 1995, pp. 30–44.

13 In summer 2001, the Women's Center in Chiyoda Ward, Tokyo, canceled a public seminar to be given by Matsui Yayori, in order to "maintain [the political] neutrality of Chiyoda Ward." This is because: "since Matsui Yayori has publicly called attention to Japan's war responsibility, the cancellation is seen to be a result of sustained objections and pressure by right-wing organizations on the ward government." Y. Matsumoto, "Internet and Women, Public Women's Center, Freedom of Speech," KNOWHOW. Online posting. Available e-mail: KNOWHOWCONF@NIC. SURFNET.NL (10 August 2001).

14 *Will Aichi* provides a mailing list and bulletin board service as well as useful links to the local women's groups.

15 For example, Hachinohe Women's Action (HWA) and AURORA are local women's groups. HWA is partially entrusted by the Ministry with the task of promoting gender equality in the city of Hachinohe. AURORA stemmed from a group of women who attended workshops held by the Dawn center, Osaka women's community center. Hachinohe Women's Action. Online. Available HTTP: *http://www.hachinohe-u.ac.jp/women97/* (27 April 2002); AURORA. Online. Available HTTP: *http://plaza27.mbn.or.jp/~aurora/AURORA* (27 April 2002).

16 New Japan Women's Association (NJWA, Shin Nihon Fujin no Kai), founded in 1962. Online. Available HTTP: *http://www.shinfujin.gr.jp* (30 October 2001); Femin (*Women's Democratic Journal, Fujin Minshu Shinbun*), founded in 1942. Online. Available HTTP: *http://www.jca.apc. org/femin* (30 October 2001).

17 *Japan Women's Movement Network (fem-net)*. Online. Available HTTP: *http://www.jca.apc.org/ fem* (30 October 2001); WOM (Women's Online Media). Online. Available HTTP: *http://wom-p.org/index.htm* (30 October 2001).

18 H. Inoue, *Josei ni yasashii Intānetto no hon* (A simple Internet book for women) Tokyo, CQ Shuppan, 1999, pp. 164–75.

19 Asia Japan Women's Resource Center (AJWRC). Online. Available HTTP: *http://www.awrc.org/ org/ajwrc/ajwrc.html* (28 September 2001). M. Tomomasa, *Building electronic networks of activist women in Asia and in Japan – experiences of Fem-net and VAWW-Net*. Online. Available HTTP: *http://www.jca.apc.org/aworc/ict/notes/ajwrc9804en.html* (28 February 2000). After the Beijing Conference, Machiko Tomomasa, a coordinator of the Asia-Japan Women's Resource Center (Ajia Josei Shiryō Sentā), inaugurated *fem-net* as an electronic network for the women's movement in Japan.

20 Y. Matsumoto, "Netto de tsunagaru onna no undō" (Women's movements linked up by the Net), *Nihon Joseigaku Kenkyūkai Nyūsu* (Japan Women's Studies Research Group Newsletter), no. 217, 2001, pp. 6–10.

21 T. Yasutanaka, "Give us our half! Women in the elections." Online. Available HTTP: *http://www1.jca.apc.org/fem/senkyo/index.html* (8 June 2000).

22 Y. Enishi, "Give us our half! Women in the elections." Online. Available HTTP: *http://www1.jca.apc.org/fem/senkyo/index.html* (8 June 2000).

23 M. Takai, "Give us our half! Women in the elections." Online. Available HTTP: *http://www1.jca.apc.org/fem/senkyo/index.html* (9 June 2000).

24 "Give us our half! Women in the elections." Online. Available HTTP: *http://www1.jca.apc.org/fem/senkyo/index.html http://www1.jca.apc.org/fem/greetings/2001_2002.html* (27 April 2002)

25 "Give us our half! Women in the elections." Online. Available HTTP: *http://www1.jca.apc.org/fem/senkyo/index.html* (27 April 2002)

26 *No Terrorism – No War – No Violence*. Online. Available HTTP: *http://www1.jca.apc. org/fem/* (27 April 2002).

27 *Dōji tahatsu tero hōfuku sensō hantai kinkyū apiiru* (Urgent appeal against war of reprisals against outbreaks of terrorism). Online. Available HTTP: *http://www1.jca.apc.org/fem/no-violence/NoWarPetition.html* (4 April 2002). The messages are from Saito Masami, 21 September 2001, Ishimaru Akira, 27 September 2001, and Onita Tomoko, 11 October 2001.

28 *Gohōkoku to kansha* (Reports and thanks). Online. Available HTTP: *http://www1.jca.apc.org/fem/no-violence.html* (4 April 2002).

29 *Nara Kenritsu Ikadaigaku Akahara Soshō*. Online. Available HTTP: *http://www.kcn.ne.jp/~jjj/akahara* (26 October 2001).

30 "Workplace bullying in universities in Japan." Online. Available HTTP: *http://www.kcn.ne.jp/~jjj/akahara/acahara.htm* (26 October 2001).

31 "I won!" Online. Available HTTP: *http://www.kcn.ne.jp/~jjj/akahara/ac/iwon.html* (26 October 2001).

32 K. Watanabe, "Kyanpasu sekusharu harasumento zenkoku nettowāku" (National campus sexual harassment network), *Women's Asia 21*, no. 20, 1999, p. 29. Online. Available HTTP: *http://www.jca.apc.org/shoc/* (27 April 2002).

33 K. Watanabe, "Kyanpasu sekusharu harasumento zenkoku nettowāku," p. 29.

34 Working Women's Network (WWN). Online. Available HTTP: *http://www.ne.jp/asahi/wwn/wwin/index.html* (6 June 2001). Women's Union Tokyo. Online. Available HTTP: *http://www.f8.dion.ne.jp/~wtutokyo* (6 June 2001). Women's Solidarity Foundation (Josei Rentai Kikin). Online. Available HTTP: *http://www02.so-net.ne.jp/~wsf/index.html* (6 June 2001). WINWIN (Women in New World International Network). Online. Available HTTP: *http://ww.winwinjp.org* (13 June 2000).

35 Dawn Center "Sūji de miru Josei" (Women seen by statistics). Online. Available HTTP: *http://www.dawncenter.or.jp/plbank/plsql/numjyo.asp* (30 October 2001).

36 M.A. Okawara, "The leadership role of women in information society." Online. Available HTTP: *http://www.glocom.org/debates/200110_tforum/dis1/index.html* (27 April 2002).

37 *Network for Amending the Civil Law* (*Emu Netto/Minpō kaisei jōhō nettowāku*). Online. Available HTTP: *http://www.ne.jp/asahi.m.net* (2 June 2000). *Organization for Considering Separate Married Surnames* (*Bessei o kangaeru kai*) Online. Available HTTP: *http://www/nwnet.or.jp/~nohiguti* (3 June 2000). Single Mothers' Forum (Shinguru Mazāzu Fōramu). Online. Available HTTP: *http://www7.big.or.jp/per centEsingle-m* (21 October 2001).

38 C. Akaishi, "Shinguru Mazāzu Fōramu" (Single Mothers' Forum), *Onna tachi no 21 Seiki*, no. 20, 1999, p. 28.

39 Violence Against Women in War-Network Japan. Online. Available HTTP: *http://www.jca.apc.org/~vawwjs* (26 October 2001).

40 *VAWW-NET Japan Nyūsu*, January 2001, p. 18 (200 persons from ninety-five foreign media and 105 persons from forty-eight Japanese media).

41 *VAWW-NET Japan Nyūsu*, January 2001, p.2. The VAWW-NET site continues to provide detailed information concerning the historical and political background to the Tribunal as well as the

daily developments, accompanied also by a live broadcast of the proceedings. Later, the Tribunal was the object of a TV program directed by Nippon Hōsō Kyōkai (NHK), the public broadcast network. However, VAWW-NET Japan regarded the program as promoting misunderstanding and prejudice toward the tribunal and, claiming that the content and intention were distorted, immediately began to collect signatures in protest against the broadcast. Within one month, the signatures reached 2,878, of which 1,357 were collected via e-mail.

42   Ōya. Online. Available HTTP: *http://www7.big.or.jp/~single-m/bbs-jft/light-jft.cgi* (13 December 2001).

43   The bulletin board system started from 7 December 2001. Online. Available HTTP: *http://www7.big.or.jp/~single-m/bbs-jft/light-jft.cgi* (15 February 2002).

44   Anonymous, No. 105, *Kinkyū kokuchi desu! Natsume sanchi no keijiban* (Urgent! Natume's Bulletin board). Online. Available HTTP: *http://bbs.fresheye.com/free/11214/living* (5 February 2002).

45   Machida. Online. Available HTTP: *http://www7.big.or.jp/~single-m/bbs-jft/light-jft.cgi* (8 and 10 December 2001).

46   C. Akaishi. Online. Available HTTP: *http://www7.big.or.jp/~single-m/bbs-jft/light-jft.cgi* (12 December 2001).

47   A. Kinoshita, "Onna tachi no idobata kaigi" (Women's chatter), *Shakai kyōiku* (Social Education) no. 30, 1986, p. 32.

48   A. Kinoshita, "Onna tachi no idobata kaigi," p. 33.

49   The organization that produced the textbooks, Atarashii Rekishi Kyōkasho o Tsukuru Kai, was formed on 2 December 1996.

50   K. Yamazaki, "Onnatachi ga tsukuru ōtanatibu media" (Alternative media made by women), *Onna tachi no 21 Seiki*, no. 10, 1997, p. 7. Many of the women ran to the nearest store equipped with a fax machine to send their signatures.

51   Just 21.1 per cent of people who have an annual income of under 4,000,000 yen have access to the Internet, and 63.2 per cent of Japanese women have an annual income of less than 3,000,000 yen. *Danjo Kyōdō Sankaku Hakusho* (White paper on a gender-equal society). Online. Available HTTP: *http://www8.cao.go.jp/whitepaper/danjo/plan2000/h13/1-2.html* (22 October 2001). Furthermore, as for the availability of Internet access in public libraries, as of February 2001, Tokyo prefecture (twenty-seven cities and twenty-three wards) provides access in the local community libraries of only one city and two wards.

52   The Ministry of Public Management, Home Affairs, Posts and Telecommunications, *Information and Communication in Japan, the 2001 White Paper*, Tokyo, Gyōsei, 2001, pp. 3, 100.

53   Y. Matsui, "Intānetto o Josei no te ni" (Let women take over the Internet), *Onna tachi no 21 Seiki*, no. 20, 1999, p. 7.

8 Cybermasculinities

Masculinities and the Internet in Japan

Romit Dasgupta

INTRODUCTION

In September 1999, I attended a two-day conference in the city of Kyoto. In addition to myself there were around 300 delegates, two-thirds of whom were male. This in itself was nothing unusual – conferences and conventions tend to be male-dominated affairs anyway, particularly in Japan. However, the delegates attending this conference were part of an emerging "community" of individuals who had gathered to discuss, problematize, and interrogate notions of "masculinity" and being male. The official title given to this gathering was the "Kyoto Men's Festival" (or, in Japanese *Kyōto Menzu Fesuta*), and delegates came from all over Japan.

As I learned, the gathering was the fourth such annual Men's Festival to be held in Japan.[1] Over the course of two days, participants had the choice of attending sessions and workshops on a variety of themes. These ranged from topics such as introducing works on "Men's Lib;"[2] representations of gender in men's magazines, through to topics like male menopause and domestic violence.[3] Like myself, many of the participants were meeting the main organizers as well as other individuals participating in the gathering for the first time. Conversations with some of the first-time attendees made me aware of two particularly interesting issues. First, it became apparent that what for many had until quite recently been vague, personal problems and anxieties relating to negotiating individual identity as a male in Japan at the end of the millennium were now being vocalized in a public space, with reference to the experiences of other individuals. Second, through these conversations I learned that for many of the participants, news about the event, as well as information about wider issues pertaining to masculinity, had been obtained through the Internet as much as through other information channels such as word-of-mouth, print media and television.

Thus, in my mind a series of interlinked questions started taking shape. What were some of the social, cultural, economic, and other factors underlying this interest in problematizing maleness and masculinity? Why now? After all, Japan has had a long tradition of male homosocial spaces, without "masculinity" being interrogated in the same way. What influence (if any) has the spread of the Internet and other forms of computer-mediated communications (CMC) had, both through facilitating the dissemination and exchange of information, and through opening a new "space" for debate? Moreover, has the Internet worked toward creating a sense of unity and solidarity, as has often been the case with many "minority" and single-issue groups,[4] or is it more a means for individual communication, as appears to be the case for many of the Japanese gay men discussed by McLelland,[5] or is the common denominator of "de-constructing masculinity" too broad to allow for a sense of coherent community identity – whatever that may be?

The core problematic around which all these questions revolve is the notion of "identity/ies" and the ways in which they are played out in the context of globalization, particularly in the context of global flows of information and technology – what Manuel Castells refers to as the "network society" which he sees as having succeeded modernity. According to Castells, "identity" in "network society" is crafted in a different manner from the ways it was negotiated during the onset of modernity.[6] Under the new conditions accompanying the network society, identity constructions are less based on the relatively stable, predictable institutions and practices of civil society, and are increasingly splintered, organized around collective resistances (essentially "networks of people") to perceived hegemonic discourses ("community" identities that range from ethno-cultural nationalisms and religious fundamentalisms, through to environmental, feminist, and non-heterosexual identity based communities).[7] Moreover these identities are often based on networks that may transcend physical place – what Featherstone refers to as a "psychological neighborhood."[8]

This chapter, then, through the experiences of *masculinity* as an identity, seeks to explore some of these issues surrounding identity negotiation in the context of a network society. As various researchers have noted, masculinity, rather than being constant over time and space, "means different things at different times to different people."[9] Thus masculinity as an identity has to be "crafted" out of a myriad co-existing and/or conflicting *masculinities*, and is embedded within a wider context of social, economic, cultural, and political conditions. At a particular point in time, one particular discourse – what Connell terms "hegemonic masculinity" – exerts the greatest power within that society, even though this hegemony has to be constantly negotiated and re-negotiated.[10]

I begin by discussing the background to the shaping of hegemonic masculinity in contemporary Japan, specifically the nexus between work and masculinity, which played a crucial role in the process, and channels such as popular culture which have worked towards strengthening this nexus. However, this hegemonic discourse has been subject to increasing critical scrutiny from various sectors within society, not least from men themselves. Thus, I discuss some of the factors underlying this unraveling process, some of the major players involved and draw attention to some of the channels through which these changes have been articulated – often the very same channels (men's magazines, for instance) which may simultaneously work to reinforce hegemonic masculinity. One significant channel in the process has, of course, been new communications technology, particularly the Internet. Consequently, the latter part of the discussion will focus on some of the ways in which this medium is being used in these negotiations and articulations of masculinity, specifically the extent to which the Internet, more than other channels of expression, has contributed to a questioning of hegemonic masculinity, through the facilitation of information exchange, and by opening up new spaces for vocalizing dissent.[11]

CONSTRUCTING AND DECONSTRUCTING JAPANESE MASCULINITY: THE RISE AND FALL OF THE SALARYMAN DISCOURSE

In popular imaginings, mention of Japanese masculinity generally conjures up a particular image – that of the sober, respectable, grey-suited "everyman" office-worker, commonly referred to as the "salaryman" (or, in Japanese, *sarariiman*). The origins of the salaryman can actually be traced back to at least the early decades of the twentieth century. [12] However, it was over the post-war decades of rapid industrialization and the expansion of white-collar employment that he came to embody the masculine *and* corporate ideal. Moreover, the salaryman became simultaneously the beneficiary as well as the victim of the fallout of Japan's industrial transformation over these decades. Although the term salaryman in the strictest sense – male, full-time, permanent, white-collar employees of generally large organizations offering such benefits as lifetime employment and seniority-based wages and promotions – only covered about one-third of the work-force even at the peak of the era of high economic growth in the early 1970s, the *values* associated with this discourse of corporate masculinity were far more pervasive, and in many respects came to embody the hegemonic discourse for *all* males.

This idealized masculinity, premised on specific assumptions pertaining to individual behavior and lifestyle choices, was not automatically realized the day a young male entered the workforce and formally became a *shakaijin* – literally, "member of society," – with all the accompanying privileges as well as expectations and responsibilities of adulthood. Rather, as I have noted elsewhere, this "crafting" process commences early on in childhood, continues through adolescence, becomes all the more pronounced once the individual enters the workforce and formally takes on his *shakaijin* responsibilities, and indeed continues over his entire life course.[13]

While this blueprint of masculinity did (and continues to) exert considerable influence on the lives of most men (*and* women), it was by no means an all-encompassing, stable constant. The reality is that most men *do not* or *cannot* measure up to the dominant "ideal" in masculinity.[14] First, in any society at any one point in time, underlying a seemingly powerful dominant discourse of masculinity are a myriad of co-existing and/or conflicting *masculinities* which engage with the dominant in varying ways.[15] Even during the rapid industrialization decades of the 1960s and 1970s the corporate warrior/salaryman discourse was far more tenuous than initial appearances would suggest, with alternative, often dissenting masculinities co-existing.[16] However, it was really from the mid-1980s that the dominant discourse of masculinity started to unravel in a significant way, and continued to do so into the 1990s. Underlying this process were social, economic, and cultural factors that had accompanied Japan's transition into a mature, affluent, late-capitalist society. Sectors which had previously been relatively unimportant started to occupy an increasingly influential role within the economy – areas such as tourism, fashion, media and communications, and education services.[17] For significant numbers of younger Japanese – male and female – there was now a greater range of lifestyle options than had been available to their parents' generation.

At the same time, as the Japanese economy slowed down and slid toward recession from the early 1990s, subscribing to the salaryman model of masculinity became even less appealing, as some of the old guarantees such as permanent lifetime employment and automatic promotion and salary increase tied to length of service were becoming increasingly difficult to sustain, even for large elite corporations. For large numbers of middle-aged males who suddenly found themselves without job security, the issues at stake went beyond mere loss of employment – their very identity as men, as fathers, as providers was destroyed. The ramifications ranged from rising suicide rates among middle-aged males through to magazine columns and Internet sites providing increased space for salarymen to give vent to their frustrations, and television programs focusing on the issue of *datsu-sara* (literally, "escaping a salaryman life").[18] Indeed, the very magazines that worked towards propagating and reinforcing hegemonic masculinity also provided spaces for dissent. For instance, every issue of *Big Tomorrow* features a "Hotline" column of the agony aunt variety, where readers can seek advice on a whole variety of personal and work related problems, although, as I discuss further on in this chapter, undertones of "making a special effort" to conform to the expectations of hegemonic masculinity tend to inform a lot of the advice *given* (as opposed to the advice sought) in these columns.

In Japan, the emergence of what was to crystallize as men's studies/masculinities studies and the men's movement did not occur as an isolated, one-off development.[19] Right from the onset there were considerable intersections, both at an individual and intellectual level, with various academic and community-based women's groups and feminist organizations, as well as interaction with lesbian and gay organizations, such as OCCUR and Sukotan Project which were becoming increasingly visible and vocal around the same time.[20] Moreover, just as increasing gay and lesbian visibility from the 1990s was partly attributable to the media-generated "gay boom" discussed by McLelland,[21] at an everyday level the mainstream media also contributed to the growing awareness of the issues raised by masculinity studies. For example, the popular men's magazine *Bart* ran a sympathetic feature on masculinity studies in July 1997 that was accessible to a wide readership.[22] The English-language press, too, picked up on the emergence of this new social trend fairly early on.[23]

The first men's group, *Menzu ribu kenkyū kai* (Men's Liberation Research Association) was established in Osaka in 1991 by a small group of academics and community members.[24] The Osaka group was instrumental in the subsequent establishment of Japan's first Men's Center in 1995.[25] Over the course of the 1990s, following the example of the Osaka group, men's groups were established in other parts of the country – starting with the establishment of Men's Lib Tokyo in 1995, and subsequently groups in Kanagawa, Fukushima, Okayama, Kyushu and elsewhere.[26] While the issues and concerns often reflect local conditions and the priorities of the members of the groups at a particular point in time, certain issues cut across most of the groups – the provision of a forum where men could talk about issues such as relationships, work and sexuality that were pertinent to their personal lives. Moreover, underlying the thinking of most of these groups was the recognition that while men did benefit as a group from the patriarchal dividends of hegemonic masculinity, individual males could also be unwitting "victims" of patriarchy. The fallout from this "burden of masculinity" could manifest in stress, domestic violence, alcoholism, depression, suicide, and other psycho-social problems. Thus, most of the groups were engaged in numerous activities – networking with other groups (largely in Japan, but also overseas) working in gender and sexuality related areas; providing support and information services to individual men (and depending on the group, women[27]) in need; and providing information and raising awareness within the general community. This was done through the holding of seminars and workshops, publications such as newsletters, pamphlet style *minikomi*, media releases and the like. For example, Men's Center Japan put out a number of quick, easy-to-read booklets with titles like *"Otokorashisa" kara jibunrashisa e* (a rough translation would be something like, "From *macho*-ness to being yourself") and *Otokotachi no "watashi" sagashi* (Men's search for true self), which contain articles on a variety of pertinent issues – self-reflective pieces, writings on friendship, work, and fatherhood.[28]

The visibility and activism of the men's movement, as well as the shaping of some sense of community, received a significant boost in November 1996, when the first Men's Festival was held in Kyoto, attracting over 160 male and female participants from across Japan.[29] The Festivals have continued to be held annually – the most recent was held in September 2001, in Fukuoka, Kyushu. Like other similar forums, these festivals (which are more akin to conference/workshops) were important in bringing together individuals from across the nation dealing with similar issues, who might otherwise not have had the opportunity to interact and exchange ideas and strengthen a sense of shared community. This was particularly relevant for men's groups outside the central Tokyo-Osaka-Kyoto orbit; regional groups were often established during periods of post-Festival optimism, by returning local residents. It was during these years of social and cultural instability and flux in the mid-1990s (significantly also years of growing diversity and segmentation of lifestyles) that the Internet first started to have a discernible influence on society.

MASCULINITIES ONLINE: CREATING AWARENESS AND COMMUNITY

The potential the Internet offered both as a means for information dissemination and as a tool for networking and community-building began to be appreciated by various groups in Japan from the mid-1990s.[30] Several of the other chapters in this volume discuss the ways in which women's groups, sexual minorities, persons living with AIDS and other groups involved in social and cultural activism turned to cyber technology as a means of disseminating information and networking.[31] Online ventures such as the Women's Online Media (WOM)[32] initiated in May 1995 demonstrated to the men's groups the possibilities offered by online techno-logy.[33] According to Kaneko Yasufumi, of Men's Lib Tokyo, Ikuji-ren, a men's group concerned with issues of male-parenting and child-rearing, set up a committee to explore the possibilities offered by the Internet, and approached Men's Lib Tokyo to see if the latter would be interested in launching a Web site. As it turned out, the Men's Lib Tokyo home page was launched before that of Ikuji-ren – in April 1996. This was followed in November 1996 by the launching of a home page for the Kansai group, Men's Center Japan, as a consequence of the very first Men's Festival which was held in Kyoto the same month.[34] Although the Web sites of the Tokyo and the Kansai groups are the most comprehensive, most of the regional men's groups have also gone online, often drawing upon the expertise and experience of the older/larger groups like the Tokyo one.[35]

In order to get an appreciation of the actual ways in which the Internet interacts with these projects of deconstructing masculinity, I would like to discuss selected Web sites that are concerned with issues of masculinity. Underlying my discussion will be the questions posed in the introduction – the role played by the Internet as a means for publicity and information dissemination and the extent to which there may be a potential for a sense of shared community identity being created through these Web sites.

In discussing these Web sites, there are a number of considerations (some raised in the introduction) which have a bearing. First, is the whole question of definition. Using "masculinity" as a determining category can be problematic in that we could be as inclusive or exclusive as we choose. For instance, a search using a Lycos search engine for the keyword *dansei* (male/men/masculine) yielded 170 Web sites related in one way or another to the term.[36] These included sites dealing with cosmetic/plastic surgery for men (*biyō seikei dansei no tame no WEB!*);[37] a male vocal group (@GOSPELLERS);[38] a cooking site for middle-aged/older males (*oyaji no tame no jisui kōza*);[39] the online sites of major male fashion magazines such as *Fine Boys* and *Gainer*[40] and a male astrology site (*dansei no uranai kanteishitsu*)[41] among others. This is in addition to listings of sites relating more specifically to masculinity issues such as *Otoko no seikatsu jiritsu o kangaeyō* (Let's think about male independence/self-sufficiency, a site dealing with issues relating to men taking a more proactive role within the household),[42] or the results of a survey on Gay Men's Mental Health (*gei dansei no mentaru herusu chōsa kenkyū kekka*)[43] as well as sites of the various men's groups discussed earlier.

Thus, on the one hand, we could make a case for *any* Web site that has some bearing on masculinity (for instance, a group like @GOSPELLERS) to be included. However, we could just as easily restrict the discussion to only those groups that are *specifically* concerned with deconstructing "masculinity" as a construct. But where would that leave Web sites like gay men's sites – given that heteronormativity has been integral to the hegemonic discourse of masculinity – which may not seek to directly challenge masculinity per se? Similarly, given the centrality of "work" in the construction of hegemonic masculinity, would a salaryman Web site providing a space for disgruntled salarymen to voice their frustrations not be equally relevant to the discussion? In posing these questions, my intention is not so much to provide some kind of prescription to help solve these problems of definition, but rather to draw attention to the complexities at play. For the purposes of this discussion, I will focus on a selection of Web sites that are either organized around the category of masculinity itself (such as the Web sites of some of the men's groups), or are more specific, single-issue sites that either focus on a particular aspect of masculinity, or sites which end up problematizing masculinity almost by default. Furthermore, the discussion only aims to provide a general overview of these sites – detailed ethnographic description (for instance, reference to individual postings, or discussion of interaction on chat room floors) has been deliberately avoided.[44]

Broad-based groups: creating awareness and disseminating information

The Men's Lib Tokyo home page[45] provides a well-designed introductory portal consisting of an explanation of the objectives of the group, as well as sub-headings ("Guide," "Info," "Issues," "Liaison," "Works," etc.). These provide the necessary links to information about other things, including the following: the group's electronic mailing list; upcoming events (such as seminars and public meetings); support services; and comprehensive archival lists of relevant reference materials (books, as well as print articles and television programs dealing with an extensive range of issues – categories include mental health, domestic violence, prostitution and pornography, and housework and cooking, among others). The Web site is frequently updated and contains an extremely comprehensive listing of links to the home pages of diverse groups concerned with men's and gender issues. These include men's groups in other areas of Japan, various men's health organizations, self help and support groups, groups dealing with family and children's issues, and sexual minority groups.[46]

A random listing of some of the Web sites that Men's Lib Tokyo links to gives an indication of the opportunities opened up (in terms of building networks and gaining information) through cyber technology. They include *Men's Telephone Hotline ("Otoko" nayami no hotorainu)* – the first crisis and counseling telephone service for men which commenced operating in 1995;[47] *JABIP (Nihon DV kagaisha puroguramu kyōkai)* – a domestic violence counseling and support group targeting perpetrators;[48] *Suzuran's Gate* – a site for cross-dressing/transgender/transvestite/transsexual issues;[49] *If He Is Raped* – a support/information site dealing with issues surrounding male victims of rape;[50] *Life AIDS Project* – an NGO HIV-support group;[51] *Sukotan Project (Sukotan kikaku)* – a gay and lesbian site with links to OCCUR, one of the major gay activist groups;[52] the Web site of The Tokyo International Lesbian and Gay Film Festival;[53] the *Singles Club* – a site dealing with issues relating to living as a single male in Japan;[54] *Dame-ren*, the group for "social misfits" which challenges many of the premises of society (including gender discourses and the centrality of work);[55] *Ikuji-ren*, the group discussed earlier that concerns itself with issues of gender-free parenting;[56] The Gender Information Site of the Prime Minister's Office;[57] and even overseas sites like The White Ribbon Campaign, an international anti-sexual/domestic violence group.[58]

In addition to the Web site, the Tokyo group also sends out an electronic news-letter, *Menzu Ribu Tōkyō NewsMail*, on a regular basis which updates members about changes to the Web site, as well as providing new information about relevant seminars, lectures, publications, etc. (and providing links to the relevant URLs). For example, the latest issue at the time of writing (Issue no. 74, 10 February 2002) informed members about a variety of upcoming events dealing with issues related to masculinity.[59] This included information about such events as the fourth session in a series of cooking classes for men organized by the Chiba City Women's Center;[60] another cooking class, this one aimed at imparting "basic cooking skills for men over sixty," organized by the Better Homes Association;[61] a two-day training/information workshop on domestic violence held under the auspices of the Japan YWCA in Osaka, targeting personnel working in areas like education, medical services, counseling, and local government, who may be required to deal with either victims or perpetrators of domestic violence;[62] and a seminar series organized by the Women's Center of the City of Takarazuka (near Kobe) on "Ways-of-living as a Male in the Future," covering such issues as "search for true self," being able to express yourself/communicate more effectively with people around you, and improving cooking skills![63] In some respects, perhaps more so than the actual Web site of the group, this newsletter, through its constant updating of news and relevant information, works towards strengthening a sense of common-ality amongst members – particularly significant given the difficulty in individual members (even members of the committee) meeting in "real-time," since most group members hold demanding, full-time jobs.[64]

However, what comes across from both the newsletter and the home page of Men's Lib Tokyo, and from surfing the Web sites of other groups such as the Osaka-based Men's Center Japan,[65] as well as the sites of smaller regional groups such as Men's Lib Fukushima, Men's Lib Ibaraki, or Men's Lib Kanagawa,[66] is their role in networking and information provision rather than interactive commu-nity building. Hence, postings on the bulletin boards of these groups tend to be dominated by "announcement" type postings. For instance, on the Men's Center Japan board (accessed on 17 January 2002) the overwhelming majority of the twenty-four postings (out of a total of sixty-two dating back to October 2001) were notices about upcoming forums, lectures, seminars and updates of past postings. In this regard, we could argue that as far as these broad-based groups are concerned, the Internet is simply augmenting existing networking and awareness building carried out through books, magazines, and pamphlets of the *minikomi* type.[67]

Single-issue groups: deconstructing masculinity by default?

In common with the Web sites of the broad-based groups, single-issue sites, such as those of Ikuji-ren (the male-parenting group discussed earlier), also provide information such as resources and details of up-coming events. However, the difference seems to be that for single-issue groups there may be more of a sense of an interactive community, one that may be facilitated through one specific issue around which debate can hinge. In contrast to the postings on the Men's Center bulletin board, postings on Ikuji-ren's bulletin boards conveyed an impression of a lively interaction revolving around issues of (primarily) male parenting: issues such as parental leave entitlements, the personal and work ramifications of taking leave, personal trials and triumphs of male-parenting and child-rearing, and so forth.[68] These issues allow scope for much agreement and disagreement, and hence more ongoing interaction, in contrast to the less concrete (and/or contentious) networking/information dissemination features of the broad-based groups' sites.

The other point that deserves consideration is that for single-issue groups, challenging masculinity *per se* is not really a concern. However, given that the specific issues that these groups address such as fatherhood, sexuality, or work are inextricably linked to individual male identity, interrogating these does indirectly impact on wider issues of masculinity. Web sites such as *Kaisha seikatsu no tomo* (Company lifestyle friend), *Sarariiman to OL no hiroba* (Open space for salarymen and office ladies) and *Sarariiman kyōwakoku* (Salaryman republic) provide excellent illustrations of this point.[69] These sites are primarily concerned with issues relating to the workplace and organizational culture – providing both light-hearted anecdotal information and entertainment (horoscopes, amusing work-related incidents, etc.), as well as more serious issues, related information and advice, and a space for employees to air grievances and frustrations. In this respect, these sites are not consciously concerned with "de-constructing" masculinity. However, given the centrality of work in constructing masculinity, the issues posted and discussed on, for example, *Kaisha seikatsu no tomo*'s Web site's open-access space, under categories such as "Past Dreams" (*katsute no yume*), "My Ideal/ Dream Job" (*yume no tenshoku saki*), "Listen to Salaryman Anger!" (*kike! Sarariiman no okori*), and "Advice Room for Any Problems" (*onayami nan demo sōdan-shitsu*), are often linked to expectations of hegemonic salaryman masculinity. A random pick of the archived monthly postings to the "Advice Room" between September 1996 and January 2002 gives us some indication of this underlying disenchantment.[70] The "problems" posted encompassed a range of topics at the nexus between corporate culture and masculinity, including dissatisfaction with the need to participate in obligatory company trips and the possible ramifications of refusing to go (October 1996, *Shain ryokō o kotowaritai!*); resistance to expectations of promotion from a 37-year-old employee in an electronics firm who felt that he was not suited to coping with the added stress and responsibility (January 1997, *Kanrishoku ni naritakunai*); financial and personal problems arising from changing jobs (December 1997, *Tenshoku o kurikaesu jibun*); concern about the negative impact of workaholism on a future marriage partner (April 1998,

Wākuhorikku to kekkon); disillusionment with the lack of freedom within Japanese work culture (March 1999, *Nihon no kaisha wa fujiyū da*); concerns about the financial and personal ramifications of imminent middle-age lay-off (August 2000, *Risutora saresō jibun*); among a host of similar concerns relating to the workplace.

Significantly, *masculinity* itself is seldom specified – it gets implicated almost by default. For instance, a posting from a 24-year-old male entrant (*shinnyū shain*) in the financial sector (*Eriito ginkōin no nayami*) illustrates this quite clearly. The complainant notes that he was attracted to his present job with a bank, largely due to the attractive salary, and due to the fact that he had nothing else he wanted to do. However, since commencing the job he finds his whole life revolving around work. He is unable to return home until late everyday, has no girlfriend, and has no real hobbies or interests – in short, a life without meaning (*hakkiri itte tsumaranai seikatsu desu*). While he does not actually allude to masculinity, all the expectations and problems surrounding salaryman masculinity are in fact encapsulated in his posting. Another posting, also from a new entrant in the financial sector, interrogates a basic premise of hegemonic masculinity – heterosexuality. The 23-year-old writer "confesses" that he is gay, and wonders how others (who he presumes to be largely heterosexual) would respond to a co-worker "coming out" to them. Once again, masculinity itself is not called into question, rather the expectations of masculinity are.

Advice from others (*mina sama no adobaisu*) in response to these problems/ questions was also posted; these responses encompassed a whole range of opinions. For example, there were 150 responses to the October 1996 posting complaining about the necessity to participate in company trips, mentioned above. Some of these criticized the complainant for being too sensitive about a relatively minor issue (Response no. 133: *seikaku ga komakakute ki ni shisugi ja nai no*), or advised him to "endure" (*gaman*) since he is a salaryman, or quit the company (Response no. 106: *ichinichi gurai gaman dekinai no? shosen sarariiman desho. Kaisha yametara.*). However, a significant proportion of the responses were supportive. These included expressions of empathy for the complainant, with statements like "it's a total waste of time, and waste of a life" (Response no. 143: *hontō ni jikan no muda, jinsei no muda*), or "I like travel, but hate company trips" and inviting someone to start a "movement to exterminate company trips" (Response no. 144: *boku wa, ryokō wa suki da ga, shain ryokō wa dai kirai. Dareka, shain ryokō bokumetsu undō shimasen ka?*). Many respondents also offered blunt advice about dealing with the problem, such as lying and using excuses like sickness (mentioned in a number of responses, such as nos 20, 49, and 143) or prior commitments like a wedding or memorial service (*hōji*) (Response no. 66), or "drinking lots of alcohol, and falling asleep straight away!" (Response no. 86: *gangan nonde, sassato neru*).

In some respects these spaces that allow for the airing of gripes and concerns may not be dissimilar to the advice columns of the agony aunt variety in salaryman magazines like *Big Tomorrow*, touched upon earlier. The difference lies in the space allowed for flexibility and dissent. McLelland, citing research done on the responses to problem letters sent in to a leading newspaper, notes that an often stern, didactic, correctional pose is adopted by the advice giver. The onus seems to be placed on the individual to conform to societal expectations in order to rectify whatever problem she/he faces.[71] A similar point could be made for salary-man magazines like *Big Tomorrow* – for example, the response, in a recent issue, to a complaint about having to cope with entirely new work demands as a consequence of a sudden transfer was to lay the blame on the writer for being overly "selfish" (*waga mama*) and that as long as one put in one's best effort, the experience was bound to be useful someday.[72]

Conversely, as indicated by the discussion of postings on the *Kaisha seikatsu no tomo* site (and the responses to these postings), Internet sites allow for the expression of a greater range of dissenting voices.[73] For instance, in contrast to the "preachy" tone adopted in the *Big Tomorrow* column mentioned above, responses to a not dissimilar concern posted on *Kaisha seikatsu no tomo*'s "Advice Room" (December 1997, *tenshoku o kurikaesu jibun*) from a 35-year-old male employee who was unable to hold a job for long elicited numerous sympathetic responses, including advice such as encouraging him to keep searching until he finds a suitable job (Response no. 80: *iijan, oishii shigoto ga mitsukaru made, tenshoku shinayo*), or telling him that a salaryman's reality is different to the salaryman heroes portrayed in TV serials (*TV dorama no bari bari sarariiman no yō ni wa ikanai ze!!*).[74] Even the responses to the posting from the young salaryman dealing with issues of sexuality, and what ramifications "coming out" in the workplace might have, cover a range of opinions, some negative, but many supportive and/or neutral.

CONCLUSION

In light of some of the issues raised in this chapter, how are we to assess the impact of the Internet and other CMCs on constructions and deconstructions of masculinity in Japan? This question is significant given the hype that often surrounds questions about the Internet's potential for socio-cultural transformation – both the utopian visions of the "liberating" potential of CMCs, as well as the dystopian views which highlight the policing and surveillance potential of the Internet.[75] This issue is particularly relevant when talking about Japan, given the rapid expansion of Internet use in recent years, due to the development of new technologies like the *i-mode* mobile phone Internet service.[76]

With reference to masculinity, as stressed earlier, the challenges posed to the hegemonic discourse drew upon socio-economic and cultural factors arising out of the very processes of rapid industrialization and increased affluence that had allowed for this particular discourse to become hegemonic in the first place. Moreover, despite the hype often surrounding the Internet, the unraveling process had commenced quite some years *before* the spread of the Internet. Rather, as the discussion of the various Web sites examined indicates, what online technology facilitated was the streamlining and dissemination of resources and information related to masculinity which were already in existence (and still continue to exist) – resources and information which could (and can) also be found through print and visual media, seminars, lectures and public discussion. The Internet has also enabled men who are geographically dispersed (particularly those living outside

Tokyo and Osaka) to network with each other and discuss problems and issues with male identity. The Internet and other forms of new communications technology have opened up new spaces, or made more accessible existing spaces, for questioning, debate and dissent.[77]

ACKNOWLEDGMENTS

I would like to thank several individuals involved with the Men's Movement in Japan for providing me with information and insights which have been invaluable in writing this chapter. First, Dr Taga Futoshi of Kurume University in Kyushu, Japan, who provided me with a range of material dealing with the topic, and kindly made all the arrangements for my discussions with key individuals connected to the Tokyo and Kansai men's groups during a visit to Japan in September/October 2001. Thanks to Dr. Taga, I was also able to attend the Japan Gender Association Conference (*Nihon jendā gakkai 2001 nendo taikai*) in Kyoto on 21 September 2001. Also, Mr Kaneko Yasufumi of the Tokyo group, Men's Lib Tokyo, interviewed by me on 2 October 2001, and Mr Nakamura Tadashi of the Kansai group, Men's Center Japan, whom I had discussions with during the Kyoto conference mentioned above, provided me with some extremely valuable background information pertaining to the online activities of these groups (see note 33, below). I also wish to express my appreciation to Professor Itō Kimio, of Osaka University, arguably the "public face" of men's studies in Japan, and Mr Toyoda Masayuki of Tokyo Men's Lib for their insights. The editors of this volume, Associate Professor Nanette Gottlieb and Dr Mark McLelland of the University of Queensland, also provided valuable feedback and suggestions on evolving drafts of the chapter; without their support the chapter may never have reached its final form!

NOTES

1 The first, also held in Kyoto, was in 1996.
2 In Japan the term Men's Lib (*menzu ribu*), shortened from Men's Liberation, was the term that was initially favored by many in the area of masculinities studies/activism – e.g. names of groups such as Men's Lib Tokyo or the *Menzu ribu kenkyūkai* (Men's Lib Research/Study Association). To persons in the English-speaking world the term Men's Lib may conjure up associations with the more revisionist, generally anti-feminist elements within the men's movement in the United States or Australia – groups such as those loosely termed the mythopoetics, or the Promise Keepers, and individuals such as Robert Bly or Steve Biddulph. However, the use of Men's Lib notwithstanding, the Japanese groups discussed in this chapter by and large share many common views and concerns with feminist and women's groups and could by no means be considered to be reactionary or revisionist. Moreover, although a discussion of the complex semantic issues surrounding use of the term "lib" is outside the orbit of this chapter, suffice it to say that members of these groups are not unaware of the possible implications of the term, and there has been considerable discussion of the issue. For example, see the September 2001 issue of *Menzu nettowāku* (Men's Network), a periodical published by Men's Center Japan for an excellent discussion of some of the issues surrounding the semantics of the term (*Menzu Nettowāku*, vol.

11, no. 4, 2001, p. 5), as well as K. Itō, "Nihon demo 'menzu mūbumento' hajimaru!" ('Men's Movement' starts in Japan too), in K. Itō, *Danseigaku nyūmon* (An introduction to Men's Studies), Tokyo, Sakuhinsha, 1996, pp. 310–15. For a discussion of the "appropriations" and reconfigurations of another English-language term, "gay," into the Japanese context, see M. McLelland, "Out and about on Japan's gay net," *Convergence: Journal of Research into New Media Technology*, vol. 6, no. 3, Autumn 2000, pp. 16–33 (particularly pp. 25, 26, 29, 30).

3   *Dai yon kai otoko no fesuteibaru hōkoku shū* (Proceedings of the Fourth Men's Festival), Osaka, 2001, Men's Center Press.

4   See for instance, M. McLagan, "Computing for Tibet: virtual politics in the post-Cold War era," in G.E. Marcus (ed.), *Connected: Engagements with Media*, Chicago, University of Chicago Press, 1996, pp. 159–94; Part V (pp. 251–94) of D. Trend (ed.), *Reading Digital Culture*, Malden, MA, Blackwell Publishers, 2001; and A. Mitra, "Virtual commonality: looking for India on the Internet," and M. Wilson, "Community in the abstract: a political and ethical dilemma?" in D. Bell and B. Kennedy (eds), *The Cybercultures Reader*, London, Routledge, 2000, pp. 676–94, 644–57.

5   McLelland, "Out and about on Japan's gay net."

6   M. Castells, *The Information Age: Volume II, The Power of Identity*, Malden, MA, Blackwell Publishers, 1997, pp. 6 –12. See also, *The Information Age: Volume III, End of Millenium*, Malden, MA, Blackwell Publishers, 1998, pp. 340–8.

7   Castells, *The Power of Identity*, pp. 11, 12.

8   M. Featherstone, "Localism, globalism, and cultural identity," in R. Wilson and W. Dissanayake (eds), *Global/Local: Cultural Production and the Transnational Imaginary*, Durham, Duke University Press, 1996, p. 63.

9   M. Kimmel, "Masculinity as homophobia: fear, shame, and silence in the construction of gender identity," in H. Brod and M. Kaufman (eds), *Theorizing Masculinities*, Thousand Oaks, CA, Sage, 1994, p. 120.

10   R.W. Connell, *Masculinities*, St Leonards, NSW, Allen and Unwin, 1995, pp. 76–81.

11   Another consideration, which is outside the scope of this chapter, is the "negative" fallouts created through the intersections between "masculine" identity and cyber-technology – the growth of male pornography sites, for instance.

12   For more details of the origins of the "salaryman" discourse and the ways in which it is disseminated in contemporary Japan see a previous paper written by myself, "Performing masculinities? The 'salaryman' at work and play," *Japanese Studies*, vol. 20, no. 2, 2000, pp. 189–200. For the historical origins of this discourse of masculinity see also, E.H. Kinmonth, *The Self-Made Man in Meiji Japanese Thought: From Samurai to Salaryman*, Berkeley, University of California Press, 1981; and T. Umezawa, *Sarariiman no jikakuzō* (Salaryman self-images),Tokyo, Minerva Shobō, 1997, ch. 1.

13   Dasgupta, "Performing masculinities," pp. 193–8. See also, A. Allison, *Nightwork: Sexuality, Pleasure, and Corporate Masculinity in a Tokyo Hostess Club*, Chicago, University of Chicago Press, 1994; and M. Ishii-Kuntz, "Japanese fathers: work demands and family roles," in J.C. Hood (ed.), *Men, Work, and Family*, Newbury Park, Sage Publications, 1993, pp. 45–67.

14   M. Kaufman, "Men, feminism, and men's contradictory experiences of power," in H. Brod and M. Kaufman (eds), *Theorizing Masculinities*, p. 144.

15   R.W. Connell, *Masculinities*, pp. 76–81.

16   For instance, popular culture representations like *Tora-san*, the bungling, tragic-comic hero of the series of extremely popular movies bearing the same name, as well as movie genre such as *yakuza* movies tapped into some of these alternative masculinities. See I. Buruma, *A Japanese Mirror: Heroes and Villains of Japanese Culture*, London, Jonathan Cape, 1984, chs 10 and 12 for a discussion of *Tora-san* and *yakuza* movies; and M. Schilling, "Into the heartland with Tora-san," in T. J. Craig (ed.), *Japan Pop! Inside the World of Japanese Popular Culture*, Armonk, NY, M. E. Sharpe, 2000, pp. 245–55. For an overall discussion of representations of masculinity in Japanese cinema see, I. Standish, *Myth and Masculinity in the Japanese Cinema: Towards a Political Reading of the "Tragic Hero,"* Richmond, UK, Curzon Press, 2000. Moreover, even within the territory of mainstream masculinity, among the salarymen/corporate warriors

themselves, and in the ways in which society at large regarded them, there was always a considerable degree of ambivalence – for instance, even in manga, catering to a salaryman readership, the (self-) representation was often that of a weak, spineless creature, "a middle-class everyman ... married to an ugly woman, [who] dreads going home, and hangs his head low after being scolded by his boss," F. Schodt, *Manga! Manga! The World of Japanese Comics*, Tokyo, Kodansha International, 1986, p. 112.

17 Significantly these were areas of employment where the emphasis was less on those attributes traditionally associated with solid, respectable corporate masculinity, and more on individuality, youth, creativity, flair, sensitivity (indeed attributes often associated with "femininity").

18 For example, Web sites such as *Kaisha seikatsu no tomo, http://www.waw.ne.jp/kaisha* (24 February 2002), and *Sarariiman to OL no hiroba, http://toko.pos.to/* (10 March 2002) and television programs like *Ai no bimbō dasshutsu daisakusen*, aired on TV Tokyo and affiliated networks.

19 It was around this time that "masculinity" as a focus of academic scrutiny was also gaining popularity in North America, Europe, and Australia, with authors such as Bob Connell, Michael Kaufman, Harry Brod, Michael Kimmel, and others starting to publish works on masculinities. Within the context of Men's Studies in Japan, there were (and are) some fairly complex issues relating to the correct terminology – there continues to be some debate as to whether the correct term to be used should be *danseigaku* (men's studies/masculinities studies) or *danseikenkyū* (study/research on/of men). I am grateful to Dr Taga Futoshi of Kurume University, for his paper, "Danseigaku/danseikenkyū no sho chōryū" (Cross-currents within Masculinities/Men's Studies) presented at the fifth Japan Gender Association Conference held in Kyoto on 22 September 2001, which provided much of the background information.

20 M. McLelland, *Male Homosexuality in Modern Japan: Cultural Myths and Social Realities*, Richmond, UK, Curzon Press, 2000, pp. 234–5; also W. Lunsing, "Japan: finding its way?" in B.D. Adam, J. Duyvendak and A. Krouwel (eds), *The Global Emergence of Gay and Lesbian Politics: National Imprints of a Worldwide Movement*, Philadelphia, Temple University Press, 1999, pp. 302–16.

21 McLelland, *Male Homosexuality*, pp. 32–7.

22 "Kanzennaru 'danseigaku' de otoko ni nare!" (Becoming a man through perfecting "men's studies"), *Bart*, vol. 7, no. 4, 14 July 1997, pp. 10–20.

23 For instance, even as early as 1992, the English language *Japan Times* published an article on the establishment of the earliest men's studies groups. See, "'Iron John' movement appears in Japan," *Japan Times Weekly International Edition*, 27 April–3 May 1992, p. 14.

24 The group was set up by a small number of academics and concerned citizens, including Itō Kimio of Osaka University, Nakamura Tadashi of Ritsumeikan University, and Nakamura Akira, a former salaryman who had started to question and actively challenge the premises of hegemonic masculinity. While all three have continued to play an active role in sustaining and popularizing men's studies, Professor Itō, in particular, has become something of a representative for men's studies in the eyes of the wider community. In addition to frequent media interviews and appearances, he has published a number of academic and semi-academic works dealing with masculinity. These include, *"Otokorashisa" no yukue: danseibunka no bunkashakaigaku* (Tracing "macho-ness/masculinity:" the cultural sociology of male culture), Tokyo, Shinyosha, 1997; *Danseigaku nyūmon* (An introduction to men's studies), Tokyo, Sakuhinsha, 1996; and (together with K. Muta) (eds), *Jendā de manabu shakaigaku* (The study of sociology through gender), Kyoto, Sekai Shisōsha, 1998. He was also the first academic to offer a course on men's studies in its own right at a Japanese university (Kyoto University) in 1992; see Men's Center Japan (ed.), *"Otokorashisa" kara "jibunrashisa" e* (From "macho-ness/masculinity" to "being yourself"), Kamogawa Booklet no. 95, Kyoto, Kamogawa Shuppan, 1996, p. 3.

25 T. Shibuya, "'Feminisuto danseikenkyū' no shiten to kōsō: nihon no danseigaku oyobi danseikenkyū hihan o chūshin ni" (A view and vision of feminist studies on men and mascu-linities), *Shakaigaku Hyōron* (Japan Sociological Review), vol. 51, no. 4, 2001, pp. 447–63. See also the home page for Men's Center Japan *http://member.nifty.ne.jp/yeshome/Mens*.

26 Shibuya, p. 450.

27 Policies regarding female participation vary. While most groups are open to both men and women in principle, some groups restrict some activities (e.g. support/counseling for perpetrators of domestic violence) to males.

28 Men's Center Japan (ed.), *"Otokorashisa" kara "jibunrashisa" e*, Kamogawa Booklet no. 95, 1996, and *Otokotachi no "watashi" sagashi* (Men's search for "self"), Kamogawa Booklet no. 104, 1997.

29 Shibuya, p. 450, 1. Also, see the Men's Center Japan Web site for details of the programs of this first, and subsequent Men's Festivals.

30 However, as noted by Rheingold some of the earliest virtual communities, such as COARA, a virtual community based in Oita, Kyushu, date back to the mid-1980s, when the Japanese tele-communications industry first began to be deregulated. See H. Rheingold, *The Virtual Community: Homesteading on the Electronic Frontier*, New York, HarperCollins, 1994, pp. 204–12.

31 See Junko Onosaka, Mark McLelland and Joanne Cullinane in this volume. See also, McLelland, "Out and about on Japan's gay net."

32 *http://wom-jp.org/* (16 January 2002).

33 Much of the information pertaining to the online efforts of men's groups in Japan – particularly the Tokyo-based group, Men's Lib Tokyo – was provided by Mr Kaneko Yasufumi, during an interview conducted on 2 October 2001. Mr Kaneko was responsible for setting up the home pages of the Tokyo group as well as that of the Kansai-based Men's Center Japan, and continues to maintain the Tokyo group's home page and online newsletter. At the time of the interview, he also kindly provided me with a summarized timeline highlighting the significant dates and events of the various men's groups' interaction with online communication. Additional information (pertaining to the Kansai-based group) was provided by Mr Nakamura Tadashi of Men's Center Japan, during a discussion on 21 September 2001, in Kyoto.

34 Although Mr Kaneko is heavily involved with Men's Lib Tokyo, his paid employment is in the computer industry. This equipped him with an expertise that came in handy when designing and launching the Web sites of both the Tokyo and Kansai groups.

35 These include for example such groups as, Men's Lib Fukushima, *http://village.infoweb.ne.jp/~fwkk2152/link3.htm* (24 February 2002) and Men's Lib Kanagawa, *http://member.nifty.ne.jp/yeshome/ML-kanagawa/* (24 February 2002).

36 As opposed to *matches* for the term *dansei* – these numbered over one million! Search conducted on 19 March 2002.

37 *http://www.kirei.com/men/* (19 March 2002).

38 *http://www.pbrand.ne.jp/gospellers/* (19 March 2002).

39 *http://www.ne.jp/asahi/jimihen/oyaji/* (19 March 2002).

40 *Fine Boys Net, http://www.fineboys.net/* (19 March 2002); *Web Gainer, http://www.kobunsha.com/gainer/menu.html* (19 March 2002).

41 *http://www.so-net.ne.jp/uranaimen/* (19 March 2002).

42 *http://web.kyoto-inet.or.jp/people/ryo-y/menslife/m_l_index.htm* (19 March 2002).

43 *http://www.joinac.com/tsukuba-survey/* (19 March 2002).

44 I restrict my discussion to the "open access" public spaces of the Web sites discussed (i.e. where the information is available to any browser, without the need for signing in). There are ethical issues at stake here. Given the relative newness of cyberspace as a space for researchers, guidelines and protocol pertaining to Internet research are still relatively vague and remain under-discussed. For example, many researchers are unaware of issues like the need for informed consent when using material from restricted access spaces like chat rooms. Moreover, for researchers (like myself) based at universities, parameters of online research ethics vary from country to country, and often across institutions within the same country. For a description of the ethical requirements of online ethnographic research, see, L. Kendall, "'Oh no! I'm a nerd!' Hegemonic masculinity on an online forum," *Gender and Society*, vol. 14, no. 2, 2000, pp. 256–74 (specifically pp. 257–9).

45 Men's Lib Tokyo, *http://member.nifty.ne.jp/yeshome/MensLib/index.html* (17 January 2002).

46 *http://member.nifty.ne.jp/yeshome/MensLib/* (17 January 2002).

47 *http://member.nifty.ne.jp/haruhiko/* (17 January 2002).

48   *http://www2.odn.ne.jp/~acq50230* (17 January 2002).
49   *http://www9.big.or.jp/~suzuran/* (17 January 2002).
50   *http://www.comcarry.net/~genbu/index.html* (17 January 2002).
51   *http://www.lap.jp/* (17 January 2002).
52   *http://www.sukotan.com/* (17 January 2002).
53   *http://l-gff.gender.ne.jp/* (17 January 2002).
54   *http://www.eiyus.com/Singles/single.html* (17 January 2002).
55   *http://www.ne.jp/asahi/r/s/dameren/* (17 January 2002).
56   *http://www.eqp.org/* (17 January 2002).
57   *http://www.gender.go.jp* (17 January 2002).
58   *www.whiteribbon.ca/* (17 January 2002).
59   *Menzu Ribu Tōkyō NewsMail (dai 74 go)* (Men's Lib Tokyo NewsMail, no. 74), received Sunday 10 February 2002. Although this is a restricted (registered members-only) online newsletter, I am a subscribed member, and I obtained the consent of the newsletter's editor/moderator (Kaneko Yasufumi of Men's Lib Tokyo), at the time of my interview with him in 2001, to use information from the newsletter for this chapter.
60   *http://www.chp.or.jp/event/event_detail.asp?KI_CODE=0043&CO_CODE=0036&KS_CODE= 4&ZI_NENDO=2001* (18 March 2002). This was a link to an electronic application form.
61   This – older men questioning the assumptions of hegemonic masculinity (e.g. men not needing to know how to cook) – is significant, in that for many men the crisis of masculinity seems to occur not so much in youth, but either in middle-age (precipitated by factors such as sudden unemployment) or after retirement, when men suddenly discover the price of a lifetime of successfully performing hegemonic masculinity – inability to communicate, lack of friends and hobbies, and a lack of self-reliance. Hence, the popularity of cooking classes for men over sixty. This particular cooking class organized by the Better Homes Association was held across eighteen different locations in Japan; significantly, close to half the sessions were booked-out, indicating the popularity of such classes. See Web site of the event provided by the Better Homes Association, *http://www.betterhome.jp/school/oneday/60men/60men0202.html* (19 March 2002).
62   *http://www.ywca.or.jp/osaka/ngo/vawss01.htm* (18 March 2002).
63   *http://www.city.takarazuka.hyogo.jp/josei/Kouza/otoko/otoko.htm* (19 March 2002).
64   What is especially impressive is the fact that the Tokyo group's Web page and electronic mailing list, was set up by and continues to be updated and maintained largely through the efforts of one particular individual, Kaneko-san, albeit with the support of other core members.
65   *http://member.nifty.ne.jp/yeswhome/MensCenter/index.html* (17 January 2002).
66   Men's Lib Fukushima, *http://village.infoweb.ne.jp/~fwkk2152/link3htm;* Men's Lib Ibaraki, *http:/ /www1.ttcn.ne.jp/~lovecomibaraki/men%27slib.ibaraki.top.htm;* Men's Lib Kanagawa, *http:// member.nifty.ne.jp/yeshome/ML-kanagawa/* (all accessed 24 February 2002).
67   An exception is *Dansei kaihō keijiban* (Men's Liberation Noticeboard), *http://442.teacup.com/ otoko/bbs* (10 March 2002), a bulletin board concerned with discussing aspects of "masculinity," where there is a sense of an ongoing interactive community. This site is primarily a bulletin board for the specific purpose of discussion, rather than a board for posting notices and information, as is generally the case with the other broad-based groups discussed.
68   Ikuji-ren, *http://www.eqg.org/bbs/papa/yybs.cgi* (17 January 2002).
69   *Kaisha seikatsu no tomo, http://www.waw.ne.jp/kaisha/, onayami nan demo sōdan shitsu* (24 February 2002); *Sarariiman to OL no hiroba* (Open space for salaryman and office ladies), *http://toko.pos.to/,* and *Sarariiman Kyōwakoku* (Salaryman Republic), *http://haab.pos.to/* (both sites accessed 10 March 2002).
70   *Kaisha seikatsu no tomo: http://www.waw.ne.jp/kaisha/, onayami nan demo sōdan shitsu* (24 February 2002).
71   M. McLelland, "Live life more selfishly: an online gay advice column in Japan," *Continuum: Journal of Media and Cultural Studies*, vol. 15, no. 1, 2001, pp. 103–16.
72   Anonymous, "Big tomorrow hotline," *Big Tomorrow*, vol. 21, no. 7, 2001, p. 78.
73   McLelland makes a similar assertion. In contrast to mainstream print advice columns, Internet spaces, such as the gay men's online advice page run by gay activists Itō Satoru and Yanase

Ryūta which he discusses, allow for greater freedom in challenging dominant societal discourses. See McLelland, "Live life more selfishly."

74 A prime example would be the Salaryman hero depicted in the popular manga series, *Salaryman Kintaro*, subsequently adapted for television and cinema.

75 The two views of the Internet have been discussed quite extensively in various papers, edited collections, and books; for instance, D. Bell and B. Kennedy (eds), *The Cybercultures Reader*; D. Trend (ed.), *Reading Digital Culture*; and T. Jordan, "Language and libertarianism: the politics of cyberculture and the culture of cyberpolitics," *Sociological Review*, vol. 49, no. 1, 2000, pp. 1–17. A powerful example of the dystopian potential of the Internet (particularly for marginalized/minority social groups) is the recent crackdown on gay men in Egypt which received widespread international media coverage – police monitoring of gay male interaction on the Internet led to large numbers of men being "framed," arrested, and incarcerated. See the Web site of the International Gay and Lesbian Human Rights Commission (IGLHRC) for background information; *http://www.iglhrc.org/* (24 February 2002).

76 M. McLelland, "The Newhalf Net: Japan's 'intermediate sex' on-line," *The International Journal of Sexuality and Gender Studies*, vol. 7, nos 2/3, 2002, p. 164.

77 Mark McLelland makes a not dissimilar observation with reference to transgender identity. He makes the point that the Internet, rather than creating a new sexual identity per se, has allowed "pre-existing images, narratives, and practices" to become far more accessible and visible. McLelland, "The Newhalf Net," p. 173.

9 "Net"-working on the Web

Links between Japanese HIV patients in cyberspace

Joanne Cullinane

INTRODUCTION

Print capitalism has long allowed people to imagine "community" across space and time.[1] Yet, within the past two decades the Internet has enabled geographically and socially isolated individuals to imagine communities on an even larger scale and density. Although some skeptics assume that Internet surfers passively traverse pre-ordained structures, such a view obscures the ways in which Internet users commandeer this new technology for their own needs and desires.[2] In addition, such a perspective obscures the fact that some of the people most active online would find it difficult, if not impossible, to find each other offline. In this chapter,[3] I explore how HIV-positive individuals in Japan use the Internet in order to share information, seek support, and socialize with other members of a highly stigmatized group. Like all communities, the HIV community is neither homogeneous nor monolithic. While HIV patients share the same illness, they differ in almost every other conceivable way: age, occupation, sexual orientation, and socio-economic status are but a few of the most obvious distinctions. Such diversity must be taken into account, while also recognizing that many People with AIDS (PWAs)[4] share an intense desire to meet each other outside of the hierarchical structures to which they are routinely subjected. In contrast to the traditional mass media, which usually reduces HIV/AIDS to a mere "spectacle,"[5] the Internet allows PWAs to gain some measure of control over how they are represented and to establish horizontal ties which foster a sense of solidarity, rather than exclusion. At the same time, the diversity of HIV-related sites is testimony to the different issues around which PWAs are mobilizing.

Unlike the disembodied subjects one might expect to find in cyberspace, online HIV communities share intense bodily concerns. In fact, their online conversations are dominated by discussions of symptoms, side effects, and medications.[6] At the same time, PWAs share many of the same social dilemmas, including concerns about discrimination,[7] employment, relationships, and sex. In exploring the recent growth of Internet networks among HIV patients, I will focus on several interrelated questions. First of all, what role has the mass media played in stimulating the growth of this Internet community? Second, is the Internet seen as an attractive alternative to face-to-face communication? If so, why? Next, how do NGO (non-government organization) initiatives differ from patient-led initiatives? Finally, what types of alliances are forged online and what types of divisions underscored? In order to answer these questions, I will provide a brief history of AIDS and the traditional mass media, followed by an exploration of some of the cultural and social factors making it unusual for HIV patients to "come out" in Japan. Then I will discuss the rise of AIDS-related NGOs. Finally, I will show how different Web sites serve different needs, promote different interests, and foster different degrees of inclusiveness. But first I would like to say a few words about the methodological and ethical quandaries posed by a study of this nature.

NOTES ON METHODOLOGY

No longer bold enough to (mis)represent the communities they study as bounded, homogeneous, and static entities, anthropologists have begun to examine the complexity of social experience in an increasingly global world.[8] In addition, they are beginning to pay attention to the role of technology in fostering new types of subjective and collective identities. Yet they have paid scant attention to the Internet as a site of social change, except perhaps to lament its role in accelerating the imposition of Western values around the world. Such assertions are rarely accompanied by a careful investigation of how people around the world use the Internet and how their online activities relate to offline interests. As such, they ignore the cultural, historical, and political contexts within which such activity is embedded and which it, in turn, helps to shape. Ideally, I believe that Internet research should be accompanied by ethnographic fieldwork.[9] In my case, fifteen months[10] of participant observation in HIV-related NGOs, attendance at AIDS conferences, and interviews with doctors, nurses, and HIV patients provided a broader context within which to understand the activities of Japan's online HIV community.

Like many other anthropologists, I initially considered the Web a diversion, rather than an integral part of the fieldwork experience. Yet I found myself questioning this view after a number of people insisted that the Internet was their principal means of establishing contact with other members of the HIV community. These connections, forged in private, linked them to a wider world of both online communication and face-to-face interactions. For example, Nojan,[11] a 43-year-old man who was diagnosed with full-blown AIDS after collapsing and being rushed to a Tokyo hospital in 1997, told me how his family members had used the Internet to locate an NGO with multiple branches throughout Japan.[12] Taking pains not to arouse suspicion, they ventured online for two reasons: Nojan's doctor was completely unfamiliar with AIDS, and the president of his company was pressuring him to resign. Subsequently, NGO staff members were able to help him find a more experienced doctor and to resign from his company on his own terms. In another case, the Nakatas, a young couple who decided to get married after they both tested positive for HIV, told me how they had used the Internet to find a Tokyo-based NGO and to communicate with other Japanese-speaking PWAs and their partners, some as far away as Canada. While the woman had told several friends and family members about her condition, her husband had told almost nobody except for his wife, NGO volunteers, and a distant Internet correspondent. In some cases, messages travel along even more circuitous routes as, for example, when Eiko, who is fluent in Spanish, posted a message on a Spanish Web site and received a reply from a Chilean man living in Japan. Although he is not HIV-positive, he was deeply moved by what she wrote. The two began dating and looking for ways to assist Spanish-speaking HIV patients in Japan.[13]

Despite these stories of online camaraderie (and romance), the initial feelings of shock and isolation suffered by most newly diagnosed individuals can hardly be overestimated. Nor are such feelings surprising, given the fact that only 4,519 cases of HIV and 2,236 cases of AIDS have been reported to the Japanese Ministry of Health and Welfare to date.[14] Most Japanese people still consider HIV a "fire on the opposite shore" and few people in Japan have ever met anyone living with the disease. Furthermore, hospitals willing to treat HIV are still the exception, rather than the norm, in Japan.[15] While the number of hospitals offering treatment for HIV has increased in recent years,[16] finding out where such hospitals are located can be a difficult and frustrating task. The same is true for AIDS-related NGOs, which occupy a somewhat marginal position on the social landscape.

Given the sensitive nature of my fieldwork, considerations of privacy raised serious methodological issues from the start. As a researcher, I was afforded many opportunities to meet HIV-positive individuals at a variety of different venues. On the other hand, NGO staff members often attempted to place strict conditions upon the location, timing, and content of such meetings. One condition was that I not ask interviewees for any personal information, including names and addresses. Hence, it came as a surprise when people began to volunteer their e-mail addresses and favorite Internet sites.[17] In this chapter, I hope to explore the role of the Internet in allowing PWAs to establish alliances, despite underlying divisions. At the same time, I realize that "the bright light of social science research can create an unpleasant glare for participants drawn to a dimly-lit online space."[18] Therefore, I have taken every precaution to disguise the names of the sites and individuals discussed here. In addition, I quote only those individuals who gave me permission to discuss their online activities.

One other point requires clarification. In this chapter, I have made a conscious decision to limit my discussion to the online activities of those HIV patients who contracted the virus through sexual contact rather than through contaminated blood products. In making this distinction, I do not mean to imply that hemophiliac HIV patients do not use the Web for some of the same purposes as their non-hemophiliac counterparts. There are, however, compelling reasons to focus on the non-hemophiliac community. One is the fact that new *yakugai* (medical malpractice) AIDS cases have been (virtually) eliminated since 1985. Furthermore, members of the hemophiliac community were already organized into regional branches under the umbrella of the National Friends of Hemophilia Association long before HIV entered Japan. Finally, the distinction between *yakugai* patients and *seikansensha* (those who have contracted HIV sexually) is by far the most culturally significant distinction applied to PWAs in Japan and one which is often used to separate "innocent" patients from their "guilty" and less deserving counterparts. Because AIDS is often equated with promiscuity, the phrase most commonly hurled at *seikansensha* is *jigō jitoku*, or "you reap what you sow."

THE PERILS OF PUBLICITY: HIV AND THE MASS MEDIA[19]

The history of AIDS in Japan, as elsewhere, is marked by secrecy and suspicion, panics and plateaus. Early on in the epidemic, the Ministry of Health and Welfare struggled to deny the risks posed to hemophiliacs by imported blood products.[20] Then, on 22 March 1985, the AIDS Surveillance Committee seized upon a 36-year-old gay male who resided in Los Angeles as the country's "first" AIDS patient. At the same time, they quietly admitted that 29 per cent of 163 hemophiliac patients' blood samples had tested positive for HIV and that two hemophiliac patients had already died of AIDS. This information went unremarked for almost a year because doctors and bureaucrats were loathe to accept any evidence which contradicted the view of AIDS as a "gay plague" so prevalent in the United States at the time.[21] Later, during the *Sandai sōdō* (the Three Commotions), in which the first three heterosexual HIV patients, all women, were identified in quick succession, the mass media embarked on an even more sensational series of witch-hunts which was to set the tone of HIV reporting for years to come.

In the fall of 1986, the Filipino Embassy reported that a 21-year-old woman from Quezon City had entered Japan on 17 September, unaware that she had just tested positive for HIV. Newspapers released her full name and reported that she had worked in Matsumoto City. After a nationwide manhunt, the woman turned herself in to immigration authorities at Narita airport on 4 November. Although the government refused to release the results of a follow-up test, the fact that she was hastily deported from the country was taken as proof of her HIV status and the media began seeking out men who had contact with her. Weekly magazines began demonizing Southeast Asian women, especially those *Japayuki-san*[22] working in the Japanese sex industry, calling them "AIDS prostitutes." Crude jokes began to appear in weekly magazines, such as the parody of a tourist poster for Nagano, which depicted a train called "AIDS – The Practical Joke Express" and photos of Southeast Asian women in various states of undress surrounded by pictures of famous tourist sites.[23] "Only three hours to infection," the ad proclaimed in a crude attempt at black humor.

Next, on 17 January 1987, the AIDS Surveillance Committee reported that a 29-year-old Japanese woman from Kobe had been hospitalized with AIDS and was in a critical condition. Local bureaucrats, professing a desire to prevent cases of secondary infection, interrogated her as she lay in her hospital bed, unable to speak. Newspaper headlines implied that the woman had been a "habitual" prostitute and suggested that her former boyfriend, a Greek sailor with a "tendency" toward bisexuality, was the source of infection. In a media ambush, journalists even infiltrated her funeral and published photos of the deceased in a number of popular weekly magazines. By 22 January, members of the Liberal Democrat Party (LDP) had already begun to draft AIDS legislation including mandatory testing for "suspicious" characters and punitive measures for "non-cooperative" individuals. Shiogawa Yuichi, head of the AIDS Surveillance Committee, declared that AIDS had spread from "specific groups" to "ordinary" people and dubbed 1987 the first year of the AIDS era.[24]

Following shortly on the heels of the Kobe case, newspapers reported that a 20-year-old Kochi woman was HIV-positive and due to give birth the following month. Her case was widely regarded as proof that HIV, usually associated with promiscuity,[25] could infiltrate the home, the site of procreative sex. Although this woman had contracted HIV from her former boyfriend, who was a hemophiliac, television commentators noted that she was aware of her HIV status and had been strictly warned to avoid marriage and childbearing. As such, she was maligned as a dangerous and disobedient patient. One doctor from Yamaguchi University Medical School was quoted in the *Mainichi* newspaper as saying, "In the first instance, people with antibodies (to HIV) should not get pregnant and HIV-positive women should not go ahead with childbirth."[26] Television commentators publicly urged the woman to have an abortion even though she was eight months pregnant and wanted to keep the child. In the end, her will prevailed and the child was born without any signs of HIV infection.

As the above examples demonstrate, the Japanese mass media tend to treat AIDS as a scandal and their early attempts to hunt down HIV patients and expose them to public scrutiny have created an atmosphere in which very few PWAs are willing to "come out" of their own accord. One exception to this trend are the excellent programs produced by Nippon Hōsō Kyoku's (NHK) Ikeda Eriko on *yakugai* AIDS, including a beautiful and moving documentary about the late Akase Noriyasu, the first plaintiff in the *yakugai* AIDS lawsuit and the first person in Japan to "come out" as HIV-positive in 1989.[27] However, the silence and secrecy surrounding sexually-transmitted cases of HIV makes it difficult for many *seikansensha* to find one another, whether they are separated by distance, a concern for public opinion (*seken*), or the polite restraint which keeps patients from approaching one another in hospital waiting rooms. While a handful of HIV patients have followed in Akase's footsteps, the majority of people reject this course of action, feeling that it would cause undue harm to their families. In such an environment, the Internet provides an important way for *kansensha* to find each other without having to forfeit their privacy or be vilified by an unsympathetic press.

PROPRIETY AND INTIMACY, DEPENDENCE AND RECIPROCITY

In addition to the fear of exposure, which is a legacy of early AIDS reporting, many Japanese PWAs share concerns about propriety, privacy, and reciprocity which make communicating on the Internet an attractive alternative, or at least a precursor, to face-to-face interaction. In interactions with members outside of one's close inner circle, there is a widespread tendency to mask one's inner feelings (*honne*) with outer propriety (*tatemae*) in order to insure the harmony of the group.[28] Yet, while many people do reserve *tatemae* for out-groups (*soto*) and revert to *honne* in the presence of an in-group (*uchi*), there are some situations in which just the opposite is true. Defining an in-group can be especially tricky for many HIV patients, who may prefer to conceal their illness from family and friends.[29] In some cases, the burden of concealing an HIV diagnosis leads to a temporary withdrawal from social interaction, accompanied by an intense desire to "open up" (*uchiakeru*) to someone. *Uchiakeru* differs from "coming out" in that it does not involve baring one's soul to an anonymous public; instead, like the English word "confide," it implies an element of trust and reciprocity.

More and more HIV patients are seeking this reciprocity from their fellow HIV patients. With other *kansensha*, there is no need to fear discrimination or worry about prevailing stereotypes of the "typical" AIDS patient. Furthermore, there is no need to justify one's desire to pursue satisfying and rewarding sexual relationships. In short, with other patients there is no need to worry about the strictures of public opinion (*seken*) which discourage public discussions of sex or serious illness. Although many patients feel that the norms of *seken* are unfair and in need of reform, most profess a desire to spare their families from its sting. As such, they often use nicknames or pseudonyms for their HIV-related activities and those few who have "gone public" have done so after receiving their families' permission. In some settings, the presence of individuals who have gone "public" about their HIV status is considered threatening to other patients, who do not feel they should be pressured into coming out against their will. In one extreme case, the editors of *HIVOICE*, a *minikomi*[30] founded in 1993, sought to alleviate this conflict by requiring that *all* contributors use pseudonyms, regardless of their HIV status.

In addition to considerations of propriety and reciprocity, the desire to confide in other PWAs is sometimes linked to the idea that one should not be a burden on anyone. Numerous scholars have described how relationships of indulgence (*amayakasu*) and dependence (*amae*) represent an "ideal" that is replicated in many Japanese social arrangements.[31] Yet such arrangements are difficult to sustain in the case of HIV, which is a chronic and incurable condition. Over an extended period of time, the sense of debt incurred by the recipient of support can become insufferable.[32] For this reason, many patients prefer to confide in each other rather than in volunteers, or family members, who may or may not have any source of support themselves.

THE NGO MOVEMENT

Most NGOs in Japan are aptly described as "volunteer organizations." At best, they are ignored by the government and by the public; at worst, they are considered highly subversive entities. At present, there are approximately ninety-nine HIV-related NGOs in Japan[33] and these organizations make up the bulk of an "AIDS Service Industry"[34] largely patterned on American models. However, unlike in the US, where the emphasis is on peer empowerment, most NGO employees in Japan are not HIV-positive and view themselves as the first line of defense against what is perceived to be a threatening and intrusive public. For better or for worse, most of these NGOs are based upon a medical model in which the "healthy" are charged with helping and protecting the "sick." In addition, they tend to exclude the "general public."

Needless to say, most NGOs go to great lengths to protect the privacy of their clients. The Japan HIV Alliance (pseudonym), for example, is located off a dark alleyway in a sparsely populated neighborhood. A sign on the front of the building has been partially covered because some of the clients found it unsettling to enter a building with the letters HIV emblazoned on the front. This is less of an issue at Tokyo Lifeline (pseudonym) since their name is more ambiguous. Yet both organizations still deem it necessary to maintain separate meeting facilities for PWAs in undisclosed locations. Naturally, these measures have been adopted in an effort to attract as many visitors as possible. Yet sometimes they can have unintended effects; while some visitors appreciate the secrecy, others find it disconcerting. Several people told me that they dislike NGOs because they find the atmosphere stifling. Furthermore, most NGOs are in Tokyo and it is sometimes difficult for HIV-positive individuals in rural areas to relate to the concerns of those in urban areas. While NGOs provide a much-needed lifeline for newly diagnosed patients, as time goes on it is common for patients to seek alternative sources of support and information. The ubiquity of personal computers and wireless communication means that they can easily take their search to the Internet.

VENTURING ONLINE

How exactly do Japanese PWAs find each other on the Internet? A general search of Lycos Japan yields 86,252 hits for the word *eizu*, or AIDS (22 April 2002).[35] This includes a wide array of sites maintained by hospitals, government organizations, professional associations, NGOs, and individuals. Yet such a generic Web search is the least efficient way of accessing HIV-related sites and presumably used only as a last resort by those who are the most socially isolated. Other individuals find out about promising Internet sites via gay search engines, NGO publications and, finally, by word-of-mouth. Many of the sites are linked, making it easy to jump from one to another. Security is minimal and most sites simply ask visitors to agree to such descriptions of themselves as, "I am a common sense person without prejudice or discrimination." While this certainly cannot prevent anyone from proceeding, it may provoke casual visitors to question their opinions on a host of issues, including hegemonic notions of "common sense."[36] In this section, I explore several sites created by different groups and individuals, paying careful attention to the similarities and differences among them.

An NGO initiative

One fairly active bulletin board service (BBS) is accessed via an NGO Web site. As with most BBSs around the world, users are encouraged to create handle names.[37] Because new members can join the BBS at any time, messages deal with an extraordinary range of concerns, from the shock of initial diagnosis, to medical treatment, career and employment issues, and relationships. Many of the messages concern the difficulty of telling past, present, or potential partners about one's HIV status. Some people are simply seeking "e-mail friends" and they use the BBS as an advertising mechanism. Interspersed among many lighthearted messages are desperate pleas from despondent newcomers. Contributors to this BBS are at different stages of their illness and treatment, and older members gain considerable prestige for their wealth of knowledge and experience. Newer members, for their part, are encouraged to attend NGO events and to stay in contact with the group. Some messages attest to the fact that several members have already met each other offline.

The fact that visitors enter this BBS from an NGO home page ensures that they will become somewhat familiar with the NGO and its services, which include a hotline, a Buddy service, monthly meetings, and periodic events for specific audiences (e.g. women, or gay men). However, there are clear tradeoffs involved in this kind of NGO sponsorship. For example, NGO staff members retain the power to delete messages without explanation.[38] One contributor to this BBS (not myself) was even scolded by the BBS manager for expressing curiosity about the content of deleted messages. The top-down nature of this BBS stands in stark contrast to many Internet communities in the United States, which rely either on prescreening or other methods of social control in order to define the limits of acceptable and unacceptable behavior.[39] For those who wish to escape from NGO influence, a number of different options are available.

HIV and the gay Web

Although a number of gay activists organized to combat AIDS in the 1980s, the initial response to AIDS in the Japanese gay community was one of exclusion. Assuming that foreigners were the biggest threat, some gay establishments simply decided to ban foreigners from the premises. In addition, many shop owners refused to allow AIDS literature on the premises, claiming that it might scare customers away.[40] This reluctance to confront HIV head-on has not yet disappeared and HIV-positive individuals are sometimes shunned by others in the gay community. Such was the case with Kasuga Ryōji, a gay writer and musician who "came out" as HIV-positive in 1999. His revelation that he had been harassed by others in the gay community at a human rights symposium preceding the 2001 Tokyo Gay and Lesbian Parade represented an attempt to address the strained relationship between HIV activists and the gay community.[41] Despite lingering tensions, many gay individuals have featured prominently in the struggle against AIDS. Most notably, Hirata Yutaka became the first non-hemophiliac to "come out" to the press as

HIV-positive in 1993. Then, in 1994, Ōishi Toshihiro "came out" at the Tenth International AIDS Conference in Yokohama, giving Japanese PWAs visibility on the global stage.

More recently, a search of *G-Press Index*,[42] a gay search engine, revealed forty sites relating to HIV/AIDS, including several bulletin boards for gay men living with HIV and several home pages belonging to HIV-positive individuals. Frank discussions of sex are common on these sites and contributors can create personal profiles in which they describe their appearance and hobbies, as well as how long they have been living with HIV. Some groups, like Hearty Network, now in its third year, hold monthly socials in addition to their online activities. The success of many gay sites depends upon their sensitivity to issues of privacy and the playful appropriation of different identities. Taking advantage of this culture of anonymity, some HIV-positive individuals have created elaborate personal home pages in which they reveal extremely private aspects of their lives, including diaries and photographs, but use pen names nonetheless. As such, they reveal names to be an extremely limited aspect of identity, which can limit, rather than encourage, certain forms of intimacy and expression.[43]

Still, some individuals choose to use their own names in their public activities. Hasegawa Hiroshi, a former editor of *G-MEN*, one of Japan's most popular gay magazines, has adopted an ambitious multimedia approach in his efforts to challenge prevalent stereotypes of gay men with HIV. Now 49 years old, Hasegawa was first diagnosed with HIV in 1992. When I first met him in 1999, he was distributing condoms to gay bars in the Shinjuku Ni-chōme area. He was also running a support group for gay men living with HIV called NoGAP. Previously, he had pioneered making information about HIV available in the pages of *G-MEN* magazine.[44] This is no easy task in Japan, where gay magazines rarely cover health and lifestyle issues.[45] In addition, he cooperated with the staff of Tokyo Lifeline to make a video on "coming out" as HIV-positive to one's family. When I last saw him, he was being photographed for an exhibit on HIV patients in Japan.[46]

In addition to his offline activities, Hasegawa has ventured online to write articles for gay sites. In a series for *Tokyo Gay Walker*, he wrote about his experiences with HIV in a style that was not only informative, but also humorous. He chided readers who consider AIDS "scary" and argued that "the scariest thing is not knowing about AIDS or not making an effort to know about AIDS."[47] On another site for gay educators, he decried the factionalism[48] within the gay community and discussed the need to be true to oneself (*jibunrashii*), rather than striving to meet society's standards for what is "normal." For him, this includes deciding whether or not to make one's sexuality or HIV status public. On 28 March 2002, Hasegawa founded a new network called JaNP+, or the Japanese Network of People Living with HIV/AIDS, which is open to all HIV patients, regardless of sexual orientation or mode of infection. In his founding statement, he noted that: "Information, support, and services for PWH/As continue to be concentrated in the large cities and much also depends upon the hospital and the community to which the PWH/A belongs. Only a fraction of all PWH/As access such services." Using the Internet, he aims to "improve the social standing of PWH/As and offer support and information equally to all PWH/As, regardless of where they live, how they contracted HIV, and what hospital they use."[49] So far, JaNP+ includes an information division, a self-help division, and an advocacy division. Members of JANP+ receive an e-magazine and have access to information about upcoming study groups. Already, Hasegawa has received pledges of support from a wide array of NGOs and individuals, including Kawada Ryūhei, the most famous of the *yakugai* plaintiffs to date.

The hetero connection

In contrast to the gay community, which has a relatively well established Web presence, it is somewhat more difficult for heterosexual HIV patients to find one another. After growing disillusioned with NGO support groups, which fail to attract heterosexual men in large numbers, Hayashi Ruri[50] decided to create a BBS for heterosexual HIV patients in search of potential partners. Diagnosed with HIV in 1995, she was expecting her first child at the time. After this relationship ended, she met and married an HIV-positive man whom she met on the World-wide Web, but the marriage was cut tragically short when he died of cancer. Finally, in the fall of 2000, she established a Japanese Web site for *kansensha* interested in establishing romantic relationships with other PWAs. She was already a frequent participant in *Heterochat*, an international Internet community for English-speaking PWAs, and was clearly influenced by their cause. On her group's home page, she states her vision thus: "I developed this site because I thought I'd like to build a place right here in Japan where hetero men and women can gather without feeling any stress about being HIV-positive." The site includes a BBS and information on how to find out about monthly social events. At present, Hetero Connection events are held only in Tokyo, but there are plans to expand the network to other parts of the country, if possible.

Not surprisingly, Hayashi Ruri's attempts to work with the mass media have not gone well. Asked to contribute an anonymous letter for a television piece on HIV, she agreed on one condition: that it be read in its entirety. Yet, when the program aired, she found her words distorted to convey an image of her as a tragic figure. The narrator focused on the bleakest and most discouraging parts of her letter, while ignoring the positive and optimistic notes. Finally, the show included interviews with *yamamba*[51] of the type being blamed in the press at the time for "spreading" AIDS, often quite deliberately. Hayashi was sufficiently distressed by this incident that she appeared at a public symposium and at a national AIDS conference in late 1999 in order to complain about it. Appalled by the passivity thrust upon her, she asserted that "she is not the type to be easily defeated."[52] This is not to say that online interaction is always hassle-free. Hayashi is reluctant to publicize the name of her Web site widely after having received a computer virus from an anonymous source shortly after the site was established. Such acts of sabotage, though rare, are disconcerting.

Despite Hayashi's efforts, a number of factors prevent heterosexual HIV support groups from taking root in Japan. Once again, there are concerns about privacy.

Hayashi must personally clear all new members before sending them information about social events. Another problem is the large age gap between male and female HIV patients in Japan. Many of the Japanese women who test positive for HIV in Japan are in their twenties and unmarried, whereas the majority of men are in their thirties and forties.[53] Many of the male patients, who represent over 80 per cent of all HIV cases in Japan, are either gay or married. In addition, the figures for female patients include many foreign women, who find it difficult to access support groups at all. Finally, the need for a gatekeeper imposes constraints on growth; as the mother of a young child, there is only so much time Hayashi can spend on these activities. Yet her initiative, like some of the gay sites discussed above, is important in that it recognizes the right of PWAs to establish romantic and sexual relationships.

CONCLUSION

While an HIV diagnosis is no longer considered a death sentence, it is still a chronic, rare, and unpredictable condition. In addition, it is a highly stigmatized state about which much fear and misunderstanding abound. In such an environment, newly diagnosed patients often worry about how, when, and if to confide in family and friends. They worry not only about rejection, but also about becoming a burden on others. Furthermore, they realize that public opinion renders them and their loved ones vulnerable to discrimination. Hence, many people turn to the Internet in an effort to find other PWAs, whether in Japan or overseas. In so doing, they are embracing a means of communication and expression which allows for a great deal of autonomy and spontaneity. Of all the resources available on the Web, NGO sites seem to attract the most diverse audience; however, they are also less likely to allow unfettered interaction. The alternatives, on the other hand, tend to separate along the fault lines of sexual orientation.

Although members of Japan's HIV community hold conflicting ideas about the necessity of "coming out" to the general public, they are nonetheless united in their desire to create meaningful alliances with each other. The Internet provides a relatively secure space in which they can share their thoughts, feelings, and concerns, without having to identify themselves by name. Thanks to the Internet, HIV patients in Japan are talking to each other in large numbers for the first time. In this respect, computer-mediated communication has already helped alter the experience of HIV in a society in which HIV patients once felt forced to suffer in solitude.

NOTES

1   This argument is put forward by Benedict Anderson in *Imagined Communities*, London, Verso, 1983, an examination of the emergence of nationalism.
2   In this respect Internet users share some similarities with the pedestrians who appropriate city streets in Michel De Certeau's *The Practice of Everyday Life*, Berkeley, University of California Press, 1984, pp. 97–108. However, some Web users are architects, as well as travelers, in

cyberspace. Furthermore, the topography of the Internet changes from moment to moment, making each journey through cyberspace a novel one.

3 This research was funded in part by a 1999–2000 Japan Foundation Doctoral Dissertation Fellowship.

4 In the United States, the term PWA (person with AIDS) was coined as an alternative to the earlier phrase "AIDS victim." This term is designed to call attention to the fact that people with HIV/AIDS possess personal agency. However, in Japan, the term PWA (HIV *kansensha oyobi* AIDS *kanja*) is a recent foreign import with less cultural currency than the terms [HIV] *kansensha* (literally, [HIV] infected person) and [*eizu*] *kanja* (literally, [AIDS] patient). The former refers to asymptomatic individuals who have contracted HIV, and the latter refers to people with clinically defined AIDS. Both of these terms have replaced the archaic term *hokinsha* (carrier), which was used during the early days of the epidemic in Japan, but which (mistakenly) refers to bacteria and carries connotations of extreme contagion. In this chapter, I will alternate between the use of PWA, HIV patient, and *kansensha*, although I am fully aware of the inadequacy of all of these terms.

5 Simon Watney describes the "spectacle of AIDS" as "a regime of massively overdetermined images," which in 1980s Britain took the form of a punitive gaze directed at "the homosexual body." "The spectacle of AIDS," in D. Crimp (ed.), *AIDS: Cultural Analysis/Cultural Activism*, Cambridge, MIT Press, 1988, pp. 71–86.

6 As in the rest of the developed world, Japan has benefited from recent trends in HIV treatment, including a rapid decline in mortality after the introduction of HAART (highly active anti-retroviral) therapy in 1996. This is not to say that HIV is no longer fatal. However, if detected early, there is a good chance that HIV can be managed like many other chronic conditions. At present protease inhibitors and anti-retrovirals are widely available in Japan and HIV patients have been eligible for generous medical and welfare benefits ever since HIV was classified as a handicap worthy of government recognition in 1998. Yet social barriers remain; some patients refuse to apply for services because they must register with the local government as a patient with a "compromised immune system."

7 Although it is difficult to ascertain how widespread discrimination is, NGO staff members report that it is quite frequent. In one ongoing civil suit, a Tokyo Police Department recruit was tested for HIV without consent during an orientation and then suddenly forced to sign a letter of resignation. Details of this case, and two others like it, can be found on the *AIDS Scandal* site, *http://www.t3.or.jp/~aids/* (22 April 2002).

8 See A. Appadurai, "Disjuncture and difference in the global cultural economy," *Theory, Culture and Society*, vol. 7, 1990, pp. 295–310; A. Appadurai, "The production of locality," in R. Fardon (ed.), *Counterworks: Managing the Diversity of Knowledge*, London, Routledge, 1995, pp. 204–25; R. Fox, *Recapturing Anthropology: Working in the Present*, Santa Fe, School of American Research Press, 1991; and A. Gupta and J. Ferguson (eds), *Anthropological Locations: Boundaries and Grounds of a Field Science*, Berkeley, University of California Press, 1997.

9 N. Wakefield, "New media, new methodologies: studying the Web," in D. Gauntlett (ed.), *Web.Studies: Rewiring Media Studies for the Digital Age*, London, Arnold, 2000, pp. 31–41; and A. Escobar, "Welcome to Cyberia: notes on the anthropology of cyberculture," in D. Bell and B. Kennedy (eds), *The Cybercultures Reader*, London, Routledge, 2000, pp. 56–76.

10 I conducted fieldwork on the meaning and management of AIDS in Japan between July 1999 and August 2000, returning to Japan for a follow-up trip in the summer of 2001. While in Japan, I interviewed HIV-positive individuals, doctors, nurses, government bureaucrats, NGO staff members and volunteers. Since my return from the field, I have used the Internet to stay in communication with many of the people I met in the field. Such continuity calls into question the notion of closure in fieldwork, and muddies the very definitions of "home" and the "field."

11 This is his own pseudonym, which he gave me permission to use. All names in this paper are pseudonyms, unless otherwise noted.

12 Nojan is somewhat unique in that he felt comfortable confiding in his family about his medical condition right away. He attributes his family's cohesiveness to the fact that his mother was the mistress (*omekakesan*) of a wealthy patron who later abandoned her and refused to recognize

their children. Even before this final betrayal, he and his siblings had bonded in the face of abuse from Nojan's father's wife and her children.

13   This group includes South American *nikkeijin* (people of Japanese descent) and their spouses, who have returned to Japan from South America and who find it easier to obtain working visas in Japan than other foreigners. Although these patients fare better than undocumented migrants, because they are (theoretically) eligible for health insurance, they still find it difficult to access care. One of the biggest problems is the lack of interpreters. Of course, the situation is far worse for illegal immigrants, who receive little or no medical care and who are often urged to leave the country.

14   *"Eizu dōkō iinkai hōkoku"* (Report of the committee on AIDS-related trends), AIDS Prevention Information Network, *http://api-net.jfap.or.jp/mhw/mhw_Frame.htm* (13 February 2002). Naturally, because these numbers include only reported cases of HIV and AIDS, they represent only a fraction of the actual number of HIV and AIDS patients in Japan. Yet, they should provide some indication of how difficult it is for HIV patients to find one another.

15   The widespread practice of refusing care to HIV patients (*shinryō kyohi*) was one of the factors prompting hemophiliac HIV patients to file civil lawsuits against the pharmaceutical companies and the Ministry of Health and Welfare in the first place.

16   The establishment of a hospital network was a condition of the settlement reached in the *yakugai eizu* trials in 1996. These are the civil lawsuits brought by hemophiliacs in Osaka and Tokyo against the Ministry of Health and Welfare and five pharmaceutical companies which continued to sell unheated blood products for over two years after they had been banned in the United States. For more on this, see E. Feldman, *The Ritual of Rights in Japan: Law, Society, and Health Policy*, Cambridge, Cambridge University Press, 2000; for more on other cases of medical contamination (*yakugai*) in the post-war era, including SMON disease (caused by clioquinol) and thalidomide poisoning, see K. Sonoda, *Health and Illness in Changing Japanese Society*, Tokyo, University of Tokyo Press, 1988.

17   I identified myself as a researcher and described my research interests to everyone I interviewed. In addition, I wrote about my research in those settings where I felt it was appropriate, e.g. on an e-mail list and in a *minikomi* devoted to HIV/AIDS. I did not, however, identify myself on any Bulletin Board Services (BBSs) where contributions from HIV-negative individuals were obviously unwelcome.

18   M. Smith, "Invisible crowds in cyberspace: mapping the social structure of the usenet," in M. Smith and P. Kollock (eds), *Communities in Cyberspace*, London, Routledge, 1999, p. 211.

19   The following account is based upon newspaper and magazine articles of late 1986 and early 1987. In addition to the national newspapers, I consulted the following weeklies: *Fōkasu*, *Shūkan Bunshu*, *Shūkan Gendai*, *Shūkan Shinchō*, *Shūkan Hōseki*, *Shūkan Sankei*, *Shūkan Asahi*, *Sandē Mainichi*, and *Shūkan Yomiuri*.

20   The AIDS Surveillance Committee was formed in July of 1983 under the leadership of Dr Abe Takeshi. Although the committee immediately reviewed the cases of two hemophiliac patients, they rejected the opinions of Center for Disease Control, Atlanta, Georgia (CDC) officials and refused to attribute either death to AIDS. They obtained definitive proof in the summer of 1984 that these two patients had died of AIDS, but kept the results from the public until March of the following year.

21   S.C. McCombie, "AIDS in cultural, historical, and epidemiological context," in D. Feldman (ed.), *Culture and AIDS*, New York, Praeger, 1990, pp. 9–28.

22   *Japayuki-san* is a play on the word *karayuki-san*, a term used for the impoverished Japanese women who were sent abroad to work as prostitutes in the Meiji era (1868–1912). Nowadays, most *Japayuki-san* come to Japan from Thailand and the Philippines, although some come from as far away as Colombia.

23   *Fōkasu*, 5 December 1986, p. 66.

24   The phrase he used was *eizu gannen*. This refers to the system of marking a dynastic change in the imperial line, making clear that this was considered a cataclysmic, epoch-marking event.

25   S. Watney, *Policing Desire: Pornography, AIDS, and the Media*, Minneapolis, University of Minnesota Press, 1989 [1987] demonstrates how British AIDS discourse simply equated

homosexuality with promiscuity, rather than addressing the complexity of sexual desire and practice.

26 "Boshi kansen no osore genjitsu ni: hasshōritsu takai akachan" (The hard facts on the dangers of mother-to-child infection: babies with high transmission rates), *Mainichi Shinbun*, 17 February 1987, p. 23.

27 The NHK special featuring Akase was called *Inochi aru kagiri: aru eizu kansensha no hibi* (While there's life: the daily life of a person with AIDS) and it aired on 6 February 1989 to critical acclaim.

28 Countless works have noted the importance of these distinctions in Japan. For more on the concepts of in-group and out-group, see T. Lebra, *Japanese Patterns of Behavior*, Honolulu, University of Hawaii Press, 1976; J. Bachnik and C. Quinn (eds), *Situated Meaning: Inside and Outside in Japanese Self, Society, and Language*, Princeton, Princeton University Press, 1994; and D. Kondo, *Crafting Selves: Power, Gender, and Discourses of Identity in a Japanese Workplace*, Chicago, University of Chicago Press, 1990.

29 As pointed out by W. Lunsing, *Beyond Common Sense: Sexuality and Gender in Contemporary Japan*, London, Kegan Paul, 2001, pp. 220–1, gay, lesbian, and transgendered individuals often find it difficult to "be themselves" around their closest acquaintances.

30 The word *minikomi* is a pun on the word *masukomi* (mass-communication media). *HIVOICE* was founded by an HIV-positive hemophiliac and it has a circulation of about 500. However, it is common for readers to share copies and readership is probably much larger than this figure would suggest. It is sustained by a small group of dedicated volunteers who have debated whether or not to move their activities online. Interestingly, some of the volunteers have resisted this move, arguing that the printed word has more immediacy than the Internet and cannot be "tuned out" quite as easily. It remains to be seen what course they will take.

31 See T. Doi, "Amae: A key concept for understanding Japanese personality structure," in T. and W. Lebra (eds), *Japanese Culture and Behavior: Selected Readings – Revised Version*, Honolulu, University of Hawaii Press, 1986, pp. 121–9; see also H. Kumagai, "A dissection of intimacy: a study of 'bipolar posturing' in Japanese social interaction – 'amaeru' and amayakasu, indulgence and deference," *Culture, Medicine, and Psychiatry*, vol. 5, no. 3, 1981, pp. 249–72.

32 T. Lebra, *Japanese Patterns of Behavior*, Honolulu, University of Hawaii Press, 1976, p. 92.

33 "NGO katsudō no goshōkai" (Introduction to NGO activities), Japan AIDS Prevention Information Network Home Page, *http://api-net.jfap.or.jp/ngo/ngo_Frame.htm* (16 October 2001). Many of these organizations are small, understaffed, and under-funded, and meet irregularly, if at all.

34 C. Patton, *Inventing AIDS*, New York, Routledge, 1990, pp. 5–23. Here Patton describes how the "coming out" model of gay community-based AIDS groups became hegemonic in American NGOs in the 1980s. It is this very model which most HIV patients in Japan reject, although it has influenced Japanese NGOs in other ways, including the adoption of the early categories of "victims," "experts," and "volunteers."

35 A search on Lycos USA on the same day yielded 6,016,505 hits.

36 Once again, see W. Lunsing's *Beyond Common Sense*. His definition of "common sense" is similar to that of hegemony as articulated by J. and J. Comaroff in *Of Revelation and Revolution: Christianity, Colonialism, and Consciousness in South Africa*, Chicago, University of Chicago Press, 1991, p. 23.

37 J. Donath, "Identity and deception in the virtual community," in M. Smith and P. Kollock (eds), *Communities in Cyberspace*, pp. 29–59. In this piece, Donath argues that pseudonymity is preferable to anonymity because it allows for the persistence of identity over time.

38 According to an NGO staff member, they delete messages which: (1) are slanderous/libelous; (2) contain advertisements; and (3) are posted by non-PWAs. This last condition applies even to messages of a positive nature. Fortunately, they have never received any discriminatory messages to date (personal communication).

39 On moderated lists, messages are usually reviewed for content before being posted. On unmoderated lists, the following methods of maintaining order and demarcating boundaries online have been documented: sarcasm, ridicule, and flaming (C. Surratt, *Netlife: Internet Citizens*

and their Communities, Commack, NY, Nova Science Publishers Inc., 1998); trolling (M. Tepper, "Usenet communities and the cultural politics of information," in D. Porter (ed.), *Internet Culture*, New York, Routledge, 1997, pp. 39–54); and even virtual torture (E. Reid, "Hierarchy and power: social control in cyberspace," in M. Smith and P. Kollock (eds), *Communities in Cyberspace*, pp. 107–33).

40 To this day, some gay business owners still embrace this philosophy. In the summer of 2001, participants at a meeting of the newly formed MASH-TOKYO (Men's Alliance for Sexual Health) raised the issue of discrimination against foreigners, but the organizers of the event argued that it is better to work with gay businesses than to alienate them by publicly attacking their stance.

41 The official Web site for the Tokyo Gay and Lesbian Parade 2001 can be found at: *http://www.tlgp.org/frame.html* (22 April 2002).

42 *http://www.gpress.com* (16 October 2001).

43 As Kollock and Smith point out, on the Internet "the poverty of signals is both a limitation and a resource, making certain kinds of interaction more difficult but also providing room to play with one's identity," *Communities in Cyberspace*, p. 9.

44 An overview of the series, "Basic information about HIV," dating back to April of 2000, can be found at: *http://www.gproject.com/hiv/hivindex.html* (22 October 2001).

45 As other researchers have noted, few of these magazines offer the types of political or lifestyle stories one might expect from gay "communities" in the West. *Adonis*, the only magazine to have attempted this in the 1980s, promptly went out of business (see M. McLelland, *Male Homosexuality in Modern Japan: Cultural Myths and Social Realities*, Richmond, UK, Curzon Press, 2000, pp. 123–41; and W. Lunsing, "Lesbian and gay movements: between hard and soft," *Mitteilungen der Gesellschaft für Natur und Volkerkunde Ostasiens*, vol. 128, 1998, pp. 279–310).

46 The photographer admitted that he was having trouble finding anyone else who would agree to be photographed. As a result, most exhibits on AIDS in Japan feature photos of American HIV/AIDS patients, thereby reinforcing the notion that HIV/AIDS is an American disease and that discrimination is never a problem in the West.

47 H. Hasegawa, unnamed articles, *Tokyo Gay Walker*, 23 January 2001. The *Gay Walker* site, which has since closed, was located at: *http://www.gaywalker.com*.

48 H. Hasegawa, "*Reinbō-furaggu* (Rainbow flag)" (10 October 2000), The Dozan Kyoshi. Online. Available HTTP: *http://www.alpha-net.ne.jp/users2/sarang1/iibana.html*, (22 April 2002). The factionalism of Japan's gay community has been discussed by W. Lunsing in "Japan: finding its way?" in B. Adam, J. Duyvendak, and A. Krouwel (eds), *The Global Emergence of Gay and Lesbian Politics: National Imprints of a Worldwide Movement*, Philadelphia, Temple University Press, 1999, pp. 293–325.

49 H. Hasegawa, *http://www.janpplus.jp/* (2 April 2002).

50 This is the stage name (*katsudōmei*) she uses for all of her HIV-related activities.

51 *Yamamba* refers to a group of young girls, usually of high-school age, who have platinum hair, wear distinctive makeup, clothing, and platform shoes. As a symbol of socially and economically marginalized youths who trade sex for money, they embody all of society's worst fears about social reproduction gone awry.

52 In her own words, "*tataitemo kantan ni wa tsuburenai taipu desu.*"

53 T. Umeda *et al.*, "Nihon no iseikan seiteki sesshoku ni yoru eizu no tokuchō-eizu sābeiransu ni yoru eikoku oyobi beikoku to no hikaku kenkyū" (Characteristics of AIDS contracted through heterosexual sexual contact in Japan: a comparative study with America and the United Kingdom by AIDS surveillance), *Nihon Koeishi* (Japanese Journal of Public Health), vol. 3, no. 38, 2001, pp. 200–7. The authors of this article speculate that many heterosexual men contract HIV from visits to sex workers and that older Japanese women have been protected from HIV by the prevalence of condom use among married couples. However, it is rare for older women to seek an HIV test, especially if they are beyond reproductive age.

10 Private acts/public spaces

Cruising for gay sex on the Japanese Internet

Mark McLelland

INTRODUCTION

This chapter critiques orientalist assumptions about the position of sexual minorities in Japan by questioning the Western gay rights discourse through which minority sexual identities in "other" societies can only be understood in terms of exclusion and impossibility. Arjun Appadurai has been a critic of the tendency among some Western researchers to assume that these "other" societies are "behind" the West on some kind of evolutionary path. He comments that it is no surprise that "this linkage of the infancy of individuals and the immaturity of groups is made with the greatest comfort about the nations of the non-Western world."[1] Rejecting this evolutionary model, Appadurai argues that it is necessary to study "the cosmopolitan ... cultural forms of the contemporary world without logically or chronologically presupposing either the authority of the Western experience or the models derived from that experience."[2] The Internet provides a useful tool for doing just this since it is one of the cosmopolitan cultural forms that has rendered more visible previously marginalized groups and communities in Japan. The Net offers a window onto the world of Japan's sexual minorities, providing one important means of access to a large number of individuals, groups and communities whose sexual practices fall outside the heteronormative roles endorsed by mainstream media, and, importantly, offers us direct access to individuals speaking in their own voice.[3]

THE INTERNET AND THE GLOBAL QUEER

There has been considerable research done into the ways in which the Internet has facilitated community identification, particularly among Western gay men, but less work exists on lesbians, transgender individuals, sex workers and other sexual minorities in both Western and non-Western societies.[4] The focus on identity and community building, although important, is but one way in which the advent of the Internet has affected the lives of sexual minorities whose access to public media has hitherto been extremely circumscribed. Of equal importance is the way in which the Internet has impacted upon individuals' sex lives but researchers have tended to steer clear of this sensitive topic. Yet, Daniel Tsang is surely not alone when he asserts that "I can't tell you how long it has been since I have been to a gay bar, except to pick up gay magazines; like numerous others, electronic cruising has replaced bar hopping."[5] For many non-heterosexual individuals, as for many heterosexuals,[6] the Internet has had a significant impact on the ways in which sexual activities are negotiated and organized. This is particularly the case in Japan which is one of the world's most wired nations.[7] However, little attention has been paid to the very wide range of online sexual activities and discourses in which people participate.[8] The few references that do exist in English-language media highlight Japan's perceived role as a major distributor of child pornography. *Time* magazine, for instance, under the headline "Japan's Shame" confidently asserts that "[Japan] is awash in child porn, and there's little attempt at hiding it."[9]

This perception of Japan as a sexually perverse society has been heightened due to negative reporting in western media about the sex and violence, often featuring depictions of young girls, which can occur in Japanese manga and anime. Yet, as Sharon Kinsella points out, "images from Japanese animation and manga presented in television shows and magazine reports on the phenomenon in Europe and America have ... generally been edited to emphasize violence, sex and strangeness" which fit snugly into "pre-existing notions of Japan as a cruel, sexist, strange and repressed society."[10] So pervasive is the western perception of what Hammond[11] terms "Japanese weirdness" that truly astonishing (but inaccurate) reports about Japanese sexual license repeatedly find their way into print.[12] Arthur Golden's surprise best seller *Memoirs of a Geisha,*[13] and the plethora of reprints of other books about geisha that its success has generated, is evidence for the continuing popularity among westerners of exotic sexual tales about Japan that emphasize its barbarity and strangeness.

Japan is therefore fertile ground for queer representations but if two recent English-language books about homosexuality in Japan are to be believed, it is hardly a place that supports queer identities. Summerhawk *et al.*'s collection *Queer Japan*,[14] which contains life stories of Japanese lesbian, gay, bisexual and transsexual individuals, details the difficulty that Japanese sexual minorities face in establishing the kind of clear-cut sexual identities that Summerhawk and her coeditors claim as ideals. For instance, the book's back copy informs us (incorrectly) that "same-sex love between men has never been legally nor socially accepted in Japan." Inside it is asserted that "a majority of Japanese gay men live in contradiction, a constant struggle with the inner self, even to the point of cutting off emotions and the denial of their own oppression."[15]

This apparently bleak situation for gays is reinforced by Francis Conlan, the Australian translator of Itō Satoru and Yanase Ryūta's *Coming Out in Japan.*[16] In the book's preface, Conlan explains that Japan's "Confucianist traditions"[17] and "feudal values"[18] are so strong as to make "the always difficult job of bringing about social change ... even more difficult in Japan than in the West."[19] These opinions are characteristic of those held by many Westerners about the political rights of disenfranchised groups in Japan. The Western characterization of Japanese women's participation in politics, which Robin LeBlanc says is "marked by exclusion [and] impossibility,"[20] seems to have set the tone for much other writing on the (lack of) politicized sensibilities among Japan's minorities.

Both representations of Japan outlined above, as a place that is simultaneously sexually perverse *and* repressive, are characteristic of orientalist discourse.[21] The failure to find the same forms of individuality and social organization characteristic of the Western onlooker's own society can result in a perception of the "other" society as both exotic and deviant. Chandra Mohanty critiques this position when she asks, "What is it about cultural Others that makes it so easy to analytically formulate them into homogeneous groupings with little regard for historical specificities?"[22]

Lunsing, in his work on Japanese constructions of homosexuality, points to the tendency of some researchers to analyze Japanese society solely in terms of its "lack" of sexual categories developed in Anglophone countries. He critiques the assumption that "gay identity among the Japanese must be strengthened, which implies that Japanese gay people must become like Americans,"[23] adding that in Japan "sexual preference is generally not seen as a feature that determines one's personhood more than partially."[24] Lunsing argues that rather than judge Japan's sexual minorities in terms of sexual identities developed in the West, it is necessary to look at specific local meanings that situate sexual acts and identities within the wider context of Japanese society.[25]

Drawing upon perspectives developed in my earlier work, this chapter looks at the particularities of a specific kind of sexual interaction within Japanese gay culture: men who cruise public spaces for sex with other men. My focus is on how the Internet has impacted upon these men's ability not only to seek out and negotiate sexual encounters but also to develop a communal discourse about the etiquette of cruising that relies upon concepts of negotiation and harmony that are considered important aspects of all social interaction in Japan.

PARK PARADISE

Park Paradise[26] is just one of many Japanese sites on the Internet that provides information about *hattenba* or "cruising grounds" where men interested in sex with men[27] come together in search of partners. As Castells points out, "New electronic media do not depart from traditional cultures: they absorb them;"[28] in this case the Internet has made information about Japan's well-established cruising sites easier to come by for a wider variety of men. Previously, men either stumbled upon these places by accident, or found out about them through word-of-mouth at gay bars, through the regular *hatten* reports in gay magazines, through commercially produced *hatten* guides[29] available in gay bookshops or through recorded message services (*dengon daiyaru*) advertised in the gay press.[30] The Internet has made this information available to a wider range of men who do not engage with the gay scene and are not part of its informal information networks. However, the increased accessibility of information about cruising sites and practices is double-edged and has sparked off concern about security in the minds of some men who use the Internet to look for sex.

To allay security fears, *Park Paradise* requires the browser to first pass a "gay check" which involves answering questions (in Japanese) about the meaning of specific gay lingo, identifying brand items popular among gay men or spotting the odd one out in a list of titles of Japanese gay magazines. This provides a certain amount of security by weeding out casual browsers (including most Western academics) and granting access only to those acculturated into Japan's gay scene. Since the site owner is obviously attempting to restrict entry to those sympathetic to the site's contents, for ethical reasons I have chosen not to reveal the URL and have also disguised the site's name.

The main purpose of the site is to provide up-to-date information about various outdoor spots in the Tokyo and Kanagawa region where men gather in the hope of meeting other men for sex, usually to be consummated on site. It does this by providing an alphabetical listing of places with information on how to get there, the type of men who visit and the time that the spot is most active. An interactive notice board allows men to post information about themselves and their availability, suggest new places to visit or modify, expand on or contradict information provided by other users. Given the popularity of accessing the Internet via mobile phones in Japan, the site also has a specially formatted *i-mode* version that enables users to browse, create and respond to personal ads via their phones as well as download other material.

Other features include a "soliloquy" (*hitorigoto*) page where the site owner, a veteran cruiser, discusses issues related to the history, practice, etiquette and politics of outdoor sex. This section contains more philosophical discussions about why some men find outdoor sex so attractive as well as the correct terminology for referring to regular partners whom one meets at the cruise spots. The Web master considers the term "sex friend" (*sekkusufurendo*) to be too impersonal for a regular cruising partner and instead proffers the coined term *kōensai* or "park wife," although this does not seem to have been taken up by other posters. This kind of coinage does, however, emphasize the role the Internet can play as a space for trying out new ideas and identities. As Miller and Slater point out, the Internet offers "*expansive potential*, the encounter with the expansive connections and possibilities of the Internet may allow one to envisage a quite novel vision of what one could be."[31] This is particularly important in Japan where neither mainstream nor gay media have offered much space for discussions of gay identities and lifestyles.[32]

Park Paradise also features an extensive links page to other sites that specialize in cruising. These include sites dedicated to saunas, train stations, parks, toilets, *onsen* (hot springs) and beaches where men frequently meet other men for sex. As Laud Humphries' pioneering ethnography *Tearoom Trade: a Study of Homosexual Encounters in Public Places*[33] showed, male-male sexual encounters in public spaces follow strictly coded procedures that render them largely invisible to the majority of men who use these facilities for their designated purposes. These Japanese cruising sites thus also fulfill the vital function of inducting inexperienced men into the codes and etiquette that will enable them to successfully negotiate sexual encounters without drawing negative attention to themselves.[34] The sites give a strong sense of the multiple opportunities that urban spaces offer gay men to come together, a point made by Bech in his up-beat survey of gay men's urban lives.[35]

In Japanese *hatten suru* means "to develop" and *hattenba* are places where sexual interaction is likely to occur. These are subdivided into free venues, including certain parks, beaches, highway stops, temple grounds and public conveniences, as well as paying sites such as saunas, health clubs, adult cinemas and hot springs. Following the trend towards "typing" prevalent on Japan's gay scene that I have discussed elsewhere,[36] gay men who cruise for sex are further subdivided into "insiders" (*shitsunaikei*) and "outsiders" (*yagaikei*) depending on their preference for indoor or outdoor venues.

In Japan, male-male sex is not mentioned in the criminal code[37] and is therefore not subject to the strict regulations that discriminate between legal and illegal sexual acts based upon the gender of participants in countries such as the UK, the US and Australia. For instance, in the UK, the Sexual Offences Act of 1967 only partially decriminalized sex between two men over the age of 21 that took place "in private."[38] Private was defined narrowly as a room where there was no chance of being observed by outsiders. Sex between more than two men, sex in a public lavatory (even in a locked stall) and sex in "public places" whether they were in a car in a deserted car park, in a park late at night or in a steam room in a gay sauna was still deemed illegal under these regulations. Considerable police resources in these countries are diverted toward interfering with sexual acts between gay men. For instance, in the UK in 1987 "Operation Spanner" spent £3 million in an attempt to bust a "ring" of sixteen gay men who regularly met up to indulge in consenting same-sex sado-masochistic practices that were videoed and later distributed among members.[39]

Although male–male sex has been partially decriminalized in most Western societies, additional restrictions apply to sex between men that do not affect sex between men and women. Furthermore, there are states in the US that still outlaw "sodomy." Currah points out that:

> Although sodomy laws are rarely enforced, they are most effective in their indirect effects on gays and lesbians: for example, sodomy laws justify solicitation and loitering laws that are used to entrap men soliciting sex from other men in public places.[40]

So far in the US, attempts to overturn state sodomy laws through appeals to the fundamental freedoms contained in the constitution have proven unsuccessful. In 1985, the US Supreme Court went so far as to brand the claim that individuals had the fundamental right to engage in homosexual sodomy as "at best, facetious."[41]

In Japan, however, public male-male sex is not singled out for specific police attention and is subject to the same "public obscenity" (*kōzen waisetsu*) laws that apply to all expressions of public sexuality, irrespective of the gender or number of the participants.[42] Discussions of police harassment and entrapment, which are common on gay cruising sites in English, are therefore mostly absent in Japanese.[43]

Indeed, some sites seem to anticipate police cooperation in preventing gay bashing, one site recommending on its entrance page that when visiting the cruising grounds to always take one's mobile phone for security reasons. Adding that, "If you witness an incident involving another person, please at least contact the police."[44]

Although Japanese gay men who choose to seek sex in public places do not generally need to worry about police entrapment, gay bashing has emerged as a concern in recent years. The Japanese public was first introduced to the concept of "gay hunting" (*homogari*) in the 1993 television soap drama *Dōsōkai* (Alumni reunion) where the series' hero, a married gay man named Fūma, was almost beaten to death by a gang of thugs patrolling a park late at night. Despite the potentially incendiary theme of gay men's use of public space for sex, the storyline was actually highly sympathetic to the young man and his plight. Indeed, Fūma was rescued by his best friend (and secret love) Atari who, appalled by the lengths that Fūma went to get sex, agreed to become his lover, albeit for one weekend only. The extremely violent and dramatic bashing scene was typical of the melo-dramatic treatment given to homosexuality throughout the series which, although not unsympathetic, did tend to stress factors that are largely irrelevant to the lived realities of most gay men in Japan.[45] It was not until February 2000, when a young man was beaten to death by a gang of youths in Yumenojima, a Tokyo park, that gay bashing was taken up as a theme in the media and began to be discussed in a systematic manner on gay Web sites.

This murder resulted in sites such as *Park Paradise* including a "bashing infor-mation" (*bashingu jōhō*) section in which men can write in and discuss incidents that they have themselves heard of, witnessed or been involved in. Indeed, the murder served as something of a catalyst for an increased sense of community among men who enjoy public sex with other men, the victim being described on *Park Paradise* as "one of us" (*ware ware no nakama*). The park became a pilgrimage site for some gay men who, even one year after the event, were still leaving behind offerings commonly made to deceased spirits such as flowers, cigarettes and soft drinks. *Park Paradise* displays pictures of the site including a letter left behind by the victim's mother, thanking his friends for their continued concern.[46]

This tragic incident of gay bashing has generated much discussion on *Park Paradise* about how exactly to deal with the potential dangers encountered when cruising for sex in outdoor (*yagai*) venues. It is often observed that the number of men, particularly young men, who regularly meet outdoors has fallen whereas the safer indoor venues have seen a boom. Some men regret this change since, as members of the "outdoor group" (*yagaikei*) they relish the "thrill" (*suriru*) that comes from transgressing public space. It is, of course, difficult to know how accurate these observations are or whether it is fear of bashing alone that has led to a reduction in the number of young men participating in outdoor cruising. One other factor may be the increased use of the Internet by young Japanese men who now find it easier to meet partners online, thus making it unnecessary for them to seek for partners in bars or cruise spots.[47]

Whether or not incidences of gay bashing are increasing is also difficult to know for sure. *Park Paradise* does include a "gay bashing questionnaire" where readers can respond to questions about their experience of and response to gay bashing. As of 29 August 2001, a total of 59 men (18 per cent) had answered that they had been bashed while cruising for sex, a further 30 men (9 per cent) replied that they had witnessed a bashing incident, whereas 240 (73 per cent) replied that they had never been involved in a bashing.

A number of discussions center on how to make cruising safer. Suggestions include advertising cruise spots only on those noticeboards on restricted sites such as *Park Paradise* which provide "gay checks" before granting entry. Other suggestions include visiting the cruise spots only at popular times when it can be guaranteed there will be a number of men around (thus deterring troublemakers). One man points out that, "If you get your timing right you should be able to leave [satisfied] within 30 minutes." In order to assist in arriving at a time when the spot is most active, *Park Paradise* includes a list of the most popular times as well as the "type" of visitors the cruise spot attracts. This very much depends upon location with spots located next to factories, office buildings and universities attracting different kinds of men. Men are particularly encouraged to avoid cruising late at night, it being pointed out that over half (51 per cent) of the bashing incidents reported in the questionnaire had taken place between 1 and 2 a.m.

However, the most noticeable effect of the murder and the increased attention paid to other incidents of bashing on sites such as *Park Paradise* has been a heightened discourse of social responsibility. There were some messages urging the responsible use of public spaces before the Yumenojima incident; one site displayed pictures of the unpleasant mess (consisting of used condoms, wrappers and tissues) left behind by users at several popular parks and beaches. Men cruising these sites were exhorted to clean up after themselves in order not to attract negative attention. It was also pointed out on this site that certain car parks, temple grounds and parking lots that were used for sex at night were in residential areas and that great care should be taken to be quiet, especially if arriving and leaving by car, so as not to upset the sleeping residents.

THE CRUISE SPOT IMAGE-UP CAMPAIGN

Since the murder, many more messages have appeared that encourage gay men to be more discreet so as not to antagonize straight (*nonke*)[48] men. The slogan "Let's cruise while observing manners" (*manā o mamotte hatten shiyō*) has become a common exhortation on these sites. In fact, the five main cruising sites in the Tokyo area are linked to each other and on their entrance pages display the message: "We are undertaking an outdoor cruise-spot image-up campaign,"[49] along with other banners including the slogans, "Let's use cruise spots in a tidy manner," and "Cruise spot beautification movement" (*hattenba bika undō*), followed by the instruction, "Put your trash in the trash can; please cooperate." Given Japanese society's stress on maintaining social harmony and avoiding situations that might create conflict, it is not surprising to find slogans such as these as well as messages criticizing gay men for being too open about their cruising practices and thus causing trouble (*toraburu*) for straight men.

One man writes in to *Park Paradise* complaining about how two men had ruined the cruising potential of a local *rotenburo* (outdoor public bath) by overtly engaging in sex and disturbing an "ordinary customer" (*ippan kyaku*) who reported them to the custodian. Now security cameras have been installed and the custodian makes regular rounds of the bath's precincts. The writer comments, "This action [by the gay men] was really terrible, imagine what would have happened if the [other] customer had brought his children with him,[50] how would he have explained what was going on?" He also complains about other unsuitable behavior such as advertising an outdoor bath or hot spring on the Web site that is not known for gay cruising in the hope that gay men will begin to gather there. He asks, "Have you considered what would happen if [gay men] come and how the [regular] customers might feel?" Creating new cruise spots is always difficult, he suggests, mentioning an old man (*oyaji*) who approached a youth in a park not known for cruising and was abused by the youth and his friends. He concludes his post with the comment, "Where and who you choose to cruise is your own choice but I'd like you to act with more good sense."

Another man writes in to *Park Paradise*'s notice board complaining about the way some gay men treat public toilets, boring glory holes[51] into the walls and covering the cubicles with graffiti. He writes, "As someone who works in the construction industry, I have helped construct more buildings than I can remember. Please think about the fact that these buildings are built for your use at taxpayers' expense before you harm them. Let's treat [toilets] more carefully."

The notion of homophobia has entered Japanese as an English loanword (*homofobia*) and has been popularized by gay activists such as Itō Satoru, both on his Web site *Sukotan Project*[52] and in his publications.[53] I have, however, never come across the term on *Park Paradise*. The animosity sometimes shown towards gay men tends to be put down to a lack of discretion on the part of gay men themselves who cause "trouble" for straight men by being too overt when cruising in public. The response to public animosity, on *Park Paradise* at least, is to encourage men who cruise to follow correct manners and avoid any action that might disturb the residents near local cruise spots who, writers point out, have an equal right to enjoy these facilities for their designated purposes.

Other cruise sites also try to engage with rather than confront straight society. For instance, one includes a message to "straight people" (*sutorēto no kata*) on its entrance page.[54] This message, written in a register of polite Japanese known as *keigo*, admits that many straight people find the use of public space by gay men to be offensive. However, the author asks that the straight reader try to imagine for a moment how difficult it is for gay men to meet each other in ordinary circumstances as society makes it impossible to adopt an openly gay identity. He writes that:

> We gay men are living right next to you. We meet with you, talk with you and laugh with you on a daily basis. We share the same space with you day to day. But when it comes to love (*ren'ai*) it isn't as easy as you might think to find a partner. We need places where we can meet together as gays (*gei ga gei toshite*).

Having provided this justification for gay men meeting together in places such as parks, the author admits that this can cause trouble for straight people who might be disturbed at the thought of men roaming around local areas late at night and who object to the rubbish left behind and the crude graffiti scrawled on walls. The writer apologizes for the lack of manners shown by some gay men but points out that sites such as his are active in trying to improve behavior. Also mentioned is the murder incident at Yumenojima, the author pointing out that some gay men live in fear of attack, and asking for the cooperation of straight people in preventing such attacks. Finally, he invites straight people to write in and leave their messages and suggestions.[55]

This conciliatory response toward anxieties routinely entertained by "straight society" about gay men and their sexual practices was also evident among some of my gay informants from an earlier project. One man in his late twenties explained:

> For some people, small things can become a huge wave and carry them away because the values that people place on different things are different…It's to be expected that you can't always understand what it is that upsets other people, especially with regard to being gay. Although I've been rejected by my parents and beaten, I don't feel bad and live a happy life.

As I have commented in earlier work,[56] my informants tended not to "other" heterosexual society in a confrontational discourse which posited society and its institutions as somehow against them. Instead, many of the men I spoke to tried to understand why it might be difficult for others to accept their homosexual inclination. Other gay voices also support this viewpoint. For instance, one of Yajima's informants points out that Japan's "gay movement" has not had to fight against legal discrimination[57] and the moral prohibitions of Christianity, as have gay groups in Europe and America. He suggests that in the face of the institutionalized persecution faced by many gays overseas, "In Japan, in terms of the social system, you could say that discrimination is slight."[58] It would be ethnocentric to suggest, as Summerhawk does, that men such as this are simply in "denial of their own oppression."[59] The discussions I had with my informants, as well as the conciliatory tone adopted in many submissions to *Park Paradise* about the importance of good manners while cruising, suggest that many gay men in Japan do not conceive of themselves in terms of an oppositional us-against-them mentality.

CONCLUSION

At the outset of this chapter I referred to two recent books about queer life in Japan that describe the lives of lesbian, gay and transgendered people in terms of constant struggle against oppression. The underlying assumption in these books seems to be that Japan is "behind" a largely unarticulated "West" in terms of gay rights. It may be the case that a relatively small percentage of affluent and well-educated gay men and lesbians who live in major cities of the Western world exert a high degree of control over their lives through their ability to choose their occupation, living space and recreation.[60] This situation is generally unparalleled in Japan where an openly gay identity would be seen as a bar to social acceptance, career advancement and even house rental. Yet a crude West/Japan juxtaposition hardly helps us to understand the place of homosexual desire and identities in Japanese society.

The small window onto the lives of gay men who cruise for sex that the Internet provides is further evidence that "oppression" is too crude a term to capture the complexities of the lives lived by many non-heterosexual men in Japan. Sites such as *Park Paradise* represent communal spaces where sexual minorities can come together and invert the usual binary of "us" and "them" that posits them as social outsiders. On these sites it is gay men who are the "us" among whom a common sensibility and code of ethics is assumed in contrast to an outside world made up of "straight society."

Yet, although it is clear that there is a distinct gay subculture in Japan (symbolized by the "gay check" that restricts entry to some sites), gay men's use of public space is not framed in terms of legal rights or political confrontation.[61] Rather, public spaces such as parks, beaches, hot springs and toilets are represented on these sites as facilities established for the use of the public as a whole. Gay men encourage each other to use these spaces for their own subcultural ends but only with discretion and following a strict code of manners. Negative responses from straight men tend to be put down to trouble caused by gay men themselves as opposed to any underlying homophobia. This would seem to support the notion of "glocalization" advanced by Appadurai,[62] among others, who argues that global technologies are put to very local uses. The "outdoor cruise-spot image-up campaign" advanced through these Internet sites seems to be a distinctively Japanese response to the threat of gay bashing and intimidation that avoids setting up an always antagonistic and persecuting heterosexual majority against a permanently oppressed gay minority.

It is clear that some Japanese gay men, whose relationships are not generally accorded public recognition,[63] use the Internet in order to meet, network with and create sexual (and other) relationships with men who share their desires. Rather than a community based upon a common "identity," however, it would be better to view these sites as a coalition of men who share an "affinity." In Appadurai's terms, *Park Paradise* and similar Web sites are "communities of imagination and interest"[64] that the Internet has enabled among men who are otherwise very diverse. The Internet has provided a public venue (in the sense of a space available to all who can access the technology) in which men from all over Japan can come together at any time of day or night and find someone else online who shares their interests. Despite the primary emphasis on the search for sexual partners on *Park Paradise* and other cruising sites, this is not incompatible with either identity or community development. In fact, discussion about these issues is taking place but largely outside the political paradigms that structure much Western gay and lesbian discourse. *Park Paradise* and other Japanese cruising sites are therefore examples of what Castells refers to as "customization," wherein "global" technologies such as the Internet are adapted to serve and reflect the "tastes, identities and moods of individuals."[65]

Different individuals' experience of sexual freedom or constraint is obviously complex and multivalent, depending, as it does, on multiple personal and social factors. However, the brief look at the community of men who use the Internet to facilitate their search for public sex should be sufficient to problematize the notion that Japan is somehow behind a more liberated West when it comes to the expression of homosexual desire, especially when the punitive measures that many Western states employ to keep homosexual behavior closeted behind closed doors are considered.

NOTES

1 A. Appadurai, *Modernity at Large: Cultural Dimensions of Globalization*, Minneapolis, University of Minnesota Press, 1996, p. 141.

2 Appadurai, *Modernity at Large*, p. 195.

3 I have discussed Internet use by Japan's sexual subcultures elsewhere. See M. McLelland, "The newhalf net: Japan's 'intermediate sex' online," *The International Journal of Sexuality and Gender Studies*, vol. 7, nos 2/3, April/July 2002, pp. 163–74; "Live life more selfishly: an online gay advice column in Japan," *Continuum*, vol. 15, no. 1, April 2001, pp. 103–16; and "Out and about on Japan's gay net," *Convergence*, vol. 6, no. 3, Autumn 2000, pp. 16–33.

4 Many edited volumes on "cyber community" or "cyber culture" include a chapter on gay (less frequently specifically lesbian, bisexual or transgendered) uses of the Internet. See for instance D. Shaw, "Gay men and computer communication: a discourse of sex and identity in cyberspace," in S. Jones (ed.), *Virtual Culture: Identity and Communication in Cybersociety*, London, Sage, 1998, pp. 133–45; R. Woodland, "Queer spaces, modem boys and pagan statues: gay/lesbian identity and the construction of cyberspace," in D. Bell and B. Kennedy (eds), *The Cybercultures Reader*, London, Routledge, 2000, pp. 416–31. The only work I have come across that has direct relevance to Queer Asia (as opposed to Queer Asian Diaspora in the West) is C. Berry and F. Martin, "Queer 'n' Asian on – and off – the Net: the role of cyberspace in queer Taiwan and Korea," in D. Gauntlett (ed.), *Web.Studies*, London, Arnold, 2000, pp. 74–81; C. Berry, F. Martin and A. Yue's (eds), forthcoming collection *Mobile Cultures: New Media in Queer Asia*, Durham, Duke University Press, will go some way to filling this gap.

5 D. Tsang, "Notes on queer 'n' Asian virtual sex," *Amerasia Journal*, vol. 20, no. 1, 1994, p.119.

6 On cyber-dating among men and women in Japan, see "Dating game: looking for prince Charming? In Japan, check your cell phone," *Time International*, vol. 157, issue 22, 4 June 2001, p. 88.

7 See, for instance, Holden and Tsuruki's chapter in this volume which discusses the popularity of *deiai* contact-sites accessed via mobile phones.

8 See, however, my articles mentioned in note 3 as well as "No climax, no point, no meaning? Japanese women's boy-love sites on the Internet," *Journal of Communication Inquiry*, vol. 24, no. 3, pp. 274–91; and "Local meanings in global space: a case ctudy of women's 'boy love' Web sites in Japanese and English," *Mots Pluriels*, no. 19, Special Issue: The Net: New Apprentices and Old Masters, October 2001. Online. Available HTTP: *http://www.arts.uwa. edu.au/MotsPluriels/MP1901mcl.html* (16 May 2002).

9 "Japan's shame: lawmakers are finally pushing legislation to help end the country's dubious distinction as the world's main source of child pornography," *Time International*, 19 April 1999, p. 34.

10 S. Kinsella, *Adult Manga: Culture and Power in Contemporary Japanese Society*, Richmond, Curzon Press, 2000, p. 14.

11 P. Hammond (ed.), *Cultural Difference, Media Memories: Anglo-American Images of Japan*, London, Cassell, 1997.

12 For instance, one journalist claims that dramas depicting rape and sodomy can be screened on Japanese TV during the dinner hour so long as no pubic hair is visible. This is, of course, a complete fiction. J. Fallows, "The Japanese are different from you and me," *The Atlantic*, vol. 258, September 1986, p. 35.

13 A. Golden, *Memoirs of a Geisha*, London, Vintage, 1998. For a critique of the orientalism embedded in this narrative, see A. Allison, "Memoirs of the Orient," *Journal of Japanese Studies*, vol. 27, no. 2, 2001, pp. 381–98.

14 B. Summerhawk, C. McMahill and D. McDonald (eds), *Queer Japan: Personal Stories of Japanese Lesbians, Gays, Bisexuals and Transsexuals*, Norwich, VT, New Victoria Press, 1998.

15 Summerhawk *et al.*, *Queer Japan*, pp. 10–11.

16 S. Itō and R. Yanase, *Coming Out in Japan* (trans. F. Conlan), Melbourne, Trans Pacific Press, 2000.

17 Itō and Yanase, *Coming Out in Japan*, p. xv.

18 Itō and Yanase, *Coming Out in Japan*, p. xvi. Ironically, male same-sex relations were common and in certain circumstances highly valued in Japan's feudal period (1600–1867). See for instance E. Ikegami, *The Taming of the Samurai: Honorific Individualism and the Making of Modern Japan*, Cambridge, MA, Harvard University Press, 1995.

19 Itō and Yanase, *Coming Out in Japan*, p. xv.

20 R. LeBlanc, *Bicycle Citizens: the Political World of the Japanese Housewife*, Berkeley, California University Press, 1999, p. 17.

21 Timon Screech argues that "orientalism" (a fixed pattern of images and assumptions that "western" researchers routinely entertain about "eastern" peoples) is characterized by "the institution-alization of structural disequilibrium, arrogation of representational power and sexual fetishization." T. Screech, review of John Treat's *Great Mirrors Shattered: Homosexuality, Orientalism and Japan*, *The Journal of Asian Studies*, vol. 59, no. 3, August 2000, p. 759.

22 C. Mohanty, "Under Western eyes: feminist scholarship and colonial discourses," *Boundary 2*, vol. XII, no. 3/vol. XIII, no. 1, Spring/Fall 1984, p. 340.

23 W. Lunsing, "Gay boom in Japan? Changing views of homosexuality," *Thamyris*, no. 4, Autumn 1997, p. 284.

24 Lunsing, "Gay boom," p. 284. In drawing attention to differences that might exist between Japan and North American or Northern European societies I do not want to reify "traditional Japanese society" and suggest that the Anglophone discourse of gay rights is, in the Japanese context, always inappropriate. However, I do want to question the certainty with which some Western observers greet the recent development of North-American models of "gay identity" in Japan as a sign of Japan's progress or advancement toward an already developed gay identity supposedly characteristic of the West. On this point, see my paper "Out on the global stage: authenticity, interpretation and Orientalism in Japanese coming out narratives," *The Electronic Journal of Contemporary Japanese Studies*, 2001. Online, Available HTTP: *http://www. japanesestudies.org.uk/articles/McLelland.html* (1 April 2002).

25 Other researchers, notably M. Manalansan IV, "In the shadows of Stonewall: examining gay transnational politics and the diaspora dilemma," *GLQ: A Journal of Lesbian and Gay Studies*, vol. 2, 1995, pp. 425–38; A. Chao, "Global metaphors and local strategies in the construction of Taiwan's lesbian identities," *Culture, Health & Sexuality*, vol. 2, no. 4, October—December 2000, pp. 377–90; and M. Chiang, "Coming out into the global system," in D. Eng and A. Hom (eds), *Q & A: Queer in Asian America*, Philadelphia, Temple University Press, 1998, pp. 374-395, have questioned the universalizing rhetoric of lesbian and gay liberation in relation to indigenous constructions of sexual identity.

26 The number of hits registered on a page can be misleading but *Park Paradise* claims to have received 321,015 visitors in the three years since its inception (2 August 1998—1 August 2001). It attracted 19,830 hits in the month of August 2001.

27 As numerous studies of men's sexual behavior have illustrated, men who have sex with men and "gay" men are not coterminous groups. Indeed, many men who cruise for casual sex with other men are married and identify themselves as heterosexual. The situation in Japan is further complicated by the fact that recreational sex for men is often spoken of as a "hobby" (*shumi*) or "play" (*asobi* or *purei*) and therefore the correlation between act and identity may not be so insistent as in Anglophone societies. For a discussion of terminology relating to same-sex desiring men in Japan, see M. McLelland, *Male Homosexuality in Modern Japan: Cultural Myths and Social Realities*, Richmond, UK, Curzon Press, 2000, pp. 7–12.

28 M. Castells, *The Information Age: Volume I, The Rise of the Network Society*, Malden, MA, Blackwell Publishers, 2000, p. 370. For a discussion of how the Internet has become a major advertising medium for Japan's long-established community of transgendered bar, cabaret and sex workers, see my paper "The newhalf net: Japan's 'intermediate sex' online."

29 These included both site maps as well as guides to cruising etiquette. An example of the latter is Kazuya's *Gei seikatsu manyuaru* (Gay lifestyle manual), Tokyo, Data House, 1998.

30 These prerecorded services anticipated the Web sites in some ways by providing constantly updated information that could be accessed according to region at the touch of a button (the advertisements for these services included the different code numbers for different areas and types of location). They were also interactive; for an extra fee, men could record their own messages and listen to messages left by others.

31 D. Miller and D. Slater, *The Internet: An Ethnographic Approach*, Oxford and New York, Berg, 2001, p. 11; emphasis in the original.

32 See my discussion in *Male Homosexuality in Modern Japan*, particularly chapters 6 and 9.

33 L. Humphries, *Tearoom Trade: a Study of Homosexual Encounters in Public Places*, London, Gerald Duckworth & Co., 1970.

34 These codes are not international but follow distinct local patterns. One example would be the manner in which sexual interactions can take place while on board Tokyo's tightly packed commuter trains. Some sites explain the conventions that exist for exhibiting one's availability along with train lines, times and carriages where gay male commuters gather.

35 H. Bech, *When Men Meet: Homosexuality and Modernity*, Oxford, Polity Press, 1997. Bech's thesis is that cities provide the optimum environments for cruising and the development of the gay "gaze."

36 Many of Japan's gay magazines and bars are aimed at certain "types" (*taipu*) of men and their admirers. These include sporty types, body builders, cute guys, foreigners, and many more; see M. McLelland, *Male Homosexuality in Modern Japan*, pp. 124–7.

37 See my discussion "Homosexuality before the law" in *Male Homosexuality in Modern Japan*, pp. 37–40.

38 In February 1994 the House of Commons voted to reduce the age of consent for gay men from 21 to 18. In 2001 the age of consent was reduced to 16 (on a par with heterosexual consent).

39 D. Bell, "Pleasure and danger: the paradoxical spaces of sexual citizenship," *Political Geography*, vol. 14, no. 2, 1995, pp. 139–53. The initial sentences of three years were, upon appeal, reduced to six months but an appeal to the European Court of Human Rights to have the sentences overturned was rejected.

40 P. Currah, "Searching for immutability: homosexuality, race and rights discourse," in A. Wilson (ed.), *A Simple Matter of Justice? Theorizing Lesbian and Gay Politics,* London, Cassell, 1995, p. 53.

41 Currah, "Searching for immutability," p. 52.

42 Although technically illegal, female (and male) prostitution (and soliciting) is tolerated by the Japanese police so long as it is restricted to designated areas and takes place in registered establishments (*fūzoku kanren eigyō*). Guide books to these areas and the establishments within them detailing services offered and fees charged are openly sold. See for example, *SEX no arukikata: Tōkyō fūzoku kanzen gaido* (How to find your way around sex: the complete guide to Tokyo's sex scene), Tokyo, Alt.books, 1998.

43 An entry to Hibiya Park in Tokyo's central business district (that attracts older salarymen in the evening) points out that there is a police box in the park and that the policemen "are well aware

of what goes on." However, the writer does not suggest avoiding the park but simply urges greater discretion.

44 *"Moshi tajin demo jiken o mitara keisatsu ni renraku suru gurai no kimochi o motte kudasai."*

45 For a discussion of *Dōsōkai*, see S. Miller, "The (temporary?) queering of Japanese TV," *Journal of Homosexuality*, vol. 39, nos 3/4, 2000, pp. 83–109.

46 The text of the letter reads: "To all [his] friends, thank you for always leaving beautiful flowers and drinks. On behalf of the person concerned [i.e. her son] I would like to offer thanks. [His] Mother." The site owner comments that "When thinking about his mother's letter I can't help crying."

47 Many of my interviewees spoke of how the Internet had become their preferred means to meet sexual partners as it cut down on time and also on chances of rejections since photos could be exchanged and sexual acts negotiated ahead of a meeting. See my discussion in *Male Homosexuality in Modern Japan*, chapters 6 and 7.

48 *Nonke* is a hybrid Anglo-Japanese term formed by the English prefix "non" and the character *ki* (or *ke*) meaning "feeling." A *nonke* is thus a man who does not have sexual feelings for other men. This term exists alongside the loanword "straight" (*sutorēto*).

49 *Yagaikei hattenba imējiappu kyanpēn o okonatte imasu.*

50 Parents bathing with their children (both at home and at public baths) is common practice in Japan where there is less self-consciousness about nudity and where "skinship" (*sukinshippu*) is thought to create a natural bond between parent and child.

51 Small holes enable peeping whereas larger ones allow sex acts to take place between two men in separate but adjoining stalls.

52 *Sukotan Project http://www.sukotan.com* (7 March 2002).

53 See for example S. Itō, *Dōseiai no kiso chishiki* (Basic information about homosexuality), Tokyo, Ayumi shuppan, 1996, pp. 131–5.

54 Again, for ethical reasons, I have decided not to reveal the URL of this site. However, anyone with Japanese ability should be able to find this and other sites like it via *Yahoo Japan* with the judicious use of key words. There are four major sites dealing with cruising in the Tokyo area that crop up in Net searches.

55 However, I saw no submissions by self-identified straight people to the page.

56 *Male Homosexuality in Modern Japan*, p. 198.

57 "Sodomy" (*keikan*) was, in fact, criminalized in the Meiji criminal code between 1873 and 1881 but was hardly ever punished. Since homosexuality is not mentioned in Japan's current criminal code, technically there is no age of consent for male-male (or female-female) sexual acts. Some Western nations, however, routinely maintain higher ages of consent for same-sex acts and in some American states, they are still illegal.

58 M. Yajima (ed.), *Danseidōseiaisha no raifuhisutorii* (Life histories of male homosexuals), Tokyo, Gakubunsha, 1997, p. 178.

59 Summerhawk *et al.*, *Queer Japan*, p.11.

60 R. Connell, *The Men and the Boys*, St Leonards, NSW, Allen & Unwin, 2000, p. 103, points out that gay communities "have a limited class composition … [participants are] highly educated and affluent in comparison with the general population." Among Connell's sample of working-class Australian gay men, he found that "Almost all find it necessary to be cautious about disclosing their homosexuality, concealing it from neighbours as well as workmates and family" (p. 109). This is not dissimilar to most gay men in Japan.

61 It could also be argued that gay sex sites in English such as *www.cruisingforsex.com* (7 March 2002) do not frame their presentation of cruising sites in terms of a political rhetoric but instead adopt a discourse which is purely recreational, not dissimilar to *Park Paradise*. However, *cruisingforsex.com* is prefaced with an extensive legal disclaimer forbidding access to browsers of a certain age in certain locations (based on a "sodomy law map" of the US) with links to sites concerning the legalities of public sex. The rhetorical space of the site is thus circumscribed by broader political concerns than simply pleasure seeking. Furthermore, some submissions to the site do specifically address notions of gay rights, such as the article "Whose Rights, Whose

Freedoms?" which can be found at: *http://www.cruisingforsex.com/columnDEAN/columnDEAN.html* (16 November 2001). Cruising sites in English do, however, share with those in Japanese admonitions on the need to clear up the evidence of sexual activity.

62 Appadurai, *Modernity at Large*.

63 Sharon Chalmers points out that in Japan "The act of heterosex (within the private sphere) is reduced to the singular function of human reproduction within a particular family type. The result of judging this representation as the 'norm' is that all other forms of 'familying' and household structures are virtually ignored," in "Lesbian (in)visibility and social policy in Japanese society," in V. Mackie (ed.), *Gender and Public Policy in Japan*, London, Routledge, 2002. Thus, while Japanese society is relatively lenient in its attitude towards recreational sex, particularly for men, it provides no social recognition of non-heterosexual relationships.

64 Appadurai, *Modernity at Large*, p. 195.

65 Castells, *The Information Age: Volume I, The Rise of the Network Society*, Malden, MA, Blackwell Publishers, 2000, p. 2.

Part III
Politics and religion

11 The great equalizer?

The Internet and progressive activism in Japan

David McNeill

INTRODUCTION

Any political assessment of the reconfiguration of Japanese politics in the 1990s would have to note two striking and perhaps related trends. The first is the acceleration in the long-term decline of the organized left, and the return of a form of popular "soft" nationalism around the nucleus of the Liberal Democratic Party, after they briefly lost their grip on power in the mid-1990s. The second is the growth of grass-roots activism, often via the auspices of non-profit or non-government organizations, and usually focused on local single-issue campaigns such as opposition to war, nuclear power plants or public works projects.

That this potentially profound crumbling of old political realities is taking place at the same time as the rapid development of the Internet in Japan is significant, and might be seen as heralding a revival of a progressive politics hinged around networks of virtual communities and Netizens operating away from the old political center in Tokyo. Such a revival is an especially attractive notion for those on the progressive left in Japan who have watched with unease the gradual shifting of the political center of gravity toward the right.

I want to argue, however, that there is little evidence that the Internet alone will make Japanese political and social institutions more democratic or that it will single-handedly rescue the left from its long-term decline. While there has indeed been something of a revival of local and national progressive activism, often hinged around imaginative and innovative uses of the Web, these relatively small forces are ranged against an entrenched establishment as well as conservative political forces also determined to make their mark on cyberspace. Moreover, the increasing commercialization of online activity threatens the gradual marginalization of progressive online campaigns.

GOODBYE TO THE JAPANESE LEFT?

The 38-year period of unbroken political hegemony by the Liberal Democratic Party (LDP) from 1955–93 and the opposition offered by the Japan Socialist Party (JSP) for much of this period was, it has been argued, essentially window dressing for a system driven by bureaucrats in the pursuit of unlimited industrial expansion.[1] The so-called 1955 system, according to this view, masked the axis of real political power in Japan and depended on the JSP's inability to provide credible opposition, before it went into coalition with the ruling Liberal Democratic Party in June 1994 and rapidly abandoned most of the policies that made it "social democratic."[2]

The coalition followed a brief post-Cold War political spring which appeared to herald the possible demise of the LDP but which instead left the JSP in terminal decline when the government dissolved in 1996. JSP voters failed to see why they should reward a party that had learned to embrace not just the market but nuclear power, the constitutionality of the Self-Defense Forces (SDF) and the Japan-US Security Treaty, and many deserted it for alternatives like the Japan Communist Party.[3] The JSP went from sixty-six upper house seats in the House of Councilors election of 1989 to thirty-eight in 1996 and just eight by 2001.

From the ashes of that doomed experiment emerged a much more conservative political terrain. Various elements of the LDP rearranged the deck-chairs to form the Democratic Party of Japan in 1996 and the Liberal Party and New Kōmeitō, both in 1998. A further split from the Liberal Party produced the New Conservative Party in 2000.

While the LDP has been unable to recover the grip on power it had under the 1955 system and has been forced into coalition government, the left has fared much worse. The issues that defined progressives in Japan, including pacifism, opposition to the expansion of the military and to nuclear power are argued by fewer and fewer Diet lawmakers. Much of the DPJ finds itself in agreement with the hawkish policies of LDP leader Koizumi Jun'ichirō. The few progressive Diet members still around find it difficult to combat an increasing slide toward populist nationalism. As Asada Akira says, "The story of the nineties is that the LDP kept on sinking, but its opponents sank even more."[4]

Of course, most of the real energy of progressive politics in post-war Japan came from outside the Diet, particularly during the period 1950–72, but here the story is little better. The anti-Stalinist New Left (*han-nikkyōkei*), which broke with the JCP in the 1950s, helped catalyze a powerful challenge to the centrist tendencies of the state, reaching a climax with mass protests against the US–Japan Security Treaty in 1960. The Anpo struggle remains the high point of mass social action in post-war Japan and radicalized a generation of progressives who concluded that because established political institutions could simply ignore popular democratic demands when it suited them, grass-roots organization and activism was necessary.[5]

The fallout from that struggle and the organizations it helped energize, particularly the national student organization *Zengakuren* and later *Beheiren* and *Zenkyōtō*,[6] were an important legacy in the movements that followed, with many Anpo-generation activists going on to play leading roles in campaigns over the environment, Women's Liberation, educational censorship, the building of Narita Airport and the Vietnam War.[7] Protests against the war and the revision of the Security Treaty in 1970 once again locked thousands in battle with the state, but the slowing of the momentum of these movements and their failure to win mass popular support contributed to the fragmentation of the left into increasingly violent and militarized sectarian factions, of which only the most well known was the now disbanded Japanese Red Army. A number of other sects, particularly Chūkakuha and Kakumaruha, spent much of the following two decades waging internecine struggles, known as *uchi-geba*, that killed over 100 activists and injured hundreds more.[8] The violence, and the crimes of doomsday cult Aum Shinrikyō, with which these groups shared superficial organizational similarities, understandably did little to add to the attractions of the left and some sects simply abandoned the "ideological" politics of the past for NGO-type activity.[9] The radical energy of the period was increasingly dissipated by the intellectual struggles between the orthodox communism of the JCP and the romanticism of the increasingly divided New Left.

Any impetus toward political activity by young males is now quickly drained by the demands of the system. Since the 1980s, an increasing number of employees are likely to find themselves in non-unionized, white-collar employment rather than the blue-collar, well protected union strongholds of Japan's manufacturing golden age. Privatization of railways and telecommunications, which began under the Nakasone administration in the mid-1980s, and the increasing shift of manufacturing offshore to China and elsewhere during the 1990s have cut a swathe through these industries, and union membership has slumped to a post-war low of under 22 per cent, down from almost 56 per cent in 1950.[10]

Parliamentary politics increasingly dominated by parties of the right, traditional avenues of organized resistance whittled down to size after a decade of restructuring, shrinking networks of progressives from the Anpo, *Zengakuren* and *Zenkyōtō* generations – obviously the rather bleak balance sheet of the left drawn up here is a product of a particular historical juncture, both in its organizational structure and its relationship with other hierarchies of state and industrial power. Japanese capitalism has changed since the heyday of mass social democratic parties and new left movements, so lamenting the demise of the various forms of organized resistance that it threw up might be seen as rather futile. Much better, it might be suggested, would be to look for any evidence of more flexible and diverse political responses to the changing economic and social landscape.[11]

THE GRASS-ROOTS

The search leads to one of the few sources of optimism for progressives in Japan – the steady growth of local grass-roots political activity, civic action, and non-profit organizations, a fact recognized by the creation of the Non-Profit Organization Law in 1998. Estimates put the number of organizations involved in some sort of non-profit work in Japan in the region of at least 200,000.[12]

Many NPOs and NGOs, of course, are hardly "progressive" and some are heavily involved in promoting government activity. Indeed, as Tessa Morris-Suzuki points out, new-right-influenced governments have increasingly shown a liking for the "rhetoric of emancipation" espoused by these organizations "as states shift responsibility for some of their social and welfare functions to private or volunteer groups."[13] It might also be possible to quibble with the notion that this is an entirely new development, given that some can trace their genealogies back decades.

Nevertheless, by any measure it is clear there has been a significant development in the micro forces that in some way or other "challenge the existing order of things," to use Morris-Suzuki's definition of "new social movements." This process seems to have accelerated in the 1990s as doubt and disquiet increased over the stumbling of the former miracle economy and as many expressed discontent with the alternatives associated with the former progressive parties.

This is hardly a unique Japanese phenomenon but, as has often been pointed out, part of a generalized shift away from old social frameworks, rooted in central-ized political and economic institutions, toward a more fluid and fragmented political landscape.[14] If the characterization of the left summarized above emerged from an era of vertical power structures and large, monolithic political organiza-tions, we live increasingly it is said, in an era of horizontal networks, heterogeneous movements and spirited local struggles. Crucially, it is an era in which the Internet is set to play the leading technological role. Indeed, for some groups and cam-paigners that only exist in cyberspace, the Net has replaced all other forms of social action.[15]

The Net has no shortage of flag wavers in Japan or of those who argue its power to tip the balance in favor of the less powerful. Among the best known is academic and IT commentator Kumon Shumpei, Head of the Center for Global Communications, International University of Japan. While no supporter of progres-sive struggles, Kumon articulates a number of common assumptions about the Internet which are popular among Japan's intellectual and policy elite, perhaps precisely because they are so banal. Essentially these assumptions extrapolate from a property IT networks are said to embody in toto – their open democratic structure – to suggest that their development will necessarily lead to a less vertically-ordered and hierarchical society.

Kumon claims what he calls "community communication" and the information revolution "empower individuals and small groups of all kinds, acting as an equalizer between them and the 'big boys' – the state and the major corporations. In fact, it tends to give the edge to these smaller actors."[16] The Internet-driven digital economy hastens the trend "toward ever-greater empowerment of the indi-vidual in modern society" and the decline of the power of government and corporations. In the information society "netizens" will "not merely be the recipi-ents of information and persuasion" but will be able to put forward their own opinions and demands.[17] The Internet is "strengthening individual brainpower and fundamentally changing the way we think and act," leading to the intellectual empowerment of netizens. Indeed, "the power to control society that was once in the hands of governments, big businesses, and mass media is now held by netizens."[18]

Netizens, moreover, are likely to be "much more active than citizens, especially the twentieth-century citizens who came to be called 'the masses,' in searching for and acquiring products and information, and also in producing and transmitting.

They won't be content just to be passive couch potatoes. They will actively course from one network to the next, searching for information and products that they personally find interesting." Here he gives a clue as to what he thinks the real issues are: "searching for and acquiring products and information."

While Kumon articulates a fairly standard, consumer-driven model of the Internet, his vision echoes the utopian technological aspirations of commentators like Howard Rheingold who believes that the Net's virtual communities have the power to "revitalize the public sphere" and "citizen-based democracy."[19] According to Rheingold, people fleeing the growing "decay of the social fabric of existing communities" in the real world are increasingly drawn to the revitalized ideals of democracy and participation in the virtual.

Comforting as these visions are, they are inflected with the technological determinism and techno-utopianism that seems to characterize much commentary on the Net; in the belief, as Kevin Robins puts it, "that a new technology will finally and truly deliver us from the limitations and frustrations of this imperfect world."[20]

> We can all too easily think of cyberspace and virtual reality in terms of an alternative space and reality. As if it were possible to create a new reality which would no longer be open to objections like that which has been left behind. As if we could substitute a reality more in conformity with our desires for the unsatisfactory real one. The new technologies seem to offer possibilities for re-creating the world afresh."[21]

A much more difficult proposition is to position the development of the Internet, as any other technology, within the context of prevailing economic, political and cultural structures. When we do, and look at the actual state of play between the embattled forces of the left and the establishment in Japan, the view looks less promising, despite an expanding range of online activities and campaigns.

THE PROGRESSIVE NET

Progressive cyberspace in Japan has been hampered by the surprisingly late development of cheap public access to the Net and the timidity of the left in using new technology. Aside from the not inconsiderable problems of developing software to handle Chinese characters, the arcane business practices of state carrier Nippon Telegraph and Telephone, strict regulations, and early investment by NTT in the slow and expensive narrow-band Integrated Services Digital Network (ISDN) has kept Internet connectivity rates comparatively low. The tiny undernourished Japanese cable sector has also helped keep network access mostly in the hands of NTT.[22] Connectivity exploded after 1998, largely through the use of mobile telephones and increasingly cheap broadband technology such as digital subscriber lines, but it still lags behind the US, much of Europe and even South Korea.[23] Even as the technology becomes cheaper and more available, the left has been slow to use it. As Koshida Kiyokazu, Director of the NGO Pacific Asia Resource Center (PARC) explains:

Japan is dominated by the old framework – written papers rather than online information. Left wing groups here consider their main job putting out a news-paper or a journal. It's partly cultural and partly technological. Even I am a little bit suspicious of it. We use it now for gathering petitions and signatures on mailing lists but we cannot, for instance, check who is really sending us these signatures.[24]

As connectivity has increased, however, progressive groups have steadily moved online to escape the stifling centralization and cartelization of the national media. It is Rheingold's phrase, "sprawling colonies of micro-organisms," that comes most readily to mind when reviewing the burgeoning Japanese Net today, with thousands of home pages documenting environmental, consumer and women's rights issues, anti-war and anti-nuclear activity, minority and foreigners' rights campaigns, movements to stop the building of dams, reactors, roads and reclamation projects, and individual struggles against repression, discrimination or injustice.

The type of cyberactivism which has enthralled Rheingold and other Internet visionaries could be seen during the war in Afghanistan in September–December 2001. With most of the mainstream media ignoring anti-war sentiment and some of the largest demonstrations since the Vietnam conflict, including an 8,000-strong peace rally in Hibiya on 18 October, and 25,000 in Yoyogi Park on 23 October, online peace groups took the initiative.[25] An organization called Chance! spread news of protests and e-mailed an anti-war statement to the country's main media stations immediately after air strikes began. Peace Boat took to the streets of Tokyo and Osaka with computers between 26 September and 1 October and asked passers-by to e-mail Prime Minister Koizumi's office.[26] Two days after the September 11 attacks, a translation of an anti-war statement by the US group War Resisters League was circulating in Japanese e-mail accounts and on bulletin boards.[27] Global Peace Campaign collected $100,000 online from Japan and the US in the three weeks after the bombing to publish an anti-war letter in the *New York Times*.[28] Asia University lecturer Ishizaki Manabu and 150 fellow academics issued an appeal declaring new legislation allowing the Self-Defense Forces to support the war on terror unconstitutional and sent it abroad in English, Chinese and Korean. "Ordinary citizens around the world are not relying on the established media but are taking action using the Internet," said Ishizaki.[29] If nothing else, the Net allowed Japanese people to "distinguish US policy from ordinary American people when criticizing retaliation," says philosopher Morioka Masahiro.[30]

The flurry of anti-war activity – which included dozens of small demonstrations outside train stations, US army bases and the national Diet building, sometimes on the same day – illustrates, however, both the strength and weakness of progres-sive forces in Japan which tend to be pluralistic and heterogeneous, but also fragmented and uncoordinated. Although the Net allowed the pacifists, feminists, socialists and religious organizations involved, perhaps for the first time, to understand the scale of their own activity, a single large demonstration on the nation's capital of the kind that stopped the center of London in mid-November was never likely.[31] As PARC's Koshida says, "The Internet is not uniting the left in Japan but linking them."[32]

PARC, an NGO that supports struggles and citizens' movements throughout Japan and Asia, runs the non-profit portal Japan Computer Access-Net, arguably the most important progressive Web site in Japan.[33] Set up by former PARC member Inyaku Tomoya, JCA-Net is probably the most serious attempt to provide some sort of electronic hub for the progressive left, and it helps illustrate some of the problems faced by Japanese cyberactivists.

The initiative was inspired by the three years Inyaku spent in Brazil in 1991–4 working at the Ibase Institute, a non-profit think tank, which had its own Web provider.[34] After taking part in the Environmental Development Conference in Brazil in 1992 and observing the activities of the Association for Progressive Communications, which networked NGOs from all over the world, Inyaku decided Japan needed something similar. It was not until 1997, however, that he managed to get it off the ground – seven years after APC was set up – hampered by a lack of money, expertise and technical and translation skills. While NPOs in the US and other countries often have tax-exempt status, their Japanese counterparts must struggle by on subscription fees and donations and enlist the help of like-minded foreigners to help with translations. Male volunteers, exhausted by the demands of the system, are in particularly short supply.

As the main Japanese portal for the global Drop the Debt campaign, Jubilee 2000, PARC collected 200,000 signatures, many online, which helped publicize and politicize the issue of third-world debt but which ultimately failed to move the government. A range of citizens groups, pitted against the useless or harmful public works construction projects central to pork-barrel politics in Japan, have challenged the state, using a combination of online campaigns to network with national and international environmentalists, and old-fashioned efforts to forge broad-based political alliances. An NGO called Save Fujimae Association played a part in persuading the Environment Agency to scrap a plan to convert the Fujimae tidal flats in Aichi Prefecture into a garbage dump. A reclamation project in Isahaya Bay in Nagasaki Prefecture was scaled down after a campaign by citizens' groups using petitions, e-mail and a media campaign, which successfully touched on fears about the national seaweed supply.[35] Citizens' groups outraged over the use of MOX fuel in a nuclear plant in Kashiwazaki-Kariwa, and against the building of a new airport in Kobe have organized cyber-campaigns, including mass online petitions and e-mails, although with mixed results.[36]

This is progressive cyberspace at its most impressive, what McChesney calls "the enormous and mostly uncensorable soapbox," giving a voice to countless marginalized groups.[37] Nevertheless, the failure of the progressive left to success-fully challenge the state on virtually any major recent issue is quite striking. Government revisions allowing the Self-Defense Forces to expand their role in support of the war in Afghanistan sailed through the Diet in December 2001, despite the anti-war cyberactivists.[38] A law allowing the police to tap citizens' telephones (the *Tsūshin bōjuhō*) was passed in 1999, easily ignoring the campaigners' efforts to block it.[39] More legislation ordering the *hinomaru* rising-sun flag and *kimigayo* national anthem to be displayed and sung in schools was passed the same year, despite the presence of the Anti-Hinomaru and Kimigayo group.[40] There has been no Japanese equivalent of the "anti-globalization" protests of Seattle or Genoa despite a decade of economic slump and restructuring.

This lack of obvious success by online progressives in Japan might be compared to the activities of citizens' groups and networks of social activists using the Internet in Korea, which have aggressively challenged the state for reforms in recent years.[41] Technology, however, is just one factor there among many, not least the legacy of the mass protests and demonstrations against the Chun Doo-hwan regime of the 1980s, and the effects of the 1997/8 Asian currency crisis which are part of the landscape of Korean politics, providing much of the energy and drive of progressive campaigns. Japanese activists struggle against a background of more historically distant successes, the closed and restricted nature of the Japanese political system, and a much more confident political establishment. It is these kinds of very different contexts that tend to be ignored by those who see the Net as a panacea to the problems that beset the political left.

THE REGRESSIVE NET

It is nevertheless tempting to extrapolate from the advances in the work of the community of activists who have gone online in Japan in the last five years to predict that theirs will be the authentic Internet experience, indelibly molding cyberspace with non-commercial, progressive ideals. But any balanced assessment of the relative weight of political forces needs to take account of the regressive political and commercial interests that shape the Net. That the uncensorable soapbox can also host ultra right-wing groups as well as progressives is a given, as the US experience shows.[42] Japan's ultra-nationalists are active online, sometimes mimicking the tactics of intimidation they employ in the real world.[43]

As Ducke's chapter in this volume shows, the campaign by radical nationalists in the summer of 2001 to promote the use of nationalist-tinged junior high school history and civics textbooks also demonstrates that networked community activism is not the exclusive preserve of the progressive left. The organization behind the textbooks, *Atarashii Rekishi Kyōkasho o Tsukuru Kai*, or the Japanese Society for History Textbook Reform, is a group of nationalist scholars and writers that aims to end the "perverse masochistic" teaching of history which, they say, has taught a generation of school children to hate their own country. The Society used the Internet extensively and relied on networks of about 10,000 local sympathizers linked to euphemistic "concerned parents" groups to lobby local education boards to use textbooks they had written themselves.[44] The anti-textbook campaign was led by a group of about 200 citizens groups called Children and Textbooks Japan Network 21 which, during the height of the campaign, was receiving up to 20,000 hits a day on their Web site.[45] In the end, just a handful of schools adopted the textbooks, mostly under political pressure from Tokyo Governor Ishihara Shintarō.[46]

While the textbook row and other disputes may help reinforce the illusion of a vast Foucauldian terrain of micro-struggles and resistance, where the overall manipulation of power has been disconnected from broader economic and political structures, they also demonstrate that progressive citizens' groups are vastly outweighed by forces backed by the state and corporate capital. The Society for Textbook Reform, struggling to reverse years of progressive textbook campaigns, was backed by corporate donations and supported by a significant section of the Ministry of Education, the LDP and conservative national press, giving it a leverage its opponents could not hope to match.[47]

Moreover, the notion that cyberactivists can simply bypass, ignore or outwit the state will eventually come up against the reality of an increasing trend towards surveillance and control by the authorities. Old-fashioned methods cannot be ruled out, such as the police raid on PARC's offices in May 2001. This was "an information trawling exercise to see what NPOs are up to," according to Koshida.[48] Much more likely, however, are newer approaches to network surveillance, documented by Ogura Toshimaru.[49] In addition to legislation drawn up under the rubric of "anti-terrorism," these approaches are anchored by guidelines created by the country's industrial and telecommunications bureaucracies, principally MPT, the former Ministry of International Trade and Industry and its auxiliary organization, the Electronic Network Consortium, comprising online services providers. This approach, according to Ogura:

> typifies the traditional method of bureaucratic domination in Japan, which seeks to set concrete policies and laws from top-down. The Internet is supposed to be an interactive communications technology that obviates the behind-the-scenes consensus building among the few at the top of a hierarchy. But the possibility of such a new form of consensus building is being denied by the old system.[50]

The attempt, only beginning, by the state to rein in the previously anarchic development of cyberspace in Japan will pit much larger and more powerful political forces against the scattered band of cyberactivists.

THE COMMERCIAL NET

A much more important factor, however, in deciding what form the Internet takes from here is the commercial interests that increasingly dominate it. If the US experience is anything to go by, these interests will push the non-profit progressive Net to the margins of cyberspace, as they indeed have started to do since the Japanese Internet was commercialized in 1993. Broadcasters, advertisers and promoters are already focusing on the possibilities of broadband technology. Music production company Avex Inc., for instance, like many other firms has found that the Internet can be lucrative after they netcast a concert of pop singer Hamasaki Ayumi in July 2001. Maeda Harumasa, director of Avex Network Inc.'s marketing department was "astounded" by the potential he found after 24,000 people logged on to watch, many at "1,600 yen a pop."[51] Japan Net Bank, an affiliate of the Sumitomo Mitsui Banking Corp. and a number of other giant firms, also pronounced itself "surprised" by the success of its operations in winning half a million Internet banking accounts in just eighteen months.[52] Many of the top portals in Japan are sponsored by the country's electronic giants, including Fujitsu's *@Nifty*, NEC's *Biglobe* and Sony's *So-net*. Together with telecommunications giant NTT, which is behind *Goo*, Microsoft's *MSN* and *Yahoo*, these firms take up the bulk of the top ten.

As in the real world, these firms would often rather cooperate to shut out smaller competitors than compete to offer the best services. In April 2002, for example, NEC, Matsushita, KDDI and Japan Telecom formed an alliance to create an ISP with 10 million users.[53] Fujitsu and Sony also announced plans to consolidate *@Nifty* and *So-net*, with a combined seven million users, in December 2001.[54] In the 90 minutes or so the average surfer spends online every day in Japan, most limit their choices to one of these top sites, with over 50 per cent of the country's 20 million surfers visiting Yahoo's site at least once a day.[55] The content is mostly the usual tabloidized package of entertainment, sports, shopping tips and banner advertising. When questioned on what they want from the Net, most users reply motion pictures, music, TV, shopping and games, in other words, simply an extension of what they currently look for in existing media.[56] Moreover, many providers of other kinds of services must negotiate with and use these top portals if they are to have any chance of success. The notion that netizens are likely to be much more active than citizens in searching for information is hardly borne out by this image of cyberspace increasingly coming to resemble the bland shopping experience of the oligopolized real world. The evidence so far seems to show that rather than coming to replace other forms of media, the Internet in Japan, as elsewhere, is being added to the existing fare.[57] Given the limited time most people have in the day, the much richer and more heavily advertised commercial content is simply drowning out the non-profit sector.

The IT industry vision of the future growth of cyberspace is clear and expounded daily in the pages of the business press inside and outside Japan. The Net is the Trojan horse that will allow manufacturers fleeing from slumping profits in older industrial sectors to colonize new areas of what they call the "service economy," in the hope of selling equipment along with new services such as securities and entertainment software. A prime example of this strategy is Sony, which has branched out from its faltering electronic business to offer banking, insurance and other financial services under Chairman Idei Nobuyuki, who sees the Net as the key to the company's future growth.[58] Takuma Otoshi, president and chief executive officer of IBM Japan Ltd, laid out the stakes in April this year when he said,

> Numerous types of equipment, such as cellular phones and personal computers, are connected to the Net. In a decade, the global communications network will connect as many as 10 times the number of individuals and 1,000 times the equipment as current levels.[59]

All this is not to deny that the Net will continue to be important as a source of alternative information and as a data bank of historical events and struggles, particularly with news and book distribution so tied to large monopolies in Japan.[60] It will help dispel the sense of isolation that afflicts so many progressive struggles and encourage activists to link with national and international groups and forge new alliances and networks. And as the costs of connectivity fall with Japan's "broadband revolution" it will become easier for activists to go online. Like all technologies, however, the phenomenal expansion in computer power and telecommunications networks that underlies the Internet should be understood dialectically, as helping to both repressively transform the "rhythm, texture and experience" of everyday life, through for example surveillance and the further extension of the market into daily life, while also providing the tools for new forms of resistance to this transformation.[61]

CONCLUSION

I have tried to challenge here the futurologists' utopian ideal of an anarchic, free-wheeling cyberspace without gatekeepers, set to tip the balance in favor of the "little guy" because it fails to account for the way technological breakthroughs must always negotiate with existing structures of power and privilege. In Japan, these structures include an employment system that drains the time and energy of most workers and helps prevent political organization, a technological sector that has prioritized corporate users over domestic customers, a political system that has so far been able to embrace, neutralize or co-opt most serious progressive challenges, and a corporate oligopoly with years of practice in spawning new industrial sectors without altering its essential structure. For those who remain indignant at the failures of the current system and who continue to believe it can be changed for the better, the Internet might better be understood as a tool rather than a panacea.

ACKNOWLEDGMENT

Many thanks to Brian C. Folk for his help in compiling this chapter.

NOTES

1 K. Van Wolferen, *The Enigma of Japanese Power: People and Politics in a Stateless Nation*, Tokyo, Tuttle, 1993, p. 40.
2 Known as the Japan Socialist Party until February 1991 and then the Social Democratic Party of Japan until January 1996. Now called The Social Democratic Party (Shamintō).
3 While undoubtedly tainted by its past, and despite moving to the right on major issues since the fall of the Soviet Union, the JCP is seen by some as an alternative to the mostly conservative parties of the Diet because of its opposition to issues like the consumption tax or US bases. See T. Kato, "From a class party to a national party," *AMPO*, vol. 29, no. 2, March 2000, pp. 11–13.

4 A. Asada, "A left within the place of nothingness," *New Left Review*, 5, September–October 2000. Online. Available HTTP: *http://www.newleftreview.net/Issue5.asp?Article=02* (19 April 2002).

5 For an account of the Security Treaty protests and their aftermath see, W. Sasaki-Uemura, *Organizing the Spontaneous: Citizen Protest in Postwar Japan*, Honolulu, University of Hawaii Press, 2001.

6 *Beheiren*: *Betonamu ni Heiwa o! Shimin Rengō* (Citizens' Federation for Peace in Japan) was a broad-based peace movement that formed in 1965 and disbanded when the war ended whereas the *Zenkyōtō*: *Zengaku Kyōtō Kaigi* comprised independent councils of student activists that attempted to distinguish themselves from traditional left-wing sects.

7 Local farmers, later backed by radical students and other sections of the New Left, formed an anti-airport alliance after the government unilaterally announced plans in the late 1960s to build what is now the main international gateway into the country on the site of their land around Narita in Chiba Prefecture. Known as the Narita Struggle (Narita Tōsō), the dispute, which went through a particularly violent phase in the early 1970s, rumbles on.

8 Chūkakuha: *http://www.zenshin.org/*; Kakumaruha: *http://www.jrcl.org/* (3 February 2002). Both of these groups have Japanese and English Web sites.

9 The former revolutionary group Senki Kyosando (also called The Bund *http://www.bund.org/*), for instance, has abandoned Marxism for environmental and anti-war issues (personal interview with Kaoru Morishita, *Senki* editor, 12 December 2001). Aum Shinrikyō is the doomsday sect that killed twelve and injured thousands after it released nerve gas on the Tokyo subway system in March 1995. It has subsequently changed its name to Aleph and continues to recruit and operate.

10 Foreign Press Center Japan, *Facts and Figures of Japan*, Tokyo, 2000, p. 58.

11 It could be argued that the country's wealth and the economic transformation of the material lives of most citizens have naturally softened the demands of progressives and undermined the need for the sort of collective action of the past. This notion conflicts with evidence that income disparities, poverty and homelessness are once again on the rise. See Keizai Kikakuchō Kokumin Seikatsukyoku (Economic Planning Agency Citizens' Lifestyle Bureau), *Shinkokumin Seikatsu Shihyō* (People's Life Indicators), Tokyo, 2000. Also Y. Takao, "Welfare state retrenchment – the case of Japan," *International Public Policy*, vol. 19, no. 3, pp. 265–92; and S. Strom, "Japan suddenly feels widening social gap," *International Herald Tribune*, 5–6 January 2001, p. 1.

12 See *http://www.igc.org/ohdakefoundation/npo/npojp.htm* (2 February 2002). The index includes an overview of the non-profit sector in Japan.

13 T. Morris-Suzuki, "For and against NGOs," *New Left Review*, 2, March–April 2000. Online. Available HTTP: *http://www.newleftreview.net/* (23 December 2001).

14 See M. Castells, *The Information Age: Volume I, The Rise of the Network Society*, Malden, MA, Blackwell Publishers, 2000.

15 See, for example, *http://ishihara.yamero.net/* (8 February 2002), a Web site dedicated to forcing the resignation of Tokyo Governor Ishihara Shintarō, and *Tokyo Progressive*, a one-man operation run by Paul Arenson that offers mainly foreigners in Japan a left perspective on global and Japanese political issues, *http://www.arenson.org/* (8 February 2002).

16 S. Kumon, *The Internet and the Information Revolution*, Tokyo, Global Communications Platform, 1998. Online. Available HTTP: *http://www.glocom.ac.jp/lib/kumon/Toronto426.e.html* (2 May 2002).

17 Kumon, *The Internet and the Information Revolution*.

18 S. Kumon, "Netizens and the information society at a crossroads: facing the emergence of cyber-activism," Tokyo, Global Communications Platform, 2000. Online. Available HTTP: *http://www.glocom.org/opinions/essays/2000011_kumon_glocom_forum/index.html* (22 February 2002).

19 H. Rheingold, *The Virtual Community: Finding Connection in a Computerized World*, London, Secker & Warburg, 1994, p. 14.

20 K. Robins, "Cyberspace and the world we live in," in D. Bell and B. Kennedy (eds), *The Cyber-cultures Reader*, London, Routledge, 2000 p.78.

21 Robins, "Cyberspace and the world we live in," p.92.

22 See D. McNeill, "Mission impossible? Japan's broadband challengers take on NTT," *Japan Inc.*, June 2001. Online. Available HTTP: *http://www.japaninc.net/mag/comp/2001/06/jun01_mission.html* (6 February 2002). The Internet began life in Japan in 1984 as a research network linking the University of Tokyo, Keio University and the Tokyo Institute of Technology – the Japan University/Unix NETwork (JUNET), before being commercialized in 1993. As in the US, there was little if any public debate about whether the private sector should become involved.

23 See Ducke's chapter in this volume for more about the South Korean government's proactive stance toward the Internet.

24 Personal interview, PARC Office Tokyo, 9 November 2001. PARC abandoned the publication of its *AMPO* journal in 2001 in favor of a vaguely defined and still unborn "electronic journal." The result, says Mark Selden, is that there is now "no significant publication offering progressive perspectives in English on contemporary Japan except such specialized and occasional ones as those growing out of the comfort woman symposium." Personal communication, 21 April 2002.

25 Figures quoted are those given by the organizers.

26 Chance!: *http://give-peace-a-chance.jp/index.shtml*; Peace Boat: *http://www.peaceboat.org/* (3 February 2002). See T. Kawahira, "Mejia o katsuyō shita shimin undō no ugoki" (Civic movements increasingly make use of the media), *Shūkan Kinyōbi*, no. 389, 23 November 2001, pp. 22–3.

27 See M. Morioka, "How did Japanese netizens respond to the World Trade Center attack?" on *International Network for Life Studies*. Online. Available HTTP: *http://www.lifestudies.org/wtc01.html* (8 February 2002).

28 G. Nees, "Letter to President," *New York Times*, Section A, 9 October 2001, p. 23. Global Peace Campaign, *http://www.peace2001.org/* (3 February 2002).

29 Kawahira, "Mejia o katsuyō shita shimin undō no ugoki."

30 M. Morioka, "How did Japanese netizens respond to the World Trade Center attack?" A list of WTC-related Japanese Web sites can be found at *http://www.lifestudies.org/wtc02.html*, see also, Katō Tetsuro's home page at: *http://www.ff.iij4u.or.jp/~katote/Home.html* (8 February 2002).

31 See J. Vidal, "Another coalition stands up to be counted," *The Guardian*, 19 November 2001, p. 13.

32 Personal interview with Koshida Kiyokazu.

33 PARC: *http://www.jca.apc.org/parc/* (7 January 2002). Japan Computer Access-Net: *http://www.jca.apc.org/* (3 February 2002).

34 Personal interview with Tomoya Inyaku of Greenpeace Japan, Tokyo, 20 November 2001.

35 See M. Corliss, "Wetland conservation efforts gain ground," *Japan Times*, 6 September 2001, p. 2. Save Fujimae Association: *http://www2s.biglobe.ne.jp/~fujimae/english/*; Isahaya: *http://kashinomori.com/higata/higata.html* (3 February 2002).

36 Citizens' groups collected about 310,000 signatures asking that a plebiscite be held on the Kobe airport project in 1998 but it was rejected by the city assembly and the project is going ahead in the face of growing opposition. Anti Kobe Airport Web page: *http://www.kobe-airport.gr.jp/* (3 February 2002).
    The foreign community has also been active online. Debito Arudou, known as Dave Aldwinckle before he took Japanese citizenship, has used the Net to draw attention to the problems *gaijin* (foreigners) face in Japan. His home page is at: *http://www.debito.org/* (5 February 2002). Currently involved in a discrimination lawsuit against the owner of a public bathhouse, Arudou and his colleagues organized a successful e-mail and fax campaign in 1996 against Asahi Television's popular *News Station* program for its discriminatory use of the word "*gaijin*." See *Japan Observer*, *http://www.twics.com/~anzu/* (2 May 2002) which attempts to bridge activists' concerns in Japan and abroad.

37 R. McChesney, *Rich Media, Poor Democracy: Communication Politics in Dubious Times*, Urbana, University of Illinois Press, 1999, p. 175.

38 See editorial "The Diet that set a precedent," *Japan Times*, 11 December 2001, p. 7.

39 See *http://www.zorro-me.com/miyazaki/tocho-index.html* (8 February 2002).

40 See *http://tokyo.cool.ne.jp/kunitachi/* (8 February 2002). Debate over the status of the rising-sun emblem (*hinomaru*) and the song celebrating the Emperor's reign (*kimigayo*)

has continued for most of the post-war period against the background of their symbolic role during Japan's Pacific War campaign. Despite the opposition of a coalition that included teachers' unions, pacifists and religious groups who argued that they were unacceptable remnants of Japan's militarist past, the government controversially passed legislation giving them legal status as the official national flag and anthem in 1999.

41   See Laxmi Nakarmi, "The power of the NGOs," *Asiaweek*, vol. 26, no. 511, February 2000. Online. Available HTTP: *http://www.asiaweek.com/asiaweek/magazine/2000/0211/nat.korea. lobby.html* (3 May 2002).

42   See W. Harkavy, "Left behind: the radical right got wired up fast. When will progressives catch up?" *The Village Voice*, 16 May 2000. Online. Available HTTP: *http://www.flipside.org/vol3/ may00/00my16a.htm* (3 May 2002).

43   Local independent councilor Wakatake Ryōko, for instance, was forced to close her online bulletin board in early 2000 after rightists posted a stream of messages threatening sexual violence when she publicly criticized Ishihara Shintarō for remarks about "Sankokijin," a derogatory term for non-Japanese Asians living in Japan. See "Reipu suru to jyosei shigi o netto de kyōhaku" (Internet threat of rape to female councillor), *Asahi.com*, 20 April 2000. Online. Available HTTP: *http://www.asahi.com/tech/jiken/20000420b.html* (23 December 2001). For a sketch of the political connections between the ultra and mainstream right, see D. McNeill, "An unwelcome visit from the Uyoku," *New Statesman*, 26 February 2001, pp. 32–3. A list of political groups (left and right) can be found at: *http://dir.yahoo.co.jp/Government/Politics/Organizations/* (8 February 2002).

44   See D. McNeill, "Marching to war over history," *South China Morning Post* (Focus Section) 17 June 2001, p. 2. Japanese Society for Textbook Reform: *http://www.tsukurukai.com/* (8 February 2002).

45   Personal interview with Tawara Yoshifumi, Secretary General of Network 21, 19 November 2001. Children and Textbooks Japan Network 21: *http://www.ne.jp/asahi/kyokasho/net21/ index.htm* (8 February 2002).

46   See "Most schools ignore disputed text," *Japan Times*, editorial, August 2001, p. 2. A Web site dedicated to forcing the resignation of Ishihara Shintarō can be found at *http://ishihara.yamero. net/* (8 February 2002).

47   Personal interview with Tawara Yoshifumi.

48   Tawara Yoshifumi.

49   T. Ogura, "Nihon ni okeru saibasupēsu no kisei" (The regulation of cyberspace in Japan), in T. Ogura and Y. Kuihara (eds), *Shimin undō no tame no intānetto* (The Internet and civic movements), Tokyo, Shakai Hyōronsha, 1996, pp. 120–57. See also his *Kanshi shakai to puraibashi* (Privacy and the Surveillance Society), Tokyo, Impacto shupansha, 2001.

50   T. Ogura, "Nihon ni okeru saibasupēsu no kisei," p. 236. As early as 1996, in the relative infancy of the Internet in Japan, the authorities arrested the user of a Web site hosted by *Bekkoame.com*, then Japan's largest Internet service provider, on suspicion of uploading indecent material to his Web site.

51   T. Koyama, "Widening Net," *The Nikkei Weekly*, 12 November 2001, p. 3.

52   "Japan Net Bank's Accounts Reach 500,000," *Nikkei Net*, 10 April 2002. Online. Available HTTP: *http://www.nni.nikkei.co.jp/* (3 May 2002).

53   "Analysis: NEC, KDDI, Others Eyeing New ISP Services," *Nikkei Net*, 23 April 2002. Online. Available HTTP: *http://www.nni.nikkei.co.jp/* (3 May 2002).

54   "Analysis: NEC, KDDI, Others Eyeing New ISP Services."

55   See Nielsen Net Ratings at: *http://www.netratings.co.jp/press_releases/pr_261101.html* (3 May 2002).

56   See, for example, "Statistics," *Japan Inc.*, September 2001, p. 67. Statistics compiled by Chiaki Kitano.

57   "Sony sees financial future in linked online businesses," *Nikkei Net*, 15 April 2002. Online. Available HTTP: *http://www.nni.nikkei.co.jp/* (3 May 2002).

58   See "Media usage: Net users vs non Net users," *Japan Inc.*, April 2001, p. 78.

59 "Broadband era transforms traditional philosophies," *Nikkei Net*, 15 April 2002. Online. Available HTTP: *http://www.nni.nikkei.co.jp/* (3 May 2002).

60 See L. Freeman, *Closing the Shop: Information Cartels and Japan's Mass Media*, Princeton, NJ, Princeton University Press, 2000.

61 A. Ross, "Hacking away at cyberculture," in D. Bell and B. Kennedy (eds), *The Cybercultures Reader*, London, Routledge, 2000, pp. 264–5. The activities of author and academic Kogawa Tetsuo, a veteran of the 1980s "free" pirate radio movement in Japan, are worth a mention here. Kogawa has spent years trying to train his students to make and operate their own communication equipment. He currently organizes a radio program via the Internet which broadcasts once a month and deals with the challenge of state and corporate power. He says, "The Internet is popular and has potential, and that is what makes it interesting. But will it change the world? I don't think so. Better to think of it as part of a much larger struggle to change things. History is not about big events. History moves through everyday, small, sometimes boring, incremental steps." Personal interview, 16 November 2001, Goethe Gallery, Tokyo. See also, Paul Murphy, "Veteran pirate brings leftism to Net night owls," *Asahi Shinbun*, 15–16 December 2001, p. 29. His show can be found on *http://anarchy.k2.tku.ac.jp/* (5 February 2002).

12 Creating publics and counterpublics on the Internet in Japan

Vera Mackie

CYBERDEMOCRACY

One theme of writings on the Internet is the possibility of creating new publics and new political communities through the use of electronically-mediated communication.[1] It has been suggested that the Internet allows activists to overcome "the politics of location" and to form webs of solidarity which cross national boundaries. We have seen, for example, extensive use of the Internet in the form of pacifist e-mail petitions in response to the US-led "war on terrorism" in the aftermath of the September 11 terrorist attacks on New York and Washington. Indeed, in the years leading up to these events, the Internet was an important site for the dissemination of information and the mobilization of international public opinion on such issues as "ethnic" riots and sexual violence in Indonesia, political repression in East Timor, the oppression of women by the Taliban regime in Afghanistan, the destruction of world heritage sites by the Taliban in Afghanistan, the campaign for a Japanese government apology and compensation to survivors of the wartime military prostitution system in East and Southeast Asia, the campaign for a government apology for the treatment of the "stolen generation" of indigenous people in Australia, campaigns against the mandatory use of the national flag and national anthem in Japan, and in many more localized campaigns. Others, however, are more skeptical of the possibilities of the Internet as a medium for local, national and transnational political action.[2]

Other commentary on new communications technology has focused on the ways in which some governments have attempted to control the flow of information. The restrictions imposed on the News Corporation's Sky Television satellite broadcasting service in India and China, and the attempts by governments in China, Singapore and other countries to restrict access to the Internet are examples of such anxiety about the flow of information through use of the new technologies.[3] Most governments attempt some kind of regulation of the content of the Internet and the uses of electronically-mediated communications technologies.

The Internet has also been an important venue for the dissemination of scholarly information through search engines, databases and the publication of electronic journals, which can overcome some of the high costs involved in producing printed publications.[4] While some journals appear solely in electronic format, in many cases publication now takes on hybrid forms, with conventional printed publications being supplemented with the use of electronic databases, indexing and alerting services, with selected articles being placed on the Internet.[5] For others, the Internet is an important site for the production of new cultural and social identities, and the imagining of alternative realities, as explored in several chapters in this volume.[6]

In this chapter, however, I will focus on another side of the Internet. New communications technologies also provide governments with the possibility of disseminating information cheaply and effectively. Indeed, the Internet may become a tool of nation-building, as much as a tool of oppositional politics. It is likely that a government which chooses to use its resources in order to disseminate information over the Internet will be able to do so much more effectively than non-governmental organizations with limited funds, resources, and staff. This will be particularly effective in a population with high literacy rates and a high diffusion of computer hardware and technical literacy. Benedict Anderson has pointed out the importance of print capitalism, the standardization of written versions of vernacular languages, and the development of mass education in the creation of nationalist identities in an earlier age.[7] Compared with older forms of publishing, effective electronic communication requires a population with computer literacy and access in addition to conventional skills in reading and writing. In this chapter, these issues will be explored through an examination of the use of the Internet by Japan's Liberal Democratic Party (LDP) government under Prime Minister Koizumi in the latter half of 2001, with a particular focus on the Prime Minister's e-mail magazine.

KOIZUMI'S E-MAIL MAGAZINE

Koizumi Jun'ichirō became Prime Minister in April 2001, and reinforced his position through his party's convincing victory in the Upper House election of July 2001. One of the keywords of the Koizumi regime has been "reform" (*kaikaku*). He received unprecedented popularity ratings between April and December 2001, although his stated commitment to reform has antagonized the old guard of the LDP and the bureaucracy. Koizumi has presented himself as a distinctive leader who is willing to break the previous patterns of Liberal Democratic Party rule – in the composition of his Cabinet, in reform of the structure of government departments and ministries, and in his willingness to promote reform of the public corporations which administer such services as the highways and public housing.[8] Nevertheless, by the beginning of 2002, the difficulty of achieving anything more than cosmetic reforms was becoming apparent, in the wake of the removal of Tanaka Makiko, who in her brief tenure as Foreign Minister had clashed with Foreign Ministry bureaucrats.

An important and innovative feature of the Koizumi regime has been the use of the media. Koizumi has used regular appearances in the mass media, a series of town meetings, and most recently an e-mail magazine, in order to construct the image of a government which is willing to rethink the relationship between government and the citizenry.[9] The e-mail magazine may also be seen as one element of the construction of the Koizumi persona. It is possible to buy "Koizumi" products such as rice crackers, keyrings, mascots, or photographic collections. In a poll on the "buzz words" of 2001, several phrases popularized by Koizumi received attention, and the Koizumi regime was named one of the top-ten news stories of 2001 by both the *Japan Times* and the Kyōdō News Group.[10] As far as I know, little academic work has been done on the conscious management of media images by serving Prime Ministers in the Japanese system. The fact that Koizumi's use of the media is the subject of such sustained interest suggests that Koizumi's attention to his media persona has been more explicit and apparent than most of his predecessors. Most existing commentary on the relationship between politicians and the media in Japan has focused on the complicity of the Press Club with tight controls on the dissemination of information on government policies.[11]

In this chapter I will focus on the use of the weekly e-mail magazine from its trial issue in late May 2001 to the last issue of 2001. At the time of writing, further transformations of the magazine were being foreshadowed, but this chapter will concentrate on the evolution of the magazine in the first seven months of its existence. My main focus will be on the use of the e-mail magazine to construct a particular form of "public." These seven months also, however, span a period of important events in both domestic politics and international politics. In addition to the program of domestic restructuring alluded to above, these months included the shock incident where a man randomly attacked primary-school students with a kitchen knife at a primary school in Osaka Prefecture, the panic over cases of "mad cow disease" (BSE) in cattle within Japan, a series of revelations of corruption at the highest levels of the Foreign Ministry, and continued anxiety about the stability of the banking system and the economy as a whole. In the international arena, these months saw the terrorist attacks on New York and Washington DC, the war against Al Qaeda and the Taliban in Afghanistan, important meetings of APEC, G8, and the World Trade Organization, negotiations on the Kyoto Protocol on global warming, and visits by Prime Minister Koizumi to the United States, Europe and several Asian countries. Japan's relations with other countries were also brought into question through Koizumi's attendance at the Yasukuni shrine in commemoration of war dead in August, through new legislation which allowed Japanese Self-Defense Forces to participate in the US-led anti-terrorist alliance, and through an incident involving the sinking of an unidentified ship (allegedly from North Korea) in Japanese waters in December.

One of the professed features of the Internet is its potential for interactivity, and the early editions of Koizumi's e-mail magazine profess a commitment to listening to the voices of its readers. I will thus be interested in examining to what extent the promise of interactivity is fulfilled in this particular electronic publication.

FORMAT

Each edition has a brief column by Prime Minister Koizumi, columns by one or two members of the Koizumi Cabinet, brief reports on the activities of the Cabinet with links to photographs and more detailed reports, and an afterword by the editor of the e-mail magazine. Some issues also include other special reports by cabinet ministers on current issues, or special sections on issues such as the terrorist attacks on 11 September 2001,[12] or the birth of Princess Aiko to the Crown Prince and Princess Masako.[13] From the issue of 6 September each magazine also includes the explication of a "keyword" necessary to understand current political events. From this edition, the Ministers' comments also change character, and are framed as responses to concerns expressed by the public. The columns by Prime Minister Koizumi are linked to his personal profile page,[14] and Ministers' columns also provide links to their own profiles. Finally there is a link to a URL from which readers can send their opinions and messages to the editors of the magazine.[15] From 6 September each issue includes a column on some aspect of the Prime Minister's official residence. From 23 August audio is added, from 18 October video links are included, and from 6 December it becomes possible to subscribe to have the magazine sent to a mobile phone number in text message format. Visually impaired readers may also access the magazine if they have vocalization software.

The e-mail version of the magazine is plain text with hypertext links to other official home pages, but when the pages are stored as back numbers, accessible from the Cabinet home page, more use is made of colored fonts, borders and photographs. In the e-mail format the editors have chosen the format which will present the fewest problems in e-mail transmission – plain text rather than html format. Although the Internet version is more interesting in presentation, it is still relatively simple, most probably to reduce the likelihood of complaints about excessive download time. Download time is a particular concern for Internet users in Japan, where even local telephone calls are timed. The Cabinet home page and the e-mail magazine also provide a contrast with the home pages of the political parties. All, except the Japan Social Democratic Party (JSDP), have very "busy" designs, with extensive use of frames. The JSDP is the most simple and elegant in design. Most political party home pages also include links to English-language information.[16]

Koizumi's e-mail magazine is all in the Japanese language, suggesting that the major focus is the shaping of public opinion among voters in Japan. By contrast, the Prime Minister's personal profile page has an English translation, as does the list of Cabinet Ministers, Ministers of State and Cabinet Secretaries. Other government departments, in particular the Department of Foreign Affairs, have extensive English translations, and most government departments provide both English and Japanese versions of major legislation. Despite the existence of communities of speakers of Korean, Chinese and other languages within Japan, these languages rarely appear in the national government home pages, although local government areas provide printed information on immigration, alien registration, and welfare matters in Korean, Chinese, English, and other languages where necessary in a particular region. In predominantly English-speaking countries like Australia, the United States and Britain, there seems to be an implicit assumption that disseminating information in English makes it accessible to an international audience. In

Japan, information aimed at a domestic audience will be in Japanese, with English added for international dissemination.

LIONHEART

The most striking aspect of the Koizumi regime is the focus on the distinctive personality of Koizumi himself. This starts with his distinctive hairstyle, apparently achieved through a permanent wave. As the hairstyle is said to resemble a lion's mane, the Prime Minister has various nicknames which play on the English word "lion" or the Japanese word "*shishi.*" The Prime Minister's column bears the title "Lionheart" (the English word in the *katakana* script which is used for non-Japanese words). This title provides a series of associations: the distinctive lion's mane hairstyle, the Prime Minister's fearlessness in pursuing reform, and the masculinity of the Prime Minister as exemplified in the leonine quality of bravery.

The masculinity of the Prime Minister is constantly reinforced. The public image of Prime Ministers in Japan makes less use of familial imagery than comparable English-speaking countries. A comparison of the home pages of Heads of State in the United States, Britain, Australia and Japan reveals much more use of family photographs and mobilization of the image of the spouse of the leader (the so-called "first lady" when the leader is male) in the English-speaking countries.[17] Historically, the "first lady" has had very little public presence in the Japanese political system. It is thus possible for a divorced man like Koizumi to carry out his public role with little comment on his single status. Nevertheless, his personal profile mentions that he is divorced and has two sons.[18] Koizumi's singleness reinforces his image as a "lone wolf" and a bit of a "weirdo,"[19] with single-minded devotion to service to the nation,[20] while mention of his ex-wife and children provides reassurance about his heterosexuality.

Anecdotes in the Prime Minister's column reinforce his masculine persona. There are a series of references to his meetings with President George W. Bush, in an emergency visit to Washington in the aftermath of the terrorist attacks, and in further meetings in connection with the APEC meeting in Shanghai. Several columns by Koizumi and other contributors to the e-mail magazine refer to Koizumi's favorite movie, *High Noon*, and Bush's presentation to Koizumi of a movie poster showing Gary Cooper in his starring role.[21] The masculinity of the two world leaders is reinforced by a photograph of the two jeans-clad men relaxing at the President's retreat, Camp David.[22] These stories also provide an image of the homosociality of world politics, as we see photographs and stories of Koizumi shaking hands, exchanging gifts, and engaging in discussions with Britain's Tony Blair, China's Jiang Zemin, South Korea's Kim Dae Jung, and other world leaders.

The personal side of the Prime Minister is emphasized through anecdotes about his childhood summer holidays,[23] his tastes in music,[24] sport,[25] and food,[26] and his ways of dealing with the stresses of his position.[27] Koizumi is presented as someone who is equally at home with the works of the classical Chinese philosophers,[28] the important figures of modern Japanese history,[29] classical music and rock music,[30] historical novels,[31] and Hollywood cinema. All subjects, however, lead inevitably to a discussion of Koizumi's plans for structural reform. Indeed, "structural reform" may be seen as one of the keywords of the Koizumi regime, and of the e-mail magazine.

Even homely anecdotes on the Prime Minister's memories and personal tastes lead inevitably back to the theme of reform:

> Now that I come to think about it, since I was little, maybe I've always liked things which changed – like creatures which shed their skins or transform themselves. The sound of the cicadas outside my office in the official residence reminds me of when I was a child. This summer it's hot. This is the summer when we make concrete plans for structural reform, when we build the framework for next year's budget … In this time of summer rain and cicadas my thoughts turn to reform. Structural reform will involve doubts and pain. It is one necessary step towards a period of great change and growth.[32]

In similar fashion, the Prime Minister's column just after the terrorist attacks on New York and Washington DC manages both to reinforce the "Lionheart" persona and reiterate the theme of structural reform:

> There are various ways to overcome a crisis [literally: to climb over]. One can bound up steep stairs; one can go round a gentle slope; one could even rock-climb the face of the crisis. What is important is to have a strong belief that the crisis can be overcome, and to make one's own decision after consulting widely. I am proceeding with plans for Japan's structural reform with the same attitude. In the coming week I hope to unveil plans for the "Koizumi reforms." World peace and Japan's structural reform. I will face these problems with the most steadfast resolve.[33]

STRAIGHT TALK

Each edition of the e-mail magazine includes columns by one or more members of the Cabinet, on a rotating basis. The title of this column is *"Daijin no honne talk."* *"Honne"* refers to the expression of personal opinions or emotions, freed from the usual social restraints and niceties which are represented by the phrase *"tatemae."* I have translated the title of this column as "straight talk from the Minister." In the evolution of the e-mail magazine, the ministers' columns shift from very personal, anecdotal comments, to a more explicit concern with readers' questions about policy matters. Ministers' columns also, however, reinforce the two main themes of the magazine: the persona of Koizumi and the government's program of structural reform. In an early column, Finance Minister Shiokawa, an "elder statesman" of the Liberal Democratic Party, presents his knowledge of the Koizumi family. Koizumi Jun'ichirō is the third generation in a line of LDP politicians. Other columns also comment on the Koizumi "weirdo" image.

No matter what policy issue is being commented on, the phrase "*kaikaku*" (reform) appears on almost every page. In a comment on the government's Plan for Gender Equality, the responsible minister refers to the "structural reform of everyday life."[34] Columns by the Prime Minister and Cabinet Ministers present a coherent program of privatization, independence from government support, and structural reform. Reform and restructure of university education is already underway. Privatization of railways was undertaken during the 1980s, privatization of the public housing corporations is foreshadowed, and privatization of the highways is still the subject of controversy. The neo-liberal ideology behind these programs has much in common with the neo-liberalism espoused by governments from the USA, Britain and Australia in recent years.

The recurrence of similar phrases and the repetition of similar anecdotes across the different sections of the magazine suggest a strong degree of editorial inter-vention in the individual columns, and that the tone and content of the magazine is carefully controlled to provide the desired image of informality, colloquiality and accessibility. The inclusion of the column about the Prime Minister's official residence reinforces the theme of accessibility. It is significant that it is the official residence that is focused on, rather than the Diet Building itself. Readers are thus provided with the illusion of access to the daily working life of the Prime Minister, and the daily workings of Cabinet meetings and press conferences. The following description of the press room reinforces this theme of accessibility, while also providing a sense of history.

> The press room is a big window which opens up to the people. The part of the Residence which everyone sees most often on television is probably the press room. Normally, the Cabinet Secretary comes here twice a day to announce Cabinet and government decisions and policies. At other times, the Prime Minister holds press conferences here in order to inform the people directly about important matters. The current press room is on the ground floor of an annex built in 1996. It has an area of around 300 square meters. Some people may have noticed that there are two kinds of curtains used as a backdrop – beige and blue. Most press conferences use the blue curtains, but for the Prime Minister's press conferences – except for emergency situations – the beige curtains are used. The creation of the press room is relatively recent, from the time when the old annex was built in 1962. Until that time the large sitting room in the Residence was used for press conferences. The press room is also a witness to history. The inaugural press conference of each new Prime Minister is held here. This was also the place where then Cabinet Secretary Obuchi announced the name of the new era "Heisei" which succeeded the "Shōwa" period. Currently there are 104 members of the press club, including the overseas media, and there are 74 observers, so there are generally close to 200 reporters in attendance.[35]

In an unusually revealing anecdote, the editor of the e-mail magazine adds a comment in the afterword to this edition, suggesting that former Prime Minister

Nakasone Yasuhiro, like Koizumi, also engaged in some conscious manipulation of his image. This suggests that the theme of media management by serving Prime Ministers is a theme which deserves more sustained study.

> … These days the press conference is conducted with the speaker standing, but they used to be conducted seated. At the side there was a potted pine tree, giving a rather different atmosphere to today. It was Prime Minister Nakasone, who wished to present a more "Presidential" style, who made these changes, which have persisted to the present.[36]

ELECTRONIC GOVERNMENT

Several sections of the magazine reinforce the government's willingness to embrace technological reform. The magazine refers to the government's policies on Information Technology (IT), and the desirability of implementing "electronic government." It also seems, however, that "electronic government" is intimately connected with rationalization and reform of public services. It may well be that the e-mail magazine also serves the purpose of preparing the public for the gradual transfer of government services from face-to-face interaction to electronic interface:

> "Electronic government" refers to a new form of administration which makes use of electronic information, that is, electronic provision of information, electronic submission of official documents and application forms, and electronic sharing and application of information. Currently, most transactions between citizens and government departments are carried out on paper or face to face. When electronic government is implemented, it will be possible to access official information, obtain official documents and forms, and pay fees and charges by Internet from the home or workplace 24 hours a day. We can expect a great improvement in convenience for citizens and businesses. In the "e-Japan strategy" (approved by the IT Strategy Headquarters on 22 January 2001) we have made a commitment to implement electronic government by 2003. In the "Priority Program for Structural Reform," approved on 21 September, we have addressed the issue of the implementation of electronic government as one of the policies which is urgently needed for the speedy implementation of structural reform.[37]

The government has thus affirmed its commitment to the use of information technology. The Blair government in Britain has embraced "e-government" with similar enthusiasm, and has shown some sophistication through the use of the label "on-line citizen's portal." The home page of "online.gov.uk" has links to the "Citizenspace" which includes space for consultation on government proposals and archived discussion forums on current issues related to government policy.[38] By contrast, the Koizumi government makes only limited use of the interactive possibilities of the Internet.

INTERACTIVITY

Issues 6 and 7 of the e-mail magazine include a summary of the readers' responses to the magazine.[39] Subsequent issues include several changes in response to readers' comments. One of the comments was that readers would like more interactivity. From issue 12, the rotating columns by ministers take the form of responses to readers' questions. The editor reports that around 15,000 e-mails from readers are received each week (after a peak of around 20,000 in the first weeks when e-mails focused on problems in subscribing to the e-mail magazine). Of these around 80 per cent are said to be on policy matters, and about 20 per cent make comments on the e-mail magazine itself. Policy matters which invited comment included Koizumi's program of structural reform, welfare, education, the stationing of United States' military bases on Okinawa, reform of the Department of Foreign Affairs, problems associated with the redeployment of retired bureaucrats (*amakudari*), the reform of public corporations, the Hansen's disease issue, the reform of the medical insurance system, and the textbook issue. However, although a list of issues which invited comment is provided, there is no detailed exposition of the content and tone of these comments, nor any indication of whether correspondents were supportive or critical of government policies.

After the creation of the e-mail magazine in June, daily hits on the Cabinet home page averaged 1,140,000, more than five times the previous number of hits before the e-mail magazine was established and fourteen times the average for the previous year.[40] Even after the change in format, however, the magazine only deals with a maximum of three readers' questions per week (one or two replies by Ministers, with the Prime Minister's column occasionally focusing on readers' questions, particularly those from children). Thus, for most readers, there is a very small chance of their specific questions being addressed in the pages of the e-mail magazine. Even allowing for the possibility that Ministers will respond to specific queries offline, this is a long way from making full use of the interactive possibilities of the electronic medium. The institution of bulletin boards or chat rooms on specific topics would allow for a much greater degree of interactivity. The format of information provided by the e-mail magazine is thus almost totally one-directional. Information emanates from the Prime Minister and his Cabinet Ministers; only those readers' questions which are specifically dealt with in columns are responded to publicly; and readers' comments are not made available to other readers of the e-mail magazine. It should perhaps be noted, however, that the "town meetings" provide a forum for more immediate interaction with the Prime Minister and his Cabinet.

All of the Web pages of Heads of State which were examined for comparative purposes included a link to "e-mail your President," or "e-mail your PM." Most also allowed readers to subscribe for regular e-mail updates. The home page of the Australian Prime Minister was the most explicit about how e-mail inquiries and comments would be dealt with, but still failed to make use of the fully interactive possibilities of bulletin boards and chat rooms.[41]

READERS AND PUBLICS

As mentioned above, a summary of the features of the readership was presented in issue 5, and a digest of their comments was presented in issues 6 and 7 of the magazine. This presented a summary of the responses to the readership survey which readers fill out when they subscribe to the magazine, although it is admitted that not all readers respond to all questions. The summary is of the responses received up to 5 July 2001. The readership of the e-mail magazine is 2,107,000, which makes up 5.7 per cent of the 37,230,000 Internet users in Japan, and 1.7 per cent of the total population of 126,920,000.[42] Of the respondents, as many as 71,000 did not specify their gender, but of the remainder more than twice as many males (67.7 per cent) as females (32.3 per cent) subscribe. The gender difference in subscribers to the magazine is greater than that of Internet users in general. Males are 55.1 per cent of Internet users and females 44.9 per cent.

The age of subscribers, as would be expected, seems to be largely those of working age, although the Prime Minister's column in particular often refers to correspondence from children. The figures provided are teens: 88,000 (4.4 per cent); twenties: 517,000 (25.6 per cent); thirties: 625,000 (31 per cent); forties: 430,000 (21.3 per cent); fifties: 241,000 (11.9 per cent); sixties: 95,000 (4.7 per cent); seventy and over: 20,000 (1 per cent); unclear: 92,000. Those in their twenties, thirties and forties are in a greater proportion than would be expected from their overall share of the population; those in other age groups are less than would be expected as a simple proportion of the population. Occupations provided are as follows: company employee: 928,000 (46.3 per cent); management: 130,000 (6.5 per cent); self-employed: 168,000 (8.4 per cent); government employee: 14,000 (6.7 per cent); other occupation: 196,000 (9.8 per cent); student: 212,000 (10.6 per cent); no occupation: 235,000 (11.7 per cent); unclear: 104,000. It is striking that the option of housewife is not given to those who complete the survey: perhaps this is now seen as a discriminatory term? It is also striking that media and journalism are not included as options.

It is reported that 26,000 of the total 2,110,000 subscribers are overseas, but no further details are provided. Of those in Japan, 344,000 (17.3 per cent) are in Tokyo; 212,000 (10.7 per cent) are in Kanagawa; 148,000 (7.5 per cent) are in Osaka; 123,000 (6.2 per cent) are in Saitama; 116,000 (5.9 per cent) are in Chiba; 115,000 (5.8 per cent) are in Aichi; 90,000 (4.5 per cent) are in Hyogo; 64,000 (3.2 per cent) are in Fukuoka; 64,000 (3.2 per cent) are in Hokkaido; and 51,000 (2.5 per cent) are in Shizuoka. As expected, a majority of readers come from the densely urbanized belt which stretches along the Pacific coast from Tokyo and down the Kanto plain to connect with the Kyoto-Osaka-Kobe region.[43]

When readers are addressed or referred to in the e-mail magazine, a range of expressions is used. Perhaps the most often used is *mina-san* (everybody). The use of this term provides a familiar, informal sense, without any specification of gender, status or age. In written communication in Japanese, it is most unusual to address the reader directly as "you." There is no single second-person pronoun which can comfortably be used when the features of addressees are unclear.[44] Subscribers to the magazine are also at times referred to as *dokusha* (readers). This is perhaps the most precise term of reference in the situation, but manages to sidestep any discussion of the politicized relationship between readers of the magazine as voters and citizens and the government which is disseminating the magazine. Readers are also at times referred to as *kokumin*, a term which literally translates as "people of the nation," but which can variously mean "the people," or "the citizens" depending on the context. The alternative term, *shimin* ("citizen;" literally "city-person") which suggests a more active form of citizenship (as in the phrase s*himin undō*, or "citizens' movements"), rarely makes an appearance in the magazine.[45]

There is also, of course, a potential ambiguity concerning the source of the information provided through the e-mail magazine. Koizumi is the Prime Minister who chairs Cabinet Meetings which include members from several parties and Ministers of State from outside the Diet, the leader of the Liberal Democratic Party, and the Member for Yokosuka. As with all information provided by members of a ruling party, there is an ambiguity between disseminating information about government decisions, promoting the political party which the Prime Minister belongs to, and promoting the personal career of the Prime Minister as sitting member. Nevertheless the home page and e-mail magazine referred to in this article emanate from the Prime Minister's Office and the Prime Minister's official residence, and not from LDP headquarters.

PUBLICS AND COUNTERPUBLICS

Andrew Vandenberg has summarized the arguments commonly advanced in favor of the idea that the Internet has the potential to create new forms of political community, with citizens becoming "netizens" or "cybercitizens:"

> [An argument] commonly raised by enthusiasts is that the new technology will undermine the authoritarian, top-down, one-way and one-to-many communication that characterises interaction between the authors and mass readers of books, pamphlets and newspapers, between playwrights and theatre goers, and between electronic broadcasters and radio listeners or television viewers. A democratic, bottom-up, two-way, many-to-many form of broadcasting (or "net-casting") via e-mail, Usenet and the World Wide Web will complement if not supplant traditional relations of communication.[46]

Stephen Lax is skeptical about the democratic potential of the Internet, arguing that those who use the Internet for the purposes of political discussion may well be those who are already more inclined to political participation: well-educated and technically literate males in white-collar occupations. Lax further argues that Internet communication tends to supplement rather than replace conventional forms of mobilization through word-of-mouth, telephone, mail, posters and newsletters.[47]

As we have seen, the Koizumi e-mail magazine espouses the rhetoric of accessibility and interactivity. Despite the claimed commitment to "interactivity," a close examination of the Koizumi e-mail magazine reveals that this publication can best be characterized as "one-to-many," rather than "many-to-many." Other individuals and groups have attempted to use the Internet to create what has been called "many-to-many" communication.

Before considering some alternative uses of the Internet, it is useful to consider the concept of "public." In recognition of the limits of communication in older forms of "public sphere," Nancy Fraser has advanced the concept of a "counter-public" which has the potential to provide an alternative site of communication for those excluded from the mainstream public sphere which is riven by inequalities of class, gender, ethnicity, educational level, wealth, language, literacy and technical competence. It could be argued that the Internet provides the potential for the creation of such "counter-publics," through bringing together people who might otherwise be kept apart through geographical distance or lack of skill in interaction in the mass media or in public forums.

Whether we consider communication in the older forms of public meetings or publications, or whether we consider newer forms of communication over the Internet, what is at stake is the connection between the mainstream public sphere and the alternative spaces of communication (or counter-publics). If there is no means of bringing issues out of the counter-public and into the mainstream public sphere, then discussion in alternative public spaces may be no more than ghetto-ization, with little possibility of achieving the kind of social change which can only be effected through the discussion of issues in the mainstream public sphere. Given the different resources available to governments and to NGOs, the alternative forums of NGO communication may well be drowned out by the weight of official communication channels. This will be an important issue for future uses of the Internet as a site for political communication.

I would now like to briefly mention some alternative ways of using the Internet, which can be seen in non-official Internet sites. One interesting example is the site managed by the group *Issho Kikaku* (which might be loosely translated as "Project Together").[48] This group is interested in reform of legislation and social policies concerning non-Japanese residents in Japan. The group's Web site includes surveys of all the political parties on these issues, and the progress of local campaigns on issues concerning the treatment of non-Japanese residents. This group has also run an especially active list-server discussion group, particularly at the time of parliamentary discussion of proposals for extending voting rights to non-Japanese residents in local government elections (as yet still not implemented). *Issho Kikaku*'s home page is presented in English and Japanese, and achieves a form of many-to-many communication, as discussed by Vandenberg.[49] This group's discussions could also be said to take the form of discussions in a "counter-public" space. As many, but by no means all, of the contributors are non-Japanese residents who are excluded from the parliamentary process, this "counter-public" space forms a particularly vital function for these residents. The group also, however, attempts to affect discussions in the "mainstream" public, through surveys of parliamen-tarians, petitions, media monitoring, and media campaigns.

Vandenberg notes that computer-mediated communication has begun to under-mine assumptions about "the commensurability of a people, a nation and a political community."[50] The activities of *Issho Kikaku* highlight the fact that there are residents within Japan who do not have an official place in the system of parlia-mentary politics, but who still wish to find ways of gaining a political voice as members of local communities. Nevertheless, their claims for legislative protection against discrimination can only be resolved through the system of parliamentary democracy. Nancy Fraser has also called for precision in the discussion of com-munication in the public sphere.

> The idea of the "public sphere" in Habermas's sense is a conceptual resource that can help overcome such problems. It designates a theater in modern societies in which political participation is enacted through the medium of talk. It is the space in which citizens deliberate about their common affairs, hence, an institutionalised arena of discursive interaction. This arena is conceptually distinct from the state; it is a site for the production and circulation of discourses that can in principle be critical of the state. The public sphere in Habermas's sense is also conceptually distinct from the official-economy; it is not an arena of market relations but rather one of discursive relations, a theater for debating and deliberating rather than buying and selling. Thus, this concept of the public sphere permits us to keep in view the distinctions between state apparatuses, economic markets, and democratic associations, distinctions that are essential to economic theory.[51]

It is thus necessary to distinguish between the official communications of governments, the space of civil society, "counter-public" spaces of communication, and the space of the market. We need to take care not to confuse market-driven and state-driven communication over the Internet with the potentiality for the creation of alternative spaces of political discussion.

Another dimension of electronically-mediated communication is illustrated by an examination of home pages which deal with the issue of enforced military prostitution in wartime, and the NGO women's tribunal on violence against women which was held in Tokyo in December 2000.[52] The tribunal attracted over 5,000 participants, including sixty survivors from the Second World War, lawyers and scholars, and spectators from over thirty countries. The tribunal was jointly sponsored by organizations from North Korea, South Korea, China, Taiwan, the Philippines, Indonesia, Malaysia, the Netherlands, Japan and Burma. Home pages dealing with this issue are involved in networks of communication which cross national borders, and this is reflected in the versions of home pages in different languages.[53] The groups involved in this issue may be said to be part of a "transnational counter-public," which involves communication and collaboration across national borders.[54] Their campaigns are directed both at the national govern-ment of Japan and at the international machinery of the United Nations and the International Court of Justice in The Hague. Campaigns around the issue of compensation for survivors of the military prostitution system illustrate some of the dilemmas of transnational communication and political action. These NGOs have successfully used international communication, including the dissemination of information on the Internet and through e-mail petitions, in consciousness-raising exercises which have sensitized an international public to the issue of militarized sexual violence.[55] Victor Koschmann has referred to this kind of activity as "horizontal globalization," in contrast with the "vertical globalization" promoted by the United States' government and its allies, including Japan.[56] However, in their engagements with the legal system of the Japanese nation-state, they have so far been unsuccessful in their suits for government compensation.

These examples illustrate some of the potentialities and limitations of the use of electronic communication in political campaigns directed at national and international audiences, and provide some contrast with the "one-to-many" style of communication employed by the Koizumi government.

CONCLUSION

The Koizumi e-mail magazine makes only limited use of the interactive possibilities of the Internet. Rather, the Koizumi e-mail magazine is a sophisticated way of marketing the Koizumi persona, making maximum use of the ambiguity between Koizumi's roles as Prime Minister, Leader of the Liberal Democratic Party, and sitting member. Through the e-mail magazine, information can be disseminated without being mediated by journalists in the commercial mass media who would be likely to add their own critical gloss and editorial commentaries. The e-mail magazine also provides a subtle way of polling readers' opinions and attitudes, and potentially performs the function of preparing individuals for the gradual transfer of government functions from face-to-face interaction to electronically-mediated communication. In the context of Japanese parliamentary politics, the Koizumi e-mail magazine represents a relatively sophisticated use of electronically-mediated communication. However, there is a certain opacity in the interactions with the readers of the magazine, who are, after all, potential voters. The information flow is largely one-way, without making full use of the potential for "many-to-many" communication provided by the new communications technologies. Readers of the e-mail magazine are positioned as passive consumers and readers with little sense of active citizenship being facilitated through the new media. Koizumi seems to be in search of an audience rather than an active citizenry.

NOTES

1　On the political potential of the Internet, see J. Everard, *Virtual States: The Internet and the Boundaries of the Nation-State*, London, Routledge, 2000; R. Holeton, *Composing Cyberspace: Identity, Community and Knowledge in the Electronic Age*, Boston, McGraw-Hill, 1998; S. Sassen, *Globalization and its Discontents*, New York, The New Press, 1998, pp. 177–94.
2　See, for example, S. Lax, "The Internet and Democracy," in D. Gauntlett (ed.), *Web.Studies: Rewiring Media Studies for the Digital Age*, London, Arnold, 2000, pp. 159–69; V. Mackie,

"The language of transnationality, globalisation and feminism," *International Feminist Journal of Politics*, vol. 3, no. 2, 2001, pp. 180–206.

3   See, *inter alia*, D. Birch, "An 'open' environment? Asian case studies in the regulation of public culture," *Continuum*, vol. 12, no. 3, 1998, pp. 335–48; A. Vervoorn, *ReOrient*, Melbourne, Oxford University Press, 1998, pp. 230–50; D. Birch, T. Schirato and S. Srivastava, *Asia: Cultural Politics in the Global Age*, Sydney, Allen and Unwin, 2001, pp. 72–99.

4   Ariadone (ed.), *Shikō no tame no intānetto* (The Internet for academic purposes), Tokyo, Chikuma Shobō, 1999.

5   Various approaches include *Intersections, http://wwwsshe.murdoch.edu.au/intersections/* (8 May 2002), which appears only in Internet and CD-ROM format; *New Left Review, http://www.newleftreview.net/* (8 May 2002), which issues a print journal but also makes indexes, tables of contents and selected articles available on the Internet; and the Japanese journal *Sekai* (The World) *http://www.iwanami.co.jp/jpworld/top.html* (8 May 2002), which makes selected articles available in English translation on the Internet.

6   T. Shibui, *Anonimasu: Netto o tokumei de tadayou* (Anonymous: adrift on the Net with a pseudonym), Tokyo, Jōhō Sentā Shuppankyoku, 2001.

7   B. Anderson, *Imagined Communities: Reflections on the Origins and Spread of Nationalism*, London, Verso, 1983. I am indebted to Mark Selden for suggesting that we might refer to the current regime as "cybercapitalism."

8   In the period under examination, the Koizumi Cabinet included five female Ministers, the largest number of any Cabinet, and also included several Ministers of State from outside the Diet. In January 2002, the number of female ministers dropped to four when Tanaka Makiko was relieved of the position of Foreign Minister.

9   *http://www8.cao.go.jp/town* (7 January 2002); *http://www.kantei.go.jp/jp/m-magazine/backnumber/2001* (7 January 2002).

10 *Japan Times Online* (25 December 2001); *Japan Times Online* (29 December 2001); *Japan Times Online* (31 December 2001).

11 See for example, T. Hara, "Kisha kurabu mondai" (The Press Club problem), in K. Isshiki *et al.* (eds), *Shin masu komi gaku ga wakaru* (Understanding new mass media studies), Tokyo, Asahi Shinbunsha, 2001, pp. 22–5.

12 *http://www.kantei.go.jp/m-magazine/backnumber/2001/0913.html* (31 October 2001); *http://www.kantei.go.jp/m-magazine/backnumber/2001/0920.html* (31 October 2001); *http://www.kantei.go.jp/m-magazine/backnumber/2001/0927.html* (31 October 2001); *http://www.kantei.go.jp/m-magazine/backnumber/2001/1011.html* (31 October 2001).

13 *http://www.kantei.go.jp/m-magazine/backnumber/2001/1206.html* (30 December 2001).

14 *http://www.kantei.go.jp/jp/koizumiprofile/index.html* (30 December 2001).

15 *http://www.kantei.go.jp/jp/m-magazine/iken.html* (30 December 2001).

16 See *http://www.jimin.jp/* (6 February 2002); *http://www.dpj.or.jp/* (6 February 2002); *http://www5.sdp.or.jp/* (6 February 2002); *http://www.hoshutoh.com/* (6 February 2002); *http://www.jiyuto.or.jp/* (6 February 2002); *http://www.komei.or.jp/* (6 February 2002); *http://www.jcp.or.jp/* (6 February 2002).

17 See *http://www.whitehouse.gov/president/gwbbio.html* (5 December 2001); *http://www.number-10.gov.uk/default.asp?pgID=3085* (5 December 2001); *http://www.pm.gov.au/yourpm/family/index.htm* (5 December 2001).

18 *http://www.kantei.go.jp/jp/koizumiprofile/index.html* (30 December 2001). When I accessed Koizumi's profile again in March 2002, the information about his family only appeared in the English version.

19 *http://www.kantei.go.jp/m-magazine/backnumber/2001/0628.html* (30 December 2001); *http://www.kantei.go.jp/m-magazine/backnumber/2001/0802.html* (30 December 2001).

20 *http://www.kantei.go.jp/m-magazine/backnumber/2001/0614.html* (30 December 2001); *http://www.kantei.go.jp/m-magazine/backnumber/2001/0705.html* (30 December 2001).

21 *http://www.kantei.go.jp/jp/m-magazine/backnumber/2001/1004p1.html* (30 December 2001).

22 *http://www.kantei.go.jp/m-magazine/backnumber/2001/0906.html* (30 December 2001).

23 *http://www.kantei.go.jp/m-magazine/backnumber/2001/0809.html* (30 December 2001).

24 *http://www.kantei.go.jp/m-magazine/backnumber/2001/0719.html* (30 December 2001).
25 *http://www.kantei.go.jp/m-magazine/backnumber/2001/0621.html* (30 December 2001).
26 *http://www.kantei.go.jp/m-magazine/backnumber/2001/1108.html* (30 December 2001).
27 *http://www.kantei.go.jp/m-magazine/backnumber/2001/0712.html* (30 December 2001); *http://www.kantei.go.jp/m-magazine/backnumber/2001/0823.html* (30 December 2001).
28 *http://www.kantei.go.jp/m-magazine/backnumber/2001/1011.html* (30 December 2001); *http://www.kantei.go.jp/m-magazine/backnumber/2001/1018.html* (30 December 2001).
29 *http://www.kantei.go.jp/m-magazine/backnumber/2001/1122.html* (30 December 2001).
30 *http://www.kantei.go.jp/m-magazine/backnumber/2001/1101.html* (30 December 2001).
31 *http://www.kantei.go.jp/m-magazine/backnumber/2001/0726.html* (30 December 2001).
32 *http://www.kantei.go.jp/m-magazine/backnumber/2001/0906.html* (30 December 2001).
33 *http://www.kantei.go.jp/m-magazine/backnumber/2001/0920.html* (30 December 2001).
34 *http://www.kantei.go.jp/m-magazine/backnumber/2001/0823.html* (30 December 2001).
35 *http://www.kantei.go.jp/m-magazine/backnumber/2001/1025.html* (30 December 2001).
36 *http://www.kantei.go.jp/m-magazine/backnumber/2001/1025.html* (30 December 2001).
37 *http://www.kantei.go.jp/m-magazine/backnumber/2001/1004.html* (30 December 2001).
38 On the promotion of "e-government" by the Blair government, see Lax, "The Internet and democracy," pp. 159–60; Interview with Patricia Hewett, E-commerce Minister, 9 March 2001. Online. Available HTTP: *http://www.epolitix.com/data/interview/articles* (22 April 2002). For the "Citizenspace" discussion forums, follow the "Your say" links from the Prime Minister's home page at: *http://www.number-10.gov.uk* (22 April 2002).
39 *http://www.kantei.go.jp/m-magazine/backnumber/2001/0719.html* (30 December 2001); *http://www.kantei.go.jp/m-magazine/backnumber/2001/0726.html* (30 December 2001).
40 *http://www.kantei.go.jp/m-magazine/backnumber/2001/0712.html* (30 December 2001).
41 The following explanation appears on the section of the Australian Prime Minister's home page which invites e-mail correspondence. "You can e-mail the Prime Minister by following the steps set out below. After your message is read, an electronic acknowledgment will be sent to you. There will be no further electronic response from the Prime Minister. However, once your views and suggestions have been carefully considered, you may receive a further reply by Australia Post. Your e-mail may be more appropriately considered by a Federal Minister and therefore could be passed on for their attention." While not particularly encouraging, this is much more explicit than the Japanese Prime Minister's home page.
42 Statistics on Internet usage in Japan in the e-mail magazine are quoted from *Heisei 13 Jōhō tsūshin hakusho* (Sōmushō) (Information and Communications White Paper 2001); population figures are taken from the census of 2000 (Sōmuchō, former Sōmushō). At the time of the election of July 2001, there were 101,309,680 registered voters, although only 56 per cent actually voted. The number of subscribers to the e-mail magazine is thus around 2 per cent of the number of registered voters, and a greater proportion of the number of those who actually exercise their right to vote. Recent population estimates come from the Statistics Bureau and Statistics Center, Ministry of Public Management, Home Affairs, Posts and Telecommunications at *http://www.stat.go.jp/english/data/jinsui/15k2.htm* (18 April 2002).
43 See tables at: *http://www.kantei.go.jp/jp/m-magazine/hyou/0712_index.html* (30 December 2001).
44 Some broadcast commercials and printed advertisements will use *anata* (you) or an alternative second-person pronoun when there is a clear sense of the features of the target audience being addressed.
45 Contrast with the Blair government's use of the phrases "on-line citizen's portal" for access to its electronic government facilities, and "Citizenspace" for its discussion forums.
46 A. Vandenberg, "Cybercitizenship and digital democracy", in A. Vandenberg (ed.), *Citizenship and Democracy in a Global Era*, London, Macmillan, 2000, p. 294.
47 Lax, "The Internet and democracy," pp. 161–4.
48 *http://www.issho.org* (5 February 2002).
49 Vandenberg, "Cybercitizenship and digital democracy," p. 294.
50 Vandenberg, "Cybercitizenship and digital democracy," p. 293.

51 N. Fraser, "Rethinking the public sphere: a contribution to the critique of actually existing democracy," in C. Calhoun (ed.), *Habermas and the Public Sphere*, Cambridge, MA, MIT Press, 1991, pp. 56–80.

52 On this tribunal, see P. Kim, "Global civil society remakes history: 'the Women's International War Crimes Tribunal 2000'," *Positions: East Asia Cultures Critique*, vol. 9, no. 3, Winter 2001, pp. 611–17.

53 For example, the Korea Council for the Women Drafted for Military Sexual Slavery by Japan publishes in Korean and English, and includes a discussion forum at *http://www.k-comfortwomen.com/* (6 February 2002); information about the Women's International War Crimes Tribunal 2000 has been disseminated in Japanese and English at: *http://www1.jca.apc.org/vaww-net-japan/e_new/index.html* (6 February 2002), while the tribunal itself was conducted with interpreters for all of the languages represented among witnesses.

54 On this concept, see Mackie, "The language of transnationality, globalisation and feminism," pp. 188–9.

55 While much of the information on these home pages does take the form of "one-to-many" communication of testimonials and tribunal judgments, affiliated groups have also made good use of e-mail petitions, and the VAWW-NET Japan home page invites readers to "Contact Us!" with several e-mail addresses, and phone and fax numbers provided. They state their intention to "respond to all inquiries in a prompt and efficient manner, and keep everyone informed of developments." *http://www1.jca.apc.org/vaww-net-japan/e-new/FinalJudgeHague.html* (19 April 2002).

56 J. Koschmann, "Opposing US hegemony: a long-term response to Bush's war," *Japan in the World*. Online. Available HTTP: *http://www.iwanami.co.jp/jpworld/text/USHegemony0.1.html* (22 April 2002).

13 Language, representation and power

Burakumin and the Internet

Nanette Gottlieb

INTRODUCTION

The Internet has been variously discussed as an enabling technology which is an unambiguously good thing for those fortunate enough to have access, a time-wasting proliferation of mostly low-quality rubbish, a hitherto unparalleled borderless source of information and communication, and a technology which simply drives another wedge between the haves and the have-nots. Whichever of these or many other perspectives we approach it from, there is little doubt that the Net (and in particular the Web) has changed the way we approach the gathering and presentation of information, and has offered both individuals and groups a new medium for the construction of identity.

Japan came to the Internet later and initially more slowly than countries such as the United States and Australia, as we saw in the introduction to this book, but it is rapidly making up for lost time. While some might argue that the Internet is still a negligible force for the "ordinary Japanese," demographic snapshots of user statistics available from a variety of sources[1] reveal this to be changing quickly as more students, women and shoppers go online. Is the glass half-full, or half-empty? Certainly commercial, government, educational and private home pages proliferate on the Web, as any Japanese search engine will attest. Among them are sites devoted to groups with similar interests – what we might now call cybercommunities or cybercultures, although the precise meaning of those terms remains contentious.

While the process of virtual identity construction may have proceeded more slowly, in terms of scale, in Japan than in the United States and Australia, the issues involved in how selves are presented to the outside world in terms of the online community are just as absorbing. And of course, language plays a crucial part in identity construction in many different ways. In this chapter, I will examine how the Internet has enabled Japan's largest minority group, the socially marginalized Burakumin, to develop faster and potentially more productive means of community organization and communication, while at the same time opening it up to wider-reaching (though arguably not more damaging) means of vilification. Burakumin groups use the Internet to circumvent the constraints on discussion of their circumstances which prevail in mainstream Japanese society, and language is an important tool in their strategies of representation. While the majority of sites intended for use within Japan are naturally in Japanese, others adopt a strategic use of English to achieve their aims of international visibility and activism. In addition, some groups use the Internet to combat online discrimination and stereotyping. We begin with a look at who the Burakumin are and what avenues were open to them and other social protest groups before the advent of the Internet.

THE BURAKUMIN AND SOCIAL PROTEST

The Burakumin are Japan's largest minority group. Estimates of their number vary from 1.2 million in different encyclopedias to 3 million.[2] They are not an ethnic minority but are Japanese, physically indistinguishable from other Japanese. Nevertheless, for a phalanx of reasons originating in the ostracism from mainstream society of their pre-modern ancestors – a taint carried down the generations by hereditary association – because of their association with occupations involving blood, death and other impurities traditionally considered polluting, they have been and continue to be subject to status discrimination. The segregation of Burakumin during the Tokugawa Period (1603–1867) into separate villages (*buraku* – the term "*burakumin*" means "people of the village"), and the imposition of strict rules of conduct on them, saw a continuing anti-Burakumin discrimination entrenched in the minds of other Japanese. Living conditions for the Burakumin have improved as a result of twentieth-century activism and resultant government policies, but the social stigma continues to this day, particularly where marriage and employment are concerned. A person with a known Burakumin background can find marriage partners harder to find. The same is true of employment, though steps have been taken to ease this by no longer requiring applicants to give full details of where they were born after a 1970s controversy when employers were found to be circulating lists giving details of areas Burakumin were likely to come from.

Mention of the Burakumin in polite Japanese society is virtually taboo. Often certain discreet finger gestures[3] will be made if the topic should unavoidably arise, or voices will be lowered. While reports of government policy on Burakumin issues may appear in the newspapers, the voices of Burakumin themselves are seldom heard in the print or other media. One reason for the lack of access is that the existence of such a minority does not sit well with the image of Japan as a homogenous, harmonious society untroubled by dissent which was the official government line in the post-war period. Japan is claimed to be racially homogenous, despite the existence of Ainu, Okinawans and resident-Korean and Chinese minorities, among others. The Burakumin minority, of course, does not confute the racial homogeneity myth in itself – Burakumin, as we have seen, are themselves Japanese – but the dissent and protest they have sustained since they banded together into organizations in the early twentieth century sit ill with the claim of a harmonious society propagated by the consensus model of Japanese society. This model, which stems from the Neo-Confucian emphasis on social harmony of Japan's feudal period, is today best illustrated in the Nihonjinron literature of

Japanese essentialism which has dominated scholarship and social comment from the 1960s on.

A second reason for the lack of comment on Burakumin issues in the media is that since the formation of the Suiheisha (the forerunner to today's Buraku Liberation League) in 1922 the Burakumin have waged an ongoing campaign against any derogatory reference to their members in any form of media or any public place. The tactic of *kyūdan* (denunciation) involves publicly denouncing and humiliating offenders, sometimes through unpleasant and extended confrontations,[4] and has effectively resulted in self-censorship by media organizations on the topic for fear of public embarrassment. While the *kyūdan* policy has put a stop to discriminatory stereotyping in the media and public documents, then, it has also virtually muzzled open discussion of Burakumin issues and heavily restricted Burakumin access to the mainstream media.

In the 1970s, before the advent of the Internet made public Web pages possible, social protest groups who for one reason or another found access to the mainstream media difficult utilized a variety of what were known as *minikomi* (mini-communication media, as opposed to *masukomi*, mass-communication media). These were alternative media, self-produced materials which could be cheaply and easily reproduced without having to approach mainstream publishers. Small one-person publishing companies specializing in *minikomi* printing were set up, enabling individuals of all persuasions to print and distribute material in a non-regulated structure similar to that of the Internet twenty years later.[5] These *minikomi* were particularly useful for grass-roots political activism, as in the case of *Kokuhatsu* (Indictment), published by a group set up to denounce the Chisso chemical company over its role in the environmental pollution disaster of the 1960s and 1970s which led to Minamata disease. By facilitating wider debate at the level of individual contribution, they extended the commonly-held view of the public sphere, creating "an independent space for critical public discourse not readily available in the mass media."[6]

The women's movement made good use of *minikomi* in the 1970s in protest against the stereotyping of women in the mass media. These included journals such as *Feminist, Agora, Onna Eros* and *Ajia to Josei Kaihō (Asian Women's Liberation)*, which permitted both debate and information dissemination, and a range of handwritten and machine-duplicated newsletters and broadsheets often concerned with targeting discrimination in the media.[7] Over time, as personal computers became more widely available and desktop publishing software more advanced, such communications changed to more technologically sophisticated and professional formats, although the focus on personal detail in their contents remained unchanged.[8] In the area of broadcasting in the early 1980s, in contradistinction to government control of the airwaves, low-powered license-free "Mini FM" radio stations proliferated in coffee shops, stores, campuses and other private spaces, some offering music programs, others performance shows to entertain customers, and still others a chance for members of the public to join in on-air discussions in an unregulated way.[9] In the case of the Burakumin, one of the main *minikomi* activities was the newsletter published by the Buraku Liberation Research Institute, *Buraku Kaihō Kenkyū* in Japanese, published bimonthly since

1977, and the *Buraku Liberation News*, its English counterpart, published bimonthly since 1981.

Minikomi may (and do) serve the needs of a particular community of interest, but they lack the reach to address society as a whole. Nor, in most cases, are they intended to: they are a means whereby members of a certain group communicate with other members or harangue those perceived to be the opposition, rather than addressing society as a whole. They are not intended to be mass circulation media. What, then, does the advent of the Internet – the least confining of communication channels, provided one has the means of access – mean for those who have produced them? The Internet enables marginalized groups to reach a potentially much greater audience than ever before, without regard to geographical location and usually for much less outlay than that involved in producing *minikomi* or other publications. This in turn challenges the artificial construct of a harmonious society by allowing the voices of members of these groups to be "heard" rather than sidelined. The putative audience, of course, has to be wired, but this is exactly the audience such groups seek to reach: those affluent enough to afford Internet access may perhaps also be assumed to have a measure of social influence if they choose to exercise it.

From the late 1990s, Internet publications came to supplement and in some (but by no means all) instances replace print *minikomi*. McLelland's book on homosexuality in modern Japan, for example, relates an instance of such Web-based activities: a deaf-gay activist sets up an Internet *minikomi* containing articles in Japanese and English on international deaf-gay issues.[10] All issues of the *Buraku Liberation News* since issue 95 (March 1997) have been made available on the Web as well as in print, as well as back issues of *Buraku Kaihō Kenkyū* to issue 120 (February 1998), although not all of the latter have live links to articles. In addition to these electronic publications, of course, the groups which produce them maintain home pages, bulletin boards and mailing lists which enable a much greater interactivity of communication across time and space for those so inclined. I turn now to a discussion of the role these activities play in constructing the online Burakumin community in its relations with both a domestic and an inter-national audience.

LANGUAGE, REPRESENTATION AND POWER

The growth of cybercultures or cybercommunities world-wide has seen discussion of issues of representation, identity and globalization and the tension between them foregrounded in recent research. Internet users may be less prone to define themselves predominantly in terms of national or local community and more likely to emphasize what they see as particular defining interests. They depend less on their immediate physical environment for knowledge or affirmation.[11] Groups of people prevented from full participation in the life of their local communities through ostracism or illness can therefore identify through Internet use with the wider national or international[12] community of others in similar circumstances, regardless of location. The choice of language under these circumstances can become an important enabling tool for the particular purposes of the chosen interaction.

Much of the computer-mediated communication (CMC) literature seeks to define the notions of community and culture inherent in the term "cyberculture" and to find alternative explanations to essentialist notions of both. Rheingold, author of *The Virtual Community*, for example, defines online or "virtual" communities as "social aggregations that emerge from the [Internet] when enough people carry on those public discussions long enough, with sufficient human feeling, to form webs of personal relationships in cyberspace."[13] Fernback and Thompson suggest that online community consists of "social relationships forged in cyberspace through repeated contact within a specified boundary or place (i.e. a conference or chat line) that is symbolically delineated by topic of interest." They comment further that "it is widely understood, that virtual communities will be communities of interest rather than of geographical proximity or of historical or ethnic origin."[14] Appadurai expands on this, commenting that in the context of communication during periods of upheaval:

> The speed of such communication is further complicated by the growth of electronic billboard communities, such as those enabled by the Internet, which allow debate, dialogue, and relationship building among various territorially divided individuals, who nevertheless are forming communities of imagination and interest that are geared to their diasporic positions and voices ... Information and opinion flow concurrently through these circuits, and while the social morphology of these electronic neighborhoods is hard to classify and their longevity difficult to predict, clearly they are communities of some sort, trading information and building links that affect many areas of life ...[15]

"Communities of imagination and interest" – that description fits the kind of Internet community constructed by Japan's Burakumin groups, as it does many of the other groups examined in this book. The Buraku Liberation League and other groups are clearly communities of interest, using the Net in a distinctive way to advance their causes in a manner not previously permitted by the constraints of Japanese society discussed above and no longer bound by the constraints of locality. As we might expect, their online activities conform to what Garnham describes as the two critical communicative functions in the public sphere: the collection and dissemination of information and the provision of a forum for public debate.[16] Friedland, discussing the concept of electronic democracy, expands on the importance of information dissemination: "as we move from a model of public life grounded in discourse toward one that expands to include greater emphasis on social networks, the role of knowledge brokers within the communications system becomes more central."[17]

The main function of Burakumin Web sites is to disseminate information about themselves both to Burakumin and, more importantly, non-Burakumin, thus brokering the knowledge given to others. This is particularly important because of the conditions discussed above, i.e. Burakumin have been traditionally relegated to the margins of society, shunted out of sight where possible and not discussed openly. While they have occasionally achieved wider newspaper coverage by conducting those very public denunciations of organizations accused of insulting them, their smaller print-based avenues such as *minikomi* have historically been slower, more expensive and less able to reach a wider and possibly unsolicited audience, not because of social considerations of literacy but because of the nature of their contents. Castells, writing on the social effects of information technologies, proposes the hypothesis "that the depth of their impact is a function of the pervasiveness of information throughout the social structure." Revolutionary technological change, he warns, must be "located in the social context in which it takes place and by which it is being shaped." [18] The Internet, with its capacity for interaction independent of location across both national and international territories, in theory allows much greater flexibility in terms of both scope of readership and opportunities for interaction. [19] In Japan's case, however, where a highly literate population combines with a relatively low diffusion of information about Burakumin, the impact of the Internet on the wider social matrix is likely to be minimal. For the Burakumin community itself, the Internet may come to play an increasingly important role if the digital divide can be overcome. The gap between Burakumin and non-Burakumin Internet users in terms of access is estimated to be around 20 per cent, which accounts for the fact that there are fewer than 100 Burakumin-maintained sites, only one of which offers chat facilities for members. [20] Some Burakumin-supporter sites do, however, provide bulletin boards and/or discussion groups which may be assumed to provide a sense of community among users.

If equality of access could be achieved, Burakumin use of the Internet has the potential to subvert the inability of such groups to be heard in mainstream media and thereby provide evidence which refutes the hegemonic view of Japan as a harmonious homogenous whole untroubled by minority groups (a view increasingly discredited in both Western and Japanese scholarship but surprisingly resilient in both the Japanese popular imagination and official discourse, as well as in international perceptions). Burakumin Web sites explain for the wired world – or at least for those users motivated to enter the Web sites – the history of discrimination and persecution faced by their members. One site carries an account written by a sociology student from Brigham Young University, detailing how his original interest in Burakumin issues had led to his decision to write a doctoral dissertation on marriage discrimination and his surprise at finding material relating to Burakumin available on the Internet. [21] A later issue of the same publication carries reflections on a Burakumin-related research trip to Japan by a doctoral student from New Zealand. [22] The Internet is therefore clearly a factor in spreading the word, to foreign researchers as well as to international human rights groups.

What I want to do in the remainder of this chapter is look at how Burakumin groups use the Internet in relation to issues of language and community, focusing on two main issues: the strategic use of English versus Japanese on certain Web sites in order to achieve different aims in the construction of their Internet presence, and the use of those sites themselves to combat discriminatory language and linguistic stereotyping. Finding out how they do these things tells us quite a lot about how they view issues of language, power and representation, and how they view the role of the Internet in addressing those issues. Burakumin groups are adept at using English strategically on the border-crossing Internet in order to gain international recognition for their cause. In this they are not alone, of course, both within Japan and in other countries. Cultural theorist Ien Ang, writing on the Huaren Web site (*www.huaren.org*) established for diasporic Chinese during the 1998 anti-Chinese riots in Indonesia, notes that "the use of English, explicitly encouraged by the site keepers, signals a desire to have a global reach, an international hearing, even as it also means that access to the site would be limited to those who are relatively well educated and economically privileged."[23]

Not surprisingly, the focus on most Burakumin English-language sites is educational. Major groups such as the Buraku Liberation League and its international offshoot IMADR (International Movement Against Discrimination and Racism) in particular run multilayered sites aimed at providing international Web users with detailed information on the state of human rights issues, and in particular Burakumin issues, in Japan. In 1998 I conducted a survey of Burakumin groups in order to find out how they saw their use (both existing and planned) of the Internet. An e-mail from the Buraku Liberation League's Hyogo branch confirms Garnham's two functions as the major thrust of the group's Internet activities:

> The rapid spread of access to the Internet is a two-edged sword: it allows people to sit in secret rooms and send out discriminatory mail anonymously or under a pseudonym, but it is also useful for spreading information useful for dispelling discrimination, and for forming anti-discrimination networks.

An explicit belief in the power of knowledge to dispel stereotypes, long a plank in Burakumin campaigns against discrimination,[24] is coupled here with an implicit belief in the power of the Internet both to reach a wider and more varied audience and to provide a forum for collective action in order to achieve the specific aim of an end to discrimination.

Most Burakumin organizations surveyed thus saw the Internet as a means for spreading information within Japan and also attracting international attention to their cause. The Buraku Liberation and Human Rights Research Institute (BLHRRI)'s site[25] is a good example of how they do this. The English version of this site is constructed very much with a view to providing information and comment for international visitors on the Burakumin situation in particular and occasionally the wider human rights situation in Japan.[26] On 4 October 2001, for instance, it contained information about the Institute and its activities; online issues of its bimonthly Buraku Liberation News;[27] pertinent information such as English translations of the Law on the Promotion of Human Rights Education and Human Rights Awareness-Raising (enacted 6 December 2000) and articles on its likely impact; a list of English-language publications of the BLHRRI; information about the "Buraku problem" and Buraku discrimination; and other items of interest.

The thrust of the English-language version of the site is thus frankly educational and aimed at keeping international (where international is assumed to mean "non-

Japanese reading") visitors abreast of the situation relating to Burakumin issues in Japan, within the wider context of human rights in general. The Japanese site is, as we might expect, much more detailed and informative, providing specifics of local information in line with the assumed needs of its local community of users. Also on 4 October 2001, it contained details of events recent and forthcoming; online issues of various newsletters; details of the activities of associated study groups, including the texts of public lectures; a moderated bulletin board; several searchable databases; and other assorted information likely to be of interest to those accessing the page, such as texts relating to United Nations activities in the human rights area. The Japanese-language site functions as a research and support area for members and the public. It sends out periodic information updates to those who register for them, mostly in Japanese but some (about one in three) in both Japanese and English. The Japanese version lists updates to the various sections of the Web site, notifications of changes to the databases available, occasional personal information about and jokes from the Web master, appeals for help with the Web page, and annual membership solicitations. The English updates are usually restricted to notification of the contents of the Buraku Liberation News. A sample of its contents, taken from the English updates 2000–1, finds articles relating to UN sub-commission activities; a protest lodged by human rights groups over discriminatory statements about foreigners made by the Governor of Tokyo in 2000;[28] discussion of how to promote the UN Decade of Human Rights in every field of society; and occasional articles describing the realities of discrimination against Burakumin, "for our new readers."

Perhaps the most internationally visible Internet activity on Burakumin issues can be found on the Web site of the International Movement Against Discrimination and Racism (IMADR),[29] an NGO which has consultative status with the United Nations Economic and Social Council. IMADR was founded in 1988 by the Buraku Liberation League, and has since grown into a global presence with regional committees in Asia, North America, Latin America and Europe. Its work takes place within five main anti-discrimination program areas,[30] all spelled out on its main site along with its other activities in Japanese and English, as well as in French, Spanish and German on associated sites. It maintains an active information and activist profile in international fora on issues relating to Burakumin as well as to other minority groups. The top page of the site on 4 October 2001 featured links to articles on such immediately topical issues as "World conference on racism – success or failure?", "Aftermath of the attacks on September 11 for the Muslim communities in US," "IMADR's urgent statement on the attacks on the World Trade Center and the Pentagon" and "Report on the sub-commission on the promotion and protection of human rights 2001," among others.

Sites maintained at universities or high schools for discussion of Burakumin-related issues are usually Japanese-language only, as might be expected. An example is the *Michikusa*[31] site at Osaka University, which aims to build a network of people interested in Burakumin and human rights issues by providing both useful information and a forum for discussion. The mailing list and bulletin board of this group were run together until April 1998, until misuse of the bulletin board resulted in unspecified "harm" to mailing list members (perhaps linked to the kind of harassment described below); they are thus currently run separately. The home page explains that the aim of the site is to create a network for people who think about Burakumin issues and human rights issues in general, and goes on to explain just what kind of network it has in mind: one which is participatory and respects the autonomy of individuals, without a one-way-only flow of information from the site owners; where debate from a variety of viewpoints is welcome; which provides fresh new information useful to readers; and where information remains firmly grounded in the realities of discrimination.

This site is internally directed, meant for debate between "insiders" in the sense of those who speak Japanese (though presumably non-Japanese able to carry on a debate in the language would not be excluded). It lacks the outward focus implied by the use of English. The use of Japanese indicates that the place of language in perceptions of network or community is firmly grounded in geography, or at least within the cultural geography of the Japanese language if not the actual physical geography of Japan. For most Japanese, despite this age of global travel and internationalization, the cultural geography and the physical geography of their language firmly coincide. The *Michikusa* site is primarily for Burakumin and their supporters engaged in dealing with Burakumin issues within Japan. There is thus no perceived need to use English, as there would be in a site aimed at those who might be useful in helping with international campaigns.

The home page of the Arai branch of the Zenkairen,[32] a Burakumin group which split from the Buraku Liberation League in 1979, restricts itself to Japanese in all but one clearly strategically-targeted case where English is used. This site, unlike *Michikusa*, is specifically polemically-oriented and points to divisions within not only the wider community over Burakumin but within Burakumin ranks themselves. For the most part the debate is presented in Japanese, but the site uses English to make one specific point clear to international readers, namely that other Burakumin groups are not "the real thing." It posts an English-language document meant for United Nations perusal which accuses the BLL and its IMADR offshoot of brutal violations of human rights: "we have decided to publish this ... to provide accurate information for UN bodies and member countries on the violent activities of the Buraku Liberation League, the core body of IMADR, with accurate information on human rights issues in Japan." The use of English in this one document alone is purely strategic, intended to reach organizations external to Japan which are assumed to be unable to read Japanese.

There is nothing about the use of the Web in this language-specific way which has not been done before in print media, or which other groups (such as the comfort women tribunal, associations of persons with disabilities, and women's groups) have not also adopted. What is new, of course, is the speed with which the information can be disseminated world-wide by the current technology, and the potentially much wider audience it can reach. Geographic spread becomes a factor here which print media could never have hoped to replicate. It might be argued that the patterns of language use described above replicate underlying assumptions about ownership of languages and patterns of power in a somewhat static way, taking no account of the changes in language demographics which have occurred over the last twenty years. Nevertheless, they highlight the commonsensical manner in which Burakumin groups assess the nature of their audience and use English strategically to achieve certain clear objectives on the international scene.

HOW THESE WEB SITES FOREGROUND DISCRIMINATION

Despite the increased exposure the Web allows, the Internet has not been an unadulterated "good thing" for Burakumin groups, because of course, as my 1998 e-mail survey respondent pointed out, its enabling conditions of anonymity and borderlessness have provided fertile ground for hate speech in Japan as elsewhere, and Burakumin groups in particular have felt its bite. In 2001, for example, 253 separate sites reviling Burakumin were discovered.[33] A second major characteristic of Burakumin groups' use of the Internet is therefore the way in which they constitute themselves as guardians against this type of online discrimination and stereotyping. On the Net we find many instances where Burakumin sites highlight and target the role of discriminatory stereotypes and hate speech in perpetuating racism, in recognition that language constructs social practice as much as it reflects it in a two-way process which informs both perceptions and their outcomes.

A speech by Kumisaka Shigeyuki, president of the Buraku Liberation League and Director of IMADR, at a forum in Germany in 1998 confirmed that "with the widespread use of electronic communication tools ... discriminatory propaganda and agitation has been on the increase on the Internet." [34] In August 2000 IMADR also made a statement on incitement to racial discrimination on the Internet at the UN Sub-Commission on the Promotion and Protection of Human Rights (52nd session). A local collaborator organization (the Network against Discrimination and for Research on Human Rights, NDRH[35]) had recently released a draft report on the incidence of incitement of racial discrimination against Burakumin and Koreans in Japan.[36] While the 1998 revision of codes of conduct on this matter by Internet providers had reduced the open use of discriminatory language, the report found, the result had been "more creative methods of disseminating such messages that are harder to detect," facilitated by an increase in foreign servers in 1999. The IMADR report called on the UN High Commissioner for Human Rights "to collect information on good practices by Internet providers including codes of conduct, and any legal measures that effectively deal with racial discrimination on the Internet."[37]

The NDRH reported nearly 70 complaints in 2000 about Web sites containing incitements to racism. A particular focus of these complaints was Channel 2, which hosts unmoderated bulletin boards and listserves. The complaints were found to revolve around nine incidents in particular, among them three targeting Burakumin. The anonymity of the Internet loosens the restrictions of public constraints on discussing Burakumin. *Time Asia* picked up on this problem in June 2001 in an article discussing the anti-Burakumin messages on a discussion group hosted by the Channel 2 Web site.[38] Contributors had demanded that names of Burakumin working in the entertainment industry be made public and had speculated on a well-known political figure's possible Burakumin origins, as well as posting general anti-Burakumin messages. In 1975, the practice by companies of using lists of Burakumin family names and neighborhoods to weed out Burakumin when hiring, mentioned earlier, was banned. Those lists, however, had now begun to appear again on the Channel 2 bulletin boards, a trend which the Buraku Liberation and Human Rights Research Institute deplored as much more serious than the 1975 incident because of the wider number of people the information could reach, with possible serious consequences in terms of both employment and marital discrimination.

The *Time Asia* article questioned whether the Internet revolution would help or hurt Burakumin groups, and quoted the Channel 2 Webmaster as being optimistic because anonymous bulletin boards permitted the perpetrators and recipients of discrimination to engage in dialogue in a way not possible anywhere else. Dialogue does indeed enable each side to see the other's perspective; of itself, however, it solves nothing, particularly when the medium is writing and when the place is the Internet. There is no guarantee that the poster of an inflammatory message will even check back to see a response, except out of curiosity to see what he or she has provoked, in which case the focus of interest is more likely to be the fact of the response itself rather than its contents. Productive dialogue, of the kind which fosters understanding or even attitudinal change, may be an unrealistic expectation in these circumstances; it is more likely that these sorts of activities merely perpetuate already-hardened attitudes. Two months after the *Time Asia* article appeared, the NDRH reported, the ISP which had initially refused their request to delete the site changed its mind after receiving from them two documents in English: a resolution of the United Nations Human Rights Commission and a paper on discrimination against Burakumin in Japan.

The fact that these documents were sent in English rather than Japanese underlines the point I made above: English is used to achieve, with forethought, certain strategic goals. In this case the weight of international opinion, as symbolized by the use of English in a United Nations document and an informational publication on anti-Burakumin discrimination clearly meant for readers outside Japan, with accompanying overtones of the threat of possible embarrassment generated by unfavorable international publicity, was brought to bear against the provider. The summary report reflects:

> This case shows us that ISPs may not respond to a request made by just one grassroots group but may do so when they realize that this problem is so serious that even the United Nations takes it into serious consideration. As a result, the Network learned a new way to approach ISPs and got another discriminatory Web site named the "Shin Jiyū Kyōkai" (or the Association of True Liberties), which we had observed for two years, finally deleted when we sent the same documents to the ISP.[39]

Two further Japanese organizations – the International Network against Discrimination on the Internet (INDI), and the New Media Human Rights Institution – have now been set up to seek both international and domestic eradication of Internet discrimination. The INDI home page[40] highlights the fact that no real discussion has ever resulted from the posting of discriminatory messages on Web sites or bulletin boards: providers have either overlooked this or have immediately deleted the message, but without any clear policy on guidelines for the future. INDI sees its role as lobbying for cyberdiscrimination to be tackled internationally via the Internet, through NGOs and individual action as well as government and international initiatives. Its members seek to utilize the Internet's international reach and rapid transmission of information to "institute dialogue with those who allegedly have sent discriminatory messages" through the same medium, in order to work toward a resolution of the problem where possible. They also pledge to mobilize an international network of concerned individuals and groups in order to more effectively tackle cyberdiscrimination and use the Web to publicize its activities. The New Media Human Rights Institution's declaration expresses similar sentiments, pledging to contribute to the protection of human rights in the face of new threats from advances in information technology.[41]

CONCLUSION

What the activities described in this chapter show us is that Burakumin groups, marginalized in everyday Japanese society, show a clear belief in the power and reach of the Internet to present information aimed at dispelling stereotypes to a wider public than ever before available to them. Whether or not greater numbers of people unconnected by personal ties to the groups actually visit the sites to read the information is not the issue here: the point is that they *can*, whereas once an interested person would have had to make a special effort to find the information in print. In that sense this is an enabling technology where results cannot be predicted. The same point is made in other chapters in this book with regard to other groups. The Internet for these groups is a useful tool in ongoing campaigns for recognition and redress and in fostering both domestic and international networks. If virtual community may be said to rest upon a shared acknowledgment of ties that bind, whatever the nature of those ties may be, and a willingness to inform and educate whoever will listen on the issues important to the group, then Burakumin groups fit the description of online communities. In their interactions with the online world at large, they use language both to achieve the outward-looking strategic objectives outlined in this chapter and as a focus of examination and protest within Japan itself. The use of English constructs an external identity as members of the international human rights community, while their Japanese-language activities both support their local members and combat discrimination expressed through the use of derogatory terms and images. Language thus plays a crucial role in their strategies of both representation and power, where power – or empowerment – stems from an active disinclination to accept the labeling imposed by other sections of society.

NOTES

1 See, for example, NUA Internet Surveys at: *http://www.nua.ie/surveys/* (15 April 2002).
2 Y. Sugimoto, *An Introduction to Japanese Society*, New York, Cambridge University Press, 1997, p. 6.
3 Four fingers are held up to indicate that Burakumin work with dead animals (e.g. as butchers or tanners), or are considered to be like animals themselves. The word "four" (*shi*) is also a homonym for death (*shi*) in Japanese.
4 See S. Pharr, *Losing Face: Status Politics in Japan*, Berkeley and Los Angeles, California, California University Press, 1990, pp. 126–44.
5 S. Kinsella, "Japanese subculture in the 1990s: *otaku* and the amateur *manga* movement," *Journal of Japanese Studies*, vol. 24, no. 2, 1998, p. 294.
6 W. Sasaki-Uemura, "Competing publics: citizen groups, mass media and the state in the 1960s," *Positions: East Asia Cultures Critique*, vol.10, no. 1, 2002, p. 90.
7 S. Buckley and V. Mackie, "Women in the new Japanese state," in G. McCormack and Y. Sugimoto (eds), *Democracy in Contemporary Japan*, Sydney, Hale & Iremonger, 1986, p. 181. See also S. Buckley (ed.), *Broken Silence: Voices of Japanese Feminism*, Berkeley, University of California Press, 1997.
8 V. Mackie, "Feminism and the media in Japan," *Japanese Studies Bulletin*, vol.12, no.2, 1992, pp. 23–5.
9 T. Kogawa, "New trends in Japanese popular culture," in G. McCormack and Y. Sugimoto (eds), *The Japanese Trajectory: Modernization and Beyond*, Cambridge, Cambridge University Press, 1988, pp. 61–4.
10 M. McLelland, *Male Homosexuality in Modern Japan: Cultural Myths and Social Realities*, Richmond, UK, Curzon Press, 2000, pp. 189, 213.
11 R. Kluver, "Globalization, informatization, and intercultural communication," *American Communication Journal* vol. 3, no. 3, 2000, p. 8. Online. Available HTTP: *http://acjournal.org/holdings/vol3/Iss3/spec1/kluver.htm* (8 October 2001).
12 Assuming language competence in either direction.
13 H. Rheingold, *The Virtual Community: Homesteading on the Electronic Frontier*, New York, HarperCollins, 1994, p. 5. He later qualified this description on his Web site, adding a caveat to his definition of virtual communities as "people who use computers to communicate, form friendships that sometimes form the basis of communities, BUT you have to be careful not to mistake the tool for the task and think that just writing words on a screen is the same thing as real community." Online. Available HTTP: *http://www.rheingold.com/vc/book/intro.html* (8 October 2001).
14 J. Fernback and B. Thompson, "Virtual communities: abort, retry, failure?" Online. Available HTTP: *http://www.rheingold.com/texts/techpolitix/VCcivil.html* 1995 (8 October 2001).
15 A. Appadurai, *Modernity at Large: Cultural Dimensions of Globalization*, Minneapolis, University of Minnesota Press, 1996, p. 195.
16 N. Garnham, *Capitalism and Communication: Global Culture and the Economics of Information*, London and Newbury Park, CA, Sage, 1990, pp. 111–12.
17 L. Friedland, "Electronic democracy and the new citizenship," *Media, Culture and Society* 18, 1996, p. 189.
18 M. Castells, *The Information Age: Volume I, The Rise of the Network Society*, Malden, MA, Blackwell Publishers, 2000, pp. 30n and 4.
19 Though it might be argued that only those who are interested in the first place might access the information, so that the readership is not necessarily wider than before.

20 Private e-mail communication from a member of the BLHRRI, 8 March 2002.

21 C. Morgan, "My encounter with the Burakumin." *Buraku Liberation News*, 118, 2001. Online. Available HTTP: *http://www.blhrri.org/blhrri_e/news/new118/new11803.html* (18 March 2002).

22 I. Laidlaw, "Reflection on a research trip to Japan," *Buraku Liberation News*, 120, 2001. Online. Available HTTP: *http://www.blhrri.org/blhrri_e/news/new120/new120.html* (18 March 2002).

23 I. Ang, *On Not Speaking Chinese: Living between Asia and the West*, London, Routledge, 2001, pp. 57–8.

24 See N. Gottlieb, "Discriminatory language in Japan: Burakumin, the disabled and women," *Asian Studies Review*, vol. 22, no. 2, 1998, pp. 157–73.

25 Buraku Kaihō Jinken Kenkyūjo, *http://blhrri.org*. The English version is at: *http://blhrri.org/index_e.htm* (8 October 2001).

26 Issue 117 of the *Buraku Liberation News*, for example, was devoted to the problem of discrimination against foreigners in Japan.

27 A bimonthly publication in English posted on the Web site as well as being available in hard copy.

28 Governor Ishihara Shintarō caused an outcry when he remarked in April 2000 that foreigners in Japan might riot in the wake of a major earthquake and that the Self-Defense Forces should therefore be especially careful to maintain public security in emergencies of that kind.

29 *www.imadr.org* (11 October 2001).

30 Elimination of racism and racial discrimination; international protection of minority rights; empowerment of the victims of multiple discrimination; facilitation of indigenous peoples' development; and advancement of migrants' rights.

31 *http://dhva.phys.sci.osaka-u.ac.jp/mitikusa/mitikusa.html* (8 October 2001).

32 *http://village.infoweb.ne.jp/~fwga9098/index.html* (8 October 2001).

33 Private e-mail communication from a member of the BLHRRI, 8 March 2002.

34 S. Kumisaka, "The current condition of minorities in Japan and challenges – the Buraku issue," *Buraku Liberation News*, 101, 1998. Online. Available HTTP: *http://blhrri.org/blhrri_e/news/new101/new10103.htm* (8 October 2001).

35 This is the acronym given on the association's Web site.

36 An executive summary of this report in English may be found at: *http://homepage2.nifty.com/jinkenken/eng3.htm* (8 October 2001).

37 IMADR Statements at UN, August, 2000. Online. Available HTTP: *http://www.imadr.org/geneva/aug.html* (8 October 2001).

38 T. Larimer and T. Sekiguchi, "Social outcasts," *Time Asia*, 1 June 2001. Online. Available HTTP: *http://www.time.com/time/asia/news/printout/0,9788,104138,00.html* (8 October 2001).

39 *http://homepage2.nifty.com/jinkenken/eng3.htm* (8 October 2001))

40 *http://homepage2.nifty.com/INDI/* (11 October 2001)

41 *http://www.jinken.ne.jp/shuisho.html* (10 October 2001).

14 Activism and the Internet

Japan's 2001 history-textbook affair

Isa Ducke

INTRODUCTION

The Internet, which provides new, informal, and horizontal ways of communication on an unprecedented scale, is often expected to "level the playing field" between powerful political actors and the masses struggling below. Compared with many familiar tools of political engagement such as organizing demonstrations, mail-outs and faxes, maintaining an Internet presence and communicating with a wide audience is relatively cheap and easy. McNeill describes in this volume his hopes that the Internet may act as an "equalizer," enabling activists with limited power and resources to compete with larger, well-financed organizations. Although several studies have stressed the *potential* the Internet has for enhancing democracy, however, evidence that it actually does so is often slender.[1]

Furthermore, concerns have been raised both about the effectiveness of the Internet and its potentially detrimental impact on democracy. One of the key notions in this debate is the "digital divide," that is the gap between the "information-rich" and the "information-poor."[2] It is usually the case that those who are already educated, informed, and politically active have greater Internet access, and therefore reap its benefits, whereas those who are uneducated and lack training and skills are excluded from new forms of communications technology. Can increased access to information (and perhaps decision-making processes) make a society more democratic when, once again, it is those already privileged who benefit the most?

Another concern regarding the Internet's possible effect on democracy is that it would mostly generate a shift from representational to direct democracy – with all its deficiencies, such as populist tendencies and limited protection of minorities. After all, online surveys and "comment pages" are at least as easy to manipulate as public opinion polls.[3] Even on the Internet, forms and norms of communication may still be defined by traditional structures. Axford, for instance, notes that in spite of the apparent immediacy of the Internet, most communication is still screened, mediated, and sometimes even censored. Some Web sites offer comment forms and gather information from their visitors but neither assure them of privacy nor explain what will happen to their information.[4] Even in the "information society,"[5] when the volume of communication (e-mails, telephone calls, Web site hits, etc.) is rapidly increasing, due to the fragmentation of the Internet the audience reached does not multiply at the same pace. Internet transmission just makes it easier for information to reach those already interested in an issue.[6]

In order to evaluate the potential benefits and effects that Internet use may have on activism and political participation, it is necessary to look closely at actual instances where Internet deployment has proven significant. Political activism usually involves activist groups with limited resources locking horns with government or business institutions that command a far more powerful position. If the Internet does indeed level the playing field, it should improve the activists' chances for advancing their goals. In order to assess the impact of the Internet on one activist cause in Japan, this chapter focuses on the so-called "textbook affair" (*kyōkasho mondai*) that erupted in Japan in 2001.

Disputes over history, notably the history of Japanese war aggression and colonialism, have been common between Japan and its neighbors and in particular have repeatedly disrupted Japanese–South Korean relations. In 2001, official recognition of a new history textbook, seen by its detractors as whitewashing Japan's aggressive military past, once more strained relations between Japan and Korea. As with previous disputes of this nature, citizens' groups both in Japan and in Korea mobilized to try to prevent the adoption of the textbook. The 2001 campaign against the book was, however, different from earlier campaigns in that the Internet became an important tool for activist groups both opposing and supporting the textbook. This chapter examines the ways in which different players in the 2001 textbook affair, both community-based and state-sponsored, both in Japan and in Korea,[7] used the Internet to advance their causes.

THE TEXTBOOK AFFAIR

Japan and its historical consciousness

History is a sensitive issue between Japan and South Korea. In 1910, Korea was annexed by Japan and remained a Japanese colony until 1945. During that time, Koreans were forced to speak Japanese and use Japanese names. As Japanese citizens, they were drafted into the Japanese military or into forced labor. The majority of the women who were forced or lured into sexual slavery for the Japanese military, the so-called "comfort women," were Koreans.[8] After the war, the peninsula was divided between the spheres of influence of the Cold War. Not least because of the division and ensuing Korean War, numerous Koreans who had come to Japan during the colonial period remained there. However, after the Second World War, they lost their Japanese citizenship (including benefits such as veterans' pensions as well as access to Japanese welfare programs and public housing) and those who stayed constitute a discriminated minority. Although resident Koreans have successfully fought many discriminatory practices such as the requirement to carry finger-printed identity cards and the exclusion of Korean-Japanese schools from national sports tournaments, and have become highly integrated in Japanese society through education and work, many areas of discrimination remain.[9]

Although Japan established diplomatic relations with South Korea in 1965, none exist with the communist North – for the Korean residents this means that some are now South Korean citizens, while a decreasing but substantial number of Koreans identifying with North Korea remain stateless.[10] Japanese politicians, prime ministers, and even the Diet and the Emperor, have issued numerous statements expressing various degrees of regret for the past, which nevertheless have failed to satisfy the Korean demand for an apology, either because they did not include sufficient acknowledgement of guilt, or because they were not accompanied by compensation payments. On the other hand, statements denying Japanese wrongdoing are frequent, although many of these are not official – in recent years officials making such comments were often forced to resign.[11] Unsurprisingly, Koreans (including those resident in Japan) concern themselves with the contents of Japanese history textbooks. Japanese textbook policies have therefore repeatedly become issues of debate between both countries as well as between reformist and conservative groups within Japan.

The 2001 textbook affair

While Japan and South Korea were busily preparing for the 2002 FIFA World Cup hosted jointly by both nations, arguments over history again disturbed their gradually improving relationship. Emotions erupted when Prime Minister Koizumi visited Yasukuni Shrine on 13 September 2001, just ahead of the anniversary of the end of the war. People in neighboring countries take offence at official visits to the shrine, not least because the souls enshrined at Yasukuni include those of several war criminals.

Controversy over the 2001 textbook affair started several years earlier, when a group of right-wingers (none of whom was an historian) had announced their intention to write a "better" history textbook. They argued that existing textbooks endorsed views "which discredit Japan's past history" and reflected a "steady decline of national principles."[12] In 2001, the group – the "Tsukurukai" (Japanese Society for History Textbook Reform) – submitted for approval by the Education Ministry (MEXT)[13] a textbook for use in junior-high schools.

In a preliminary screening, the ministry itself found an unusually high number of 137 factual errors that had to be changed.[14] Based on the so-called "neighboring countries clause," which was the direct result of a previous textbook affair in 1982 (described below), both South Korea and China were also consulted regarding the content of the books submitted. Both governments criticized the Tsukurukai book heavily as "distorting history" and demanded numerous changes. MEXT agreed with only two of these further objections, but finally approved the book along with seven others in April 2001. Critics in Japan and abroad, including the Chinese and Korean governments, protested against the decision, but without success. A number of their objections were about misrepresentations or omissions, for example of references to the wartime sex-slaves. MEXT argued that these were "not factual errors" and agreed only with some minor points.

This was only the first round, however. Due to the somewhat complicated selection process for junior-high school textbooks, the conflict continued throughout the summer. Regional (or supra-communal) selection committees (*saitaku shingikai*), of which there are 544 in all, select one book each from the list of those approved by the Ministry, and "recommend" it to the local education committees (*kyōiku iinkai*) in their region. Usually, the latter accept this choice and notify the ministry that all schools in the municipality will use this book.[15] For the protest movements, this offered two more opportunities for lobbying, first at the regional committee level and then at the local one. They used various methods of protest in an attempt to prevent the controversial book from actually being used in schools, including demonstrations, letters, faxes, telephone calls, e-mails, local information meetings and seminars, and cyber-demos.

The committees had to make a decision by 15 August. On 13 July, the regional committee in Shimotsuga was the first to decide in favor of the book, but due to a flood of protests, one out of ten local education committees covered by this decision took the unusual step of rejecting the recommendation. Eventually, the regional committee had to decide on a different textbook. Perhaps put off by these events, no other committee selected the Tsukurukai book. Only a few private schools and special-needs schools, which are exempted from the above selection system, were to use it.[16] The Tsukurukai succeeded in having the book adopted in only twelve schools, covering a total of 601 students.[17]

Previous textbook activism

This local level of lobbying distinguished the 2001 textbook affair from a similar controversy in 1982, which disturbed Japanese–Korean relations for months. At that time, protests were directed at the Ministry's (perceived)[18] censorship of a textbook that had critically described Japanese wartime activities. Japanese media reported in June 1982 that the Ministry had ordered the term "invasion" (*shinryaku*) for Japan's move into China to be changed to "advance" (*shinshutsu*). This was then taken up by the South Korean media, and on 26 July and 3 August, respectively, the Chinese and South Korean governments launched formal protests.[19] The South Korean parliament even issued a resolution demanding a rectification of the textbook, although South Korea was struggling at that time to obtain a major loan from Japan.[20] The textbook controversy eventually subsided in the context of an improvement of relations that went along with the new Prime Minister Nakasone's visit to Korea and an agreement on the loan issue. For activist groups, there was little to do except express their dissatisfaction with the government: due to the Ministry's censorship, the book they supported never entered the regional selection process in its original form. This was different from 2001, when they were *opposing* a book which had passed the Ministry screening and moved on to the regional and local selection process, thus entering a new phase of campaigning and lobbying.

However, in those cases where it was confirmed that the Ministry's screening process involved censorship, the authors and activist groups supporting them launched lawsuits against the government in an effort to gain public attention and to force the Ministry to change its decision. Among the most famous of these lawsuits were the three Ienaga suits, which dragged on for over 30 years.[21] The support network for Professor Ienaga Saburō was to play a major role in the 2001 textbook affair.

INTERNET, ACTIVISM, AND THE STATE

"Network 21" and the traditional activists

Ienaga is a Japanese history professor whose textbooks have long proven controversial because of their critical stance on Japanese colonialism and war. His partial victories against the screening practices of the Education Ministry have changed the Japanese textbook landscape, leading to a situation in the 1990s where it was possible to mention Japanese war atrocities. However, the process of liberalization also generated a backlash by the right wing, including the Tsukurukai. When the last of Ienaga's lawsuits was resolved in 1997, his support organization renamed itself as "Children and Textbooks Japan Network 21" (*Kodomo to kyōkasho netto 21*) and continued to campaign, this time against the new textbook. During 2001, the focus of the group's protests shifted from the Ministry and the screening system to the regional and local selection committees.

Network 21 is based in Tokyo and links local organizations throughout Japan. The Web site maintained by the secretariat in Tokyo plays a major part in its campaign activities. It is devoted entirely to the textbook issue and offers extensive background information as well as links to similar sites and a newsletter. For example, the site contains an appeal signed by Network 21 and eleven other activist groups entitled "A Textbook that Treads the Path of Constitution Denial and International Isolation should Not Be Handed over to Japanese Children."[22] The news articles posted on the Web site during the summer of 2001 were generally more concerned with future protests than with descriptions of past activities: the site regularly posted urgent appeals ahead of committee meetings. One entry on 7 August concerned the planned use of the Tsukurukai book for special-needs schools in Tokyo:

> We strongly oppose the decision of the Tokyo Board of Education to select the "Tsukurukai" textbook, and urge its members to reconsider that decision. We also appeal to people throughout Japan, not just in Tokyo, to protest against the decision of the Tokyo Board of Education and to demand a review of this decision. Let's send our protest to the members of the Tokyo Board of Education!

Below the appeal were provided the full postal address, telephone and fax numbers, and a link to the comment site of the Tokyo Metropolitan Government.[23] Network 21's Secretary-General, the historian Tawara Yoshifumi, explained that documentation of activities or press conferences was sometimes slow to appear on the Web site due mostly to a lack of resources since the Web site was maintained by a group of volunteers. Urgent matters were given priority, he said, and the secretariat would advise the volunteers to deal with upcoming appeals first, before posting follow-up information on previous events.[24]

While effective in mobilizing opposition to the Tsukurukai text, there were clear limitations to the group's Internet use. Interactive pages were largely limited to online membership application and the provision of an e-mail address and offline contact information. A proper Bulletin Board System (BBS) was not even planned for fear of sabotage: "[A BBS would be] a good tool, but on the advice of an expert we decided against it when we established the home page in 1999."[25] Additional measures were planned, including a comment form and Q&A pages, but both were "under construction" during the whole period of the textbook affair. Tawara explained that compared to a BBS, such features would prove more manageable, and would be less "dangerous" (in terms of aggressive or illegal submissions as well as sabotage). Tawara also seemed concerned about technical security (hackers, viruses, cyber-demos) although this is in reality a relatively minor problem. Network 21 also maintains a mailing list that is limited to members. The secretariat promptly answers most e-mail queries sent to the e-mail provided on the home page, or forwards them to the members on the mailing list if it cannot answer them itself.

Despite the limitations mentioned above, the group judges its use of the Internet to be "very effective."[26] Campaigning became much faster and more manageable compared with earlier activities. In a similar campaign in 1996/97, for instance, appeals were made only via fax and telephone, but in 2001, detailed up-to-date information on upcoming activities was always easily available on the Web site. The number of hits per month rose to nearly 10,000 in 2001, almost ten times as many as in the previous year. Network 21 estimates that about 20 to 30 per cent of those visits were by opponents of the Web site's position. "Information you put on the Internet can be used by your opponents, too," Tawara noted. He appeared reluctant to use the Internet for wider exchanges than among members and people sympathetic to the group's cause. Like many of my interviewees, he apparently did not regard information exchange with opponents as providing a chance for improved understanding or better quality discussion.

The group's Internet presence has, of course, facilitated international access and thus enhanced the potential for links with similar organizations in other countries. Indeed, Network 21 has experienced a heightened reaction from outside Japan. Although, as Tawara concedes, "language is still a problem," an increasing number of foreign organizations have got in touch with Network 21. According to Tawara, some Korean groups could communicate with Network 21 in Japanese via outposts in Japan. Other responses, however, are in English, and this apparently poses a problem when the letters and materials received are lengthy. Network 21 only offers a very limited English-language page on its Web site which is rarely updated. This is largely due to a lack of human resources, and to some extent, like familiarity with the Internet itself, is related to age factors. Japanese textbook activists tend to be middle-aged or older, and even though aware that other groups use an international "name and shame" strategy (see Gottlieb's chapter in this volume), regularly updating an English-language page is difficult for many of them.

Overall, the Internet home page has proven useful for Network 21's campaigning and network building or at least network maintenance. For instance, many other groups link to the Network 21 home page (Google Japan listed 250 links in April 2002, and well over 1,300 references to the group). It was certainly easier for interested individuals not previously linked with the group to either join the network or get involved in campaigning. The group's organizational structure has not changed much, though, and the cost- and time-saving effects of Internet usage have so far been limited to some extent by the additional workload and technical difficulties related to maintaining an Internet presence.

Network 21 has some links with a similar group in Korea. The "Headquarters of the movement to correct Japan's textbooks" is an umbrella organization of Korean civic groups protesting against the new textbook. The group is based on traditional activist methods and is rooted in the strong culture of protests and street demonstrations in South Korea. Like Network 21, it links numerous groups concerned with history textbooks in both Japan and Korea, providing background information, news, and updates on upcoming demonstrations, meetings, symposia or boycott activities. The Web site of the group is in Korean only, but designed so as not to require a high-speed Internet connection. It offers a busy BBS and numerous links and background texts. Interestingly, they published on their home page an appeal that they received from Network 21 via fax.[27] Network 21's decision to use fax rather than e-mail to send such an appeal was probably related to concerns about encoding problems, but the incident does illustrates the difference between Japan and Korea in terms of Internet usage, in spite of the great similarity of both groups.

Internet presence by resident Koreans in Japan regarding the textbook issue was rather low. Of the two long-standing rival groups representing the resident Koreans identifying with South Korea (Mindan) and North Korea (Chōsōren), Mindan only set up its Web page in 2000, and it offered little more than an online version of its newsletter. While Mindan did conduct local campaigns using traditional methods such as faxes and demonstrations, they had not openly established links with Japanese activists for strategic reasons – it would only play into the hands of the right wing, explained a representative, if textbook activists were openly supported by "those foreigners." Additionally, they did not want to endanger their lobbying contacts with Japanese conservatives whose support they court on numerous issues.[28] The pro-North Korean organization, Chōsōren, apparently only developed its comparatively flashy home page during 2001. The earliest texts available date from November 2001, after the textbook affair had already subsided, and the textbook issue is not mentioned. Only in 2002 did it become possible to find the page via Internet search engines. Other Web sites of the Korean community like *Chaeil.net* or *Hantongryōn* mention the textbook affair on their BBSs or in online newsletters, sometimes providing background information and links. These sites are generally less politicized community sites. Addressed to a relatively young audience, they cover a wide range of interests, from the soccer World Cup to film reviews. The textbook affair came up only occasionally during the summer as a current affairs topic. [29]

Like the groups and activists mentioned above that oppose the "New History Textbook," the Tsukurukai itself, which authored it, is also an activist group. They regard themselves as a citizens' protest movement against the "deterioration of patriotic sentiment" and in appearance, they have much in common with the activists working against them. The group inhabits a small, cramped basement office in Tokyo, with a security camera outside for fear of arson attacks by opponents (there was such an attack in August 2001, they say).[30] Etoh Takahiro, one of the more computer-literate staff members, set up a Web site in December 2000 without any extra software or funds. It is a rather amateurish site with scroll-down text lines on a patterned background, providing most of the materials otherwise available as printed pamphlets. Etoh said that some of the younger members had demanded the group set up a home page for recruiting new members – but their bank account is the only contact information given, except for a telephone and fax number. The site comes up in search engines under the keywords "Tsukurukai," "history text-book," or "textbook" (in Japanese), but the group has not attempted any further promotion of the site. As Etoh pointed out, "Ideally, people who are interested in the subject should find our Web site, and if they agree with us, join the organization." This seems to work to some extent, although the Internet still results in at least as many problems as benefits for the Tsukurukai. Although the Web site offers no interactive features, not even an e-mail address, the group ranks security issues like viruses and hacker attacks as major concerns, along with the additional workload involved in putting files on the Web site.[31]

Antijapan and cyber-demos

The Tsukurukai has reason to be concerned about virtual security: their server crashed and the Web site was made unavailable for several days in August 2001 due to a Korean-sponsored cyber-demo. The group that initiated the demo was a very untraditional activist network: *Antijapan* is simply a home page run by three South Korean high-school students. They use a combination of manual means such as repeatedly pressing the browser's reload button as well as reload software in rather simple DOS (Denial of Service) attacks – overburdening the server of the target site with requests so that it becomes unable to respond. Established in protest against the Japanese textbook – hence the name – in May 2001, *Antijapan* has since managed to crash the Warner Brothers' Web site in protest against a program on Korean dog meat eating. It has also been suggested that some of their members may have been involved in a crash of the Olympic server in Salt Lake City in protest against the disqualification of a Korean speed skater.

The *Antijapan* site provides some background information and several links to similar groups, but nothing like the amount of information provided by the traditional activist groups mentioned above. During the textbook affair, the page offered a Korean BBS, and later a BBS in English was introduced. Notices on the BBS explained the strategy of "virtual sit-ins," which consist of many Internet users continually reloading the pages to be attacked. A message on 10 August 2001 detailed the strategy for an online demonstration planned for 15 August, including a timetable and the Web sites to be attacked. Apart from the Tsukurukai, the Web sites of the newspaper *Sankei Shimbun*, the Education Ministry, the ruling Liberal Democratic Party, and some other organizations were targets of the partially successful attack.[32] *Antijapan* even provided a program for performing the task. Additionally, protesters were asked to send comments to the Japanese Prime Minister's office via the form on its home page. Most of the other sites attacked did not offer a comment form (or even a BBS). Although the appeal did mention that the large majority of Japanese were "honest people" who also opposed the new textbook, most of the messages on the BBS were decidedly polemical.

The "cyber-demo" was briefly mentioned in the Japanese media but received considerably more attention in South Korea.[33] While the Tsukurukai, which was directly affected by the cyber-demo, became aware of the group's activities after the attack, people at Network 21 had only vaguely heard about the cyber-demo and did not know who was behind it. This suggests that *Antijapan*, too, had failed to exploit the full potential of the Internet since links with traditional Japan-based activists might have greatly enhanced their impact.

For the students who founded the site, cyber-demos were a convenient and less time-consuming alternative to traditional offline protests which have a particularly strong tradition in Korea. They saw their activities not as illegal hacking but as an extension of other citizens' protest movements in which they also participated. Given the importance the Internet has come to occupy in Korean society and the education system, it was not difficult for them to set up the site and find supporters; even teachers welcomed their "social commitment." While the visibility and effect of the cyber-demos in Japan may have been rather low, they certainly gained a lot of attention in Korea.[34] However, the actual impact of such activities on politics was probably limited to being an indicator to the Korean government of the public mood.

Other international activist groups did not go quite so far as to launch cyber attacks on the sites of their opponents, but some used the special transnational features of the Internet to collect signatures for an appeal against the Tsukurukai textbook. This included groups based in Japan trying to attract an international audience, usually reflected in the extensive use of English. The "Center for Research and Documentation on Japan's War Responsibility" (JWRC), for example, collected close to 400 signatories to an "International Scholars' Appeal," most based in the US, but also including Japan and other Asian countries.[35] Statements or links to the JWRC's appeal appeared on many other sites not primarily concerned with the textbook issue, including those of the YMCA and the International Movement Against Discrimination and Racism (IMADR), an offshoot of the Buraku Libera-tion League.[36]

State actors and the Internet

Apart from the Japanese Education Ministry (MEXT), the foreign ministries of both countries – MOFA (Japan) and MOFAT (South Korea) – also became involved in the textbook affair. However, MEXT mentioned the new textbook only once on its home page, stating that the Ministry would "make efforts to prevent such problems from happening again."[37] On a page entitled "public comment," MEXT never solicited comments on the issue, nor did it publish any. The foreign ministries provided more information, although the Japanese ministry, MOFA, did not even devote a section of its Web site to the new textbook. Apparently it did not regard the controversy as an "issue" meriting a dedicated presence on its Web site, unlike, for example, whaling, or the (then) controversial Foreign Minister Tanaka Makiko. Inter-ministerial considerations also played a role, as a ministry official explained:

> This textbook issue is really the responsibility of the Education Ministry; therefore it is rather difficult for us to have a specific page on the issue. If MEXT does not take the lead in presenting the government's position, we cannot interfere in that. [38]

In the English version of the Web site, the issue was rarely mentioned, but the search function of the site brought up various Japanese documents relating to the textbook affair, mostly transcripts of press conferences. Usually, officials sought to circumvent journalists' questions about the affair with promises to "continue explaining the screening system," noting that within that system, it was "difficult" to make any further changes to the book.[39] As with the traditional activists, MOFA's Web site includes mainly material that would usually be made available in print or at press conferences: official statements, treaties, press releases, etc. The actual contents of the site differ depending on the department providing them, but generally, few materials are offered in English. Materials in other languages are not available, not even on bilateral relations.

MOFA's Web site offers few interactive features, only an e-mail address and a form for submitting comments and questions. It states clearly that not all questions will be answered, nor will the questions themselves be posted on the site, and indeed, there was no reference to the textbook issue on this page. E-mails addressed to the Ministry are ideally either dealt with by one of the officials in charge of the Web site (one for the Japanese and one for the English version), or forwarded to the relevant division, but rarely answered in either case. In 2001, the Ministry received a total of about 200 e-mails regarding the textbook issue. By comparison, total e-mails received numbered about 100 per week on average, mostly chain mails and advertising – a very small number compared with the responses to Prime Minister Koizumi's highly promoted online magazine (see Mackie this volume). Only about 10 per cent of e-mails to the Ministry are questions, according to officials. They say that the Internet presence itself is "a must," however, since "a few years ago it was just a service, but now people get suspicious if you fail to put certain information online."[40] MOFA certainly has better technical resources and more staff to deal with the Internet than the Japanese activist groups mentioned earlier, but neither side has begun to exploit the full potential of the medium. In effect, this reluctance on the part of the state actors to creatively engage with Internet technology has to some extent leveled the playing field, since activists have had a chance to catch up – but Internet-specific factors such as cost efficiency and immediacy have played only a minor role in this.

By comparison, the Web site of the South Korean Foreign Ministry (MOFAT) offers many more interactive features and more information regarding the textbook issue. Throughout the summer of 2001, the home page had a direct link to the page offering information on the textbook affair – with a large, clearly visible button showing a map of Japan, a book and the word "history" in Chinese characters (thus recognizable to Japanese as well as Korean readers). This reflects the greater importance given to the issue in South Korea. While the divisions usually haggle about what issues should be highlighted on the home page, there was "no question," according to a Ministry official dealing with Japan, that the textbook affair would be provided with its own link button.[41]

MOFAT regularly updated its Korean-language page on the textbook affair, and the English-language version usually followed several days later. The texts available in this section focused on the features of the new textbook that the Korean government had criticized, stating that the approval of the book "breaches the promise made by the Japanese Minister of Education on history textbooks in 1982."[42] The Korean language site of the Ministry also offered interactive tools like a BBS, a comment site and e-mail addresses. The official pointed out that the Ministry has clear directives to consult the public via BBS or Q&A forms. Often, more than 100 postings per day appeared on the BBS regarding the textbook issue. The official pointed out, however, that "Most statements on the bulletin board are very emotional and not really useful. In fact, almost nothing is useful. Usually we don't respond." On the other hand, MOFAT does use the Internet to obtain useful information, and this is where activists may have an impact: the wealth of materials provided by Network 21, for example, was eagerly read by ministry staff. The activists thus managed to reach a potentially influential audience with their arguments. Through providing an informative and reliable Web site, Network 21 may well have achieved more than did *Antijapan*'s more aggressive activism in that they got the Korean government to back their position vis-à-vis the Japanese government – an important strategy for groups engaged in transnational activism.[43]

The official also complained about the additional workload brought about by the Internet – the division has no extra staff to deal with Web site contents and Internet communication, and dealing with the increasing number of responses offsets the time saved by the new technology. This suggests that here, too, the Internet levels the playing field somewhat, at least to the extent that the ministry has not so far dedicated a large amount of additional resources to its online operations. The ministry has never considered using the Internet as a means of protest elsewhere, for example by asking people to send e-mails about the textbook issue to Japanese government sites. Official South Korean protests continue to be lodged through traditional diplomatic channels.[44] It is likely, however, that the public statements on MOFAT's (and other official agencies') Web sites have served as one indicator of public opinion for politicians, whose strong stance on the issue was arguably due at least in part to the public outcry. The Korean parliament adopted a resolution urging Japan to reconsider the textbook decision, and MOFAT itself issued an official statement of "disappointment."[45]

While none of the ministries mentioned offered information in the language of the other country, both foreign ministries had links to their embassies abroad (although the Japanese embassy in South Korea was not always accessible via this link). They provided a rather detailed coverage of the issue, including translations of important documents and statements in their host country's language, often in addition to an English version.[46]

CONCLUSION

On the whole, the case studies in this chapter do not support the hypothesis that the Internet has acted as an equalizing force between political Davids and Goliaths. Indeed, the differences observed between Japanese and Korean community groups' utilization of Internet technology are far greater than those observed between state and non-state actors. The activist groups examined actually offer fewer interactive features than do the ministries. In Japan, the Internet has turned out to be an equalizer of sorts only to the extent that both sides encountered essentially the same problems with the new technology – a lack of dedicated skills and resources. For a number of reasons outlined above, all parties in the debate failed to realize the full potential of the Internet to recruit and mobilize a wide range of individuals behind their cause. The question remains whether and to what extent the Internet can create a new forum for heightened political action or debate.

Community activism on both sides of the textbook debate is still based on traditional campaign methods that have only been slightly enhanced by basic Internet features. The Internet is certainly a very useful tool, but so far has proven no more revolutionary than the fax machine. As other contributors to this volume have also found, communication among activist groups and other subcultures has not exponentially increased, and e-mail contacts have certainly not been substituted for face-to-face meetings. Networking by all the groups considered was still largely initiated via personal contact, even in the transnational arena where the Internet could have proven useful in contacting distant supporters.

Age has been a factor in the development of Internet activism around the textbook affair since most activists in Japan are middle-aged or older (as those most concerned tend to be the parents of teenage children – the intended audience of the textbooks). The more proactive stance of Korean groups, however, is to be explained not only by the wider adoption of Internet technology in Korea but also by the larger number of young people concerned about the textbook issue. In Korea, arguments about Japanese history textbooks are tied up with wider concerns about Korean nationalism and appeal to a broad audience including both young and older people. Yet, when primarily younger activists did use innovative Net-based methods of campaigning, they were not taken very seriously. This was partly due to the biased and polemical nature of their statements but also due to their lack of experience in lobbying (for instance, flooding selection committees with e-mails may have proven a more successful strategy than jamming the Web sites of textbook supporters). In order for the Internet to prove a useful campaign tool, the experience of younger and older activists needs to be integrated. As noted above, traditional activists have largely failed to grasp the opportunities offered by the Internet, while younger Internet activists are lacking in campaign experience and strategy. Although the potential is there, a powerful, large-scale, transnational Internet-based campaign for textbook reform has not materialized.

In some respects, it is still impossible to effectively use the Internet in a campaign for textbook reform in Japan since many of the *target* organizations themselves lack the necessary features. For instance, "virtual sit-ins" are most effective on BBS, which are rare on Japanese Web sites related to the textbook affair. Indeed, regional and local selection committees often had no Internet presence that could become a target for online protest. Effective Internet campaigning depends not only on the activists' own approach to the Internet, but on that of their targets as well.

Even the expressly stated efforts of ministries to collect public comments either via BBS or via comment forms and e-mail did not have a great effect. They could only provide one, partial indication of public opinion trends, albeit a very immediate one. In this context it is interesting to note that officials did not pay much attention to comments sent to them because they found them polemical and useless (*yaku ni tatanai*) – although they did use informative materials provided online by activists even when not directly addressed to them. The biased nature of much Internet-based interaction also increases the fragmentation of the Net – the activist groups all admit that their sites mostly address people already interested in their activities. At the same time, polemical (and sometimes aggressive) usage of the Internet has reinforced security fears, which are particularly widespread in Japan. Activist groups in both camps showed concern about virtual attacks or harassment from opponents. This has hampered further development of interactive features on their sites and is, in turn, probably utilized as a pretext for adopting a timid approach to new technologies. Rather than seeking a virtual dialogue with opponents, activists preferred to offer information to their supporters while keeping their opponents out.

While the Internet has proven a useful tool for the activists campaigning for textbook reform, the new technology alone has not radically empowered them. If it has leveled the playing field, it is only because neither state nor non-state actors have grasped the full potential of the medium, making access to greater or lesser Internet resources largely irrelevant. The Internet does have the potential to empower activists – not only progressive ones – but only when used efficiently. For this to take place, however, a change of attitude and strategy is necessary, including greater coordination and networking with other groups, especially across age as well as national and language borders.

NOTES

1   See, for example, P. Norris, *A Virtuous Circle: Political Communications in Postindustrial Societies*, Cambridge, Cambridge University Press, 2000. H. Geser, *Auf dem Weg zur Neuerfindung der politischen Öffentlichkeit* (On the way to re-inventing the political public), 1998. Online. Available HTTP: *http://socio.ch/intcom/t_hgeser06.htm* (23 April 2002). S. Moog and J. Sluyter-Beltrao, "The transformation of political communication," in B. Axford and R. Huggins (eds), *New Media and Politics*, London, Thousand Oaks, New Delhi, Sage Publications, 2001, p. 55. K. Grönlund, *Do New ICTs Demand more Responsive Governance*, paper presented at European Consortium for Political Research, Workshops, Grenoble: 6–11 April 2001. Online. Available HTTP: *http://www.essex.ac.uk/ecpr/jointsessions/grenoble/papers/ws3.htm* (26 April 2002).

2   T. Morris-Suzuki and P.J. Rimmer, *Cyberstructure, Society and Education: Possibilities and Problems in the Japanese Context*. Online. Available HTTP: *http://www.nime.ac.jp/conf99/pre/Morris-Suzuki.paper/Morris-Suzuki.html* (23 April 2002).

3   See J. Åström, *Digital Democracy: Ideas, Intentions and Initiatives in Swedish Local Governments,* paper presented at European Consortium for Political Research, Workshop, Grenoble, 6–11 April 2001. Online. Available HTTP: *http://www.essex.ac.uk/ecpr/jointsessions/grenoble/papers/ws3.htm* (26 April 2002). S. London, *Electronic Democracy – A Literature Survey*, 1994. Online. Available HTTP: *http://www.scottlondon.com/reports/ed.html* (23 April 2002).

4   B. Axford, "The transformation of politics or anti-politics?" in B. Axford and R. Huggins (eds), *New Media and Politics*, London, Thousand Oaks, New Delhi, Sage Publications, 2001, p. 15. M. Taylor, M. Kent, and W. White, "How activist organizations are using the Internet to build relationships," *Public Relations Review*, vol. 27, no. 3, 2001, p. 266.

5   M. Castells, *The Information Age: Volume I The Rise of the Network Society*, Malden, MA, Blackwell Publishers, 2000.

6   See R. Huggins, "The transformation of the political audience," in B. Axford and R. Huggins (eds), *New Media and Politics*, London, Thousand Oaks, New Delhi, Sage Publications, 2001, p. 134.

7   Internet use is more widespread in Korea than in Japan. A very proactive government policy has resulted in high access rates and a majority of households are linked to broadband services. Internet users amount to 51.5 per cent in Korea as of end 2001; numbers for Japan are certainly lower. Statistics often use two overlapping categories of "subscribers" instead of "users:" about 20 per cent dial-up and broadband subscribers, and 40 per cent mobile phone Internet subscribers. Many of of the latter do not use the Internet at all, or use only e-mail functions). For an overview of the history of the Internet in Korea, see W. Kang, "The engine for the next economic leap: the Internet in Korea," in S. Rao and B. Klopfenstein (eds), *Cyberpath to Development in Asia: Issues and Challenges*, Westport, CT and London, Praeger, 2002, pp. 111–36; S. Yang, "Information Ministry looking to upgrade Korea's info-tech infrastructure," *Korea Herald*. Online. Available: HTTP: *http://www.koreaherald.co.kr/SITE/data/html_dir/2002/04/22/200204220003.asp* (30 April 2002); S.Y. Uhm and R. Hague, "Electronic governance, political participation and virtual community: Korea and UK compared in a political context," paper presented at European Consortium for Political Research, Workshop, Grenoble: 6–11 April 2001. Online. Available HTTP: *http://www.essex.ac.uk/ecpr/jointsessions/grenoble/papers/ws3.htm* (30 April 2002); and Sōmushō, "Number of Internet users (as of 28 February 2002)." Online. Available HTTP: *http://www.soumu.go.jp/joho_tsusin/eng/Statistics/number_users020329.html* (30 April 2002).

8   See Y. Tanaka, *Japan's Comfort Women: Sexual Slavery and Prostitution During World War II and the US Occupation*, London, Routledge, 2002, p. 31; and G. Hicks, *The Comfort Women*, St Leonards (Aus), Allen & Unwin, 1995.

9   See, for example, "Chōsen kōkyū gakkō nado e monko kaihō" (Opening the door for Korean high schools), *Asahi Shimbun*, 23 February 1998, p. 10; N. Iwamoto, "*Nikkan kankei no mirai o kanjisaseru Shinjuku rittoru koria*" (Little Korea in Shinjuku, Where You Feel the Future of

Japanese–South Korean Relations), *Jitsugyō no Nihon*, May 1997, pp. 50–2. Some Koreans become naturalized, although nationality is still based on origin, and naturalization criteria remain rather tight. See C. Kashiwazaki, "The politics of legal status," in S. Ryang (ed.), *Koreans in Japan: Critical Voices from the Margin*, London and New York, Routledge, 2000, p. 26.

10 For an overview of the issue of Korean residents, see S. Ryang (ed.), *Koreans in Japan*.

11 See I. Ducke, *Status Power: Japan's Foreign Policy toward Korea*, New York, Routledge (forthcoming).

12 Tsukurukai, *The Restoration of a National History*, Tokyo (pamphlet) 1998, p. 3.

13 MEXT is the official abbreviation of the Ministry of Education, Culture, Sports, Science and Technology. Other abbreviations used are also those officially endorsed by the organizations.

14 J. Conachy, "Japanese history textbook provokes sharp controversy," *World Socialist Web Site*, 7 June 2001. Online. Available HTTP: *http://www.wsws.org/articles/2001/jun2001/text-j07.shtml* (26 April 2002).

15 S. Saaler, "Grassroots initiative against historical revisionism," *DIJ Newsletter*, February 2002, pp. 1–2. Online. Available HTTP: *http://www.dijtokyo.org/dij-e/berichte/pdfs/newsletter15-e.pdf* (26 April 2002).

16 The system described here applies to the majority of schools, those under the authority of municipal governments. However, private schools can select any of the approved books. For special-needs schools, the decision is made by their prefecture's education boards.

17 Embassy of Japan in Korea, 12 September 2001, *Yōksa gyogwasō munjae > 2002 nyōndo junghakgyo yōksa gyogwasō ui saengdoyong suyosōg (chung: chaetaeg yul) e daehayō* (History textbook issue: on the demand ranking [adoption rate] for middle-school history textbooks). Online. Available HTTP: *http://www.japanem.or.kr/textbook/textbook_18.htm* (26 April 2002).

18 It was only later discovered that the change required by the Ministry (then called Monbushō) had never taken place. Apparently, a journalist comparing two versions of a book had made a mistake (B. Bridges, *Japan and Korea in the 1990s*, Cambridge, Cambridge University Press, 1993, p. 61). For a detailed description of the issue, see A. Ortmanns-Suzuki, "Japan und Südkorea: die Schulbuchaffäre" (Japan and South Korea: the textbook affair), *Japanstudien*, 1989, pp. 135–82.

19 Ortmanns-Suzuki, p. 139.

20 See House of Representatives, *Shūgiin gaimu iinkaigiroku* (Records of the House of Representatives Foreign Policy Committee), 96-24-1-4, 9 August 1982, p. 12; and R. Kang, "A comparison of the foreign policy making process in Japan and South Korea, in the case of the loan negotiations 1981–3," Newcastle University, PhD dissertation, 1994, p. 133.

21 Canada Association for Learning and Preserving the History of WWII in Asia, *Ienaga Textbook Screening Lawsuits, 1965–1997: An Overview*, 1997. Online. Available HTTP: *http://www.vcn. bc.ca/alpha/overview.htm* (26 April 2002).

22 *http://www.ne.jp/asahi/kyokasho/net21/e_010403seimei_1.htm* (26 April 2002).

23 Network 21, 7 and 8 August 2001, *Tōkyōto kyōiku iinkain no "tsukurukai" kyōkasho no yōgo gakkō e no saitaku ni kōgi suru* (Protest against the adoption of the "Tsukurukai" textbook by Tokyo government for schools for the disabled). Online. Available HTTP: *http://www.ne.jp/ asahi/kyokasho/net21/seimei_03-04.htm* (25 April 2002).

24 Tawara Yoshifumi, personal interview, 6 March 2002.

25 Tawara.

26 Tawara.

27 *Ilbon gyogwasō barojapgi undong bonbu* (Headquarters of the movement to correct Japan's textbooks), *http://www.japantext.net/* (26 April 2002).

28 Chung Mong Joo, vice-secretary and director of the education section of Mindan headquarters in Tokyo, personal interview, 1 November 2001. Mindan, *http://www.mindan.org/* (26 April 2002).

29 Chōsōren, *http://www.chongryon.com/* (26 April 2002). *Chaeil.net*, *http://www.chaeil.net/* (26 April 2002). *Hantongryōn*, *http://korea-htr.com/* (26 April 2002).

30 Tsukurukai, *"Tsukurukai" jimusho hōkai! Hiretsu na tero o yurusanai!* (Arson on "Tsukurukai" office! Don't permit mean terror!). Online. Available HTTP: *http://www.tsukurukai.com/*

houka.html (8 August 2001). See Network 21, *Daiwa: "Tsukurukai" jimusho kazan o meguru sankei shinbun no bōriakuteki hōdō ni genjū ni kōgi suru* (Talk: vehement protest against the scheming coverage by Sankei Shinbun of the fire at the "Tsukurukai" office). Online. Available HTTP: *http://www.ne.jp/asahi/kyokasho/net21/seimei_03-04.htm* (23 April 2002).

31 Etoh Takahiro, personal interview, 25 February 2002, Tokyo. Tsukurukai: *http://www. tsukurukai.com/* (23 April 2002).

32 *Antijapan*, 10 August 2001, *815 ssaibō shiui e gwanhan jilyong ibnida* (Directives for the cyber-demo on 15 August). Online. Available HTTP: *http://board2.hanmir.com/blue/Board.cgi? path=db48&db=antijap&page=14&cmd=view&no=2* (25 April 2002). *Antijapan*'s home page, *http://www.antijap.wo.ro/*, has since become inaccessible.

33 *Asahi.com*, 13 August 2001, *Kankoku no gurūpu, 15 nichi ni "saibā demo" yobikake* (ROK group calls for "cyber-demo" on 15th). Online. Available HTTP: *http://www.asahi.com/ international/update/0812/005.html* (13 August 2001). *SBS*, 15 August 2001, *"'Ilbon gyutan' saibō shiui"* ("Japan denunciation" cyber-demo). Online. Available HTTP: *http://news.sbs.co.kr/ society/news_society_IndexDetail.jhtml?news_id=N0311115701* (15 August 2001). *Hangyore*, 6 February 2002, *Saibō shiuidae 'anti jaepaen' ui gogyosaengdūl* (The cyber-demo students of "anti-Japan"). Online. Available HTTP: *http://worldcup.hani.co.kr/section-005000000/2002/02/ 005000000200202061430617.html* (6 February 2002). *Korea Herald*, 21 January 2002, "Koreans launch Internet campaign against US, French networks over dog meat reports." Online. Available HTTP: *http://www.koreaherald.com* (21 January 2002).

34 Ishibashi Hideaki, personal interview, 26 March 2002. Ishibashi, an *Asahi* journalist, held extensive interviews with the students of *Antijapan* and their teachers. *Asahi Shimbun*, "17sai, rekishi toi netto kōgeki" (At 17, Net attack on history question), 2 April 2002, p. 34.

35 JWRC (*Nihon no sensō sekinin shiryō sentā* [Center for Research and Documentation on Japan's War Responsibility]), *http://www.jca.apc.org/JWRC/index-j.html* (7 August 2002).

36 IMADR (International Movement Against Discrimination and Racism), 19 April 2001, *Contemporary Forms of Japanese Nationalism and Racism*. Online. Available HTTP: *http:// www.imadr.org/attention/news2001.1.html* (24 April 2002). YMCA World Index, September 2001, *Korea: YMCA protests Japan's history distortion*. Online. Available HTTP: *http:// www.ymca.int/Publications/YMCAWorld/Sept2001/3_2001korea.htm* (24 April 2002).

37 MEXT, 13 July 2001, *Chūgaku rekishi kyōkasho shūsei yōkyū ni kakawari kentō kekkara ni kansuru monbu kagaku daijin komento* (Comments of the Education Minister regarding inquiry results on the demands for changes in junior-high school history textbooks). Online. Available HTTP: *http://www.mext.go.jp/b_menu/houdou/13/07/010799.htm* (23 April 2002).

38 MOFA official, Domestic Public Relations Division, personal interview, 18 October 2001.

39 MOFA, 10 July 2001, *Jimu jikan kaiken kiroku* (Protocol of the Vice Minister's Press Conference). Online. Available HTTP: *http://www.mofa.go.jp/mofaj/press/kaiken/jikan/j_0107.html* (28 April 2002).

40 MOFA official, Overseas Public Relations Division, personal interview, 18 October 2001.

41 MOFAT official, North East Asia Division I, personal interview, 19 November 2001.

42 MOFAT, *Review of the Japanese History Textbooks: Notes on Necessary Corrections Related to Korea*, 10 July 2001. Online. Available HTTP: *http://www.mofat.go.kr/main/etop.html* (10 July 2001).

43 See M. Keck and K. Sikkink, *Activists Beyond Borders: Advocacy Networks in International Politics*, Ithaca, NY, Cornell University Press, 1998, p. 23.

44 See also *Napsnet*, 3 July 2001, "Daily report." Mailing list/news clipping service. Online. Available HTTP: *http://www.nautilus.org/napsnet/dr/0107/jul03.html* (27 April 2002).

45 *Asahi.com*, 20 July 2001, "Koizumi's diplomacy means single-minded devotion to the US" Online. Available HTTP: *http://www.asahi.com/english/politics/K2001072000288.html* (25 April 2002). MOFAT, 9 July 2001, "Statement by MOFAT spokesman on the Japanese Government's response to the ROK's request on history textbook corrections." Online. Available HTTP: *http://www.mofat.go.kr/en/rel/e_rel_view_dip.mof?seq_no=77&b_code=~diplomatic* (7 August 2002).

46 Embassy of Japan in Korea, July 2001, *Junghak yŏksa gyogwasō sujŏng-yogu gomto gyŏlgwa e gwanhan munbu gwahak daeshin komentū* (Comments of the Education Minister regarding inquiry results on the demands for changes in junior high school history textbooks). Online. Available HTTP: *http://www.japanem.or.kr/textbook/textbook_14.htm* (23 April 2002).

15 Self-representation of two new religions on the Japanese Internet

Jehovah's Witnesses and
Seichō no Ie

Petra Kienle and Birgit Staemmler

INTRODUCTION

Japanese new religions, that is religious organizations founded since about the middle of the nineteenth century, are not represented on the Internet as much as one might expect given their propensity towards proselytization. While organizations such as Sōka Gakkai, Tenrikyō or Aleph (the former Aum Shinrikyō) admittedly run several extensive Web sites, other new religions equally active on an international level, such as Sekai Kyūseikyō, Kōfuku no Kagaku or Sekai Mahikari Bunmei Kyōdan, have but few sites which contain – or contained when research for this chapter was done in 2001 – little more than most of their pamphlets.[1]

Despite considerable differences among individual organizations, new religions share a number of characteristics such as a simple, often syncretic doctrine focusing on this world rather than life after death; a clear hierarchic structure centering on the founders and his or her descendants; and easy-to-grasp religious practices and activities often believed to help gain this-worldly benefits. Several hundred new religions are active in Japan, claiming between one hundred and a few million members. Apart from new religions founded in Japan with doctrinal backgrounds drawing primarily on Shintō, Buddhism and Confucianism, there are also imported new religions based on Christian or Hindu teachings.

Although many new religions unobtrusively fulfill religious demands and social functions, political involvement or excessive missionizing by some new religions and especially the sarin-gas attack in 1995 have kindled public criticism of new religions in general and in particular of well-known organizations such as Aum Shinrikyō, Sōka Gakkai and the Unification Church.[2]

As examples for this chapter we chose two important new religions whose usage of the Internet is entirely dissimilar: the Watch Tower Bible and Tract Society, commonly known as "Jehovah's Witnesses," and Seichō no Ie, one of Japan's largest new religions. We chose these groups because they both operate on an international basis, place high emphasis on written materials and engage actively in proselytization – with written materials and word-of-mouth as the two most important means of attracting new members. Although they emerged from different backgrounds and in different countries, they thus have several significant features in common. Our research focuses on two aspects of Internet usage: external communication between the organizations or groups involved and society at large and internal communication among group members. Our purpose is to discover how the Internet may modify forms of communication and proselytization.[3]

Web sites consist of multiple elements such as written texts, pictures, links and sometimes acoustic sequences, video streams and animations. To consider all these elements we developed a detailed encoding system which incorporates information about both the organization maintaining the site and the site itself. Information gathered included:

- details of the organization (address, affiliation, further references)
- formal information about the site (name of Web master, date of publication, etc.)
- list of contents
- presentation (colors, graphic elements, frames, etc.)
- possibilities of interaction in real life and online (road map, e-mail, games, etc.)
- miscellaneous (mobile phone version, external links, hit counter)
- technical data (date of download, etc.).

We were thus able to provide comprehensive descriptions of the sites' relevant features. The analysis ensuing from the description of a site is intratextual, considering not only a site's written texts but its entirety, including textual, visual and acoustic elements as well as Internet-specific features such as frames and links. Additional information about religious Web sites was derived from interviews with those in charge and comparison with other sources of self-representation such as books or pamphlets.[4]

In order to better evaluate sites maintained by religious organizations we distinguished between official and private sites. Official sites are those published by a religious organization's management or their authorized staff, whereas private sites are published by individual members or groups and are not necessarily authorized by the organization. "Official" and "private" are not, however, opposites, but two ends of a continuum. One indication that a site is official is mention of copyright, but distinguishing between the two is not always easy or unambiguous and is sometimes impossible. Yet, it is often the only information available about a site's author – apart perhaps from the Web master's personal name.

JEHOVAH'S WITNESSES

The Japanese history of Jehovah's Witnesses began in 1911 when Charles T. Russell (1852–1916), the founder of what was later to be called the Watch Tower Bible and Tract Society, visited Japan. It was not until 1926, however, that a Japanese branch was established by Akashi Junzō (1889–1965), known as *Tōdaisha* or "The

Lighthouse." As in many other countries, Jehovah's Witnesses were also persecuted under Japan's military regime for their reserved stance toward secular government, their refusal to serve in the army, to salute the flag or to worship the Japanese emperor. Although disbanded in 1940, the *Monomi no Tō Seisho Sasshi Kyōkai*, as today's Watch Tower Bible and Tract Society is called in Japanese, recovered and expanded rapidly after the war. They claim more than 200,000 adherents in Japan today, although the figures have been decreasing in recent years according to the Jehovah's Witnesses' own statistics. Their avowed goal is to establish God's Kingdom, which they believe will emerge following Armageddon. The traditional concept of hell is rejected, as is that of eternal life. Preaching and publishing are regarded as much more valuable than praying and studying the Bible.

Although the Internet could be regarded as another very convenient medium for publishing and although "since their early days, the Witnesses have proudly availed themselves of the latest technology for the propagation of their message,"[5] there is only one official Web site for this organization in Japanese on the Web. This Web site is not independent, but consists of several Japanese pages included within the official site of the American parent organization. According to the description in the source code, this Web site is "the authoritative source about beliefs, teachings and activities."[6] The first priority of these pages is to communicate to non-members information about various aspects of Jehovah's Witnesses' beliefs and moral standards and to resolve common misunderstandings, since "it is the desire of Jehovah's Witnesses that you become better acquainted with them."[7] These Japanese pages are literal translations from the English pages and are available in eighteen other languages. It is notable that these pages – as well as their numerous and diverse print publications – have been translated into various languages without having been adapted to the particular cultural and social circum-stances of the respective societies.

The private sites of Jehovah's Witnesses found on the Japanese Internet may be divided into three major groups: members' sites, supporters' sites, the so-called *yōgoha* (a group of non-members protecting and defending the Jehovah's Witnes-ses), and sites of opponents and apostates. There are approximately the same number of members' and *yōgoha* sites as there are sites of opponents. According to the dates given on the Web sites, the opponents' sites were created first and members' and sympathizers' sites followed and are characterized by their defensive positions. One such site claims that:

> In recent years many people wishing to learn about the Jehovah's Witnesses through the Internet have been harmed by mistaken information about the Jehovah's Witnesses because of propaganda activities by people opposing them in dishonest ways, especially through the Internet.[8]

All of these sites, both pro and contra the Watch Tower Bible and Tract Society, aim to inform a wider public about beliefs and to resolve mutual misunderstandings.

Because people are unable to distinguish correct information about the Jehovah's Witnesses from mistaken information, they take lies for the truth and are led to the wrong conclusion that the Jehovah's Witnesses are a harmful cult.[9]

As regards content, private sites – whether for or against Jehovah's Witnesses – focus mainly on those social issues which were and are most prominently discussed in public: Jehovah's Witnesses' refusal of blood transfusion, difficulties faced by their children, and problems occurring when spouses are not both members of the organization.

Jehovah's Witnesses' members' and sympathizers' sites are addressed primarily to non-members, have some evangelistic elements, and advertise their way of life and religious practices. None of the sites, however, contain any announcements about events and meetings, addresses or other information for making contact offline. The main purpose and aim of the Jehovah's Witnesses' Web presence is to defend themselves against criticism.

The Web master of *Ehoba no Shōnin Kisha Kurabu*, who aims to provide "correct and useful information" for researchers, journalists, Bible students and other Christians, asserted that due to his Web site prejudices against Jehovah's Witnesses had been overcome.[10]

Among the private sites opposing the organization are Web sites of children of Jehovah's Witnesses – referring to themselves as *nisei* or "second generation" – suffering because of their parents' faith. The main purpose of these sites is to create and provide a forum for apostatized children of Jehovah's Witnesses – not only through the Web sites themselves, but also via linked mailing lists and bulletin boards:

> This exchange forum is a forum to share with others one's wounds and pain. The wounds and pain we members of the second generation suffered from heal other people bearing the same wounds and pain ... Please, let this home page be a lively forum to heal and to be healed.[11]

Due to the disproportionately high representation of housewives among Japanese Witnesses, there are also Web sites of unsatisfied and unhappy husbands whose spouses are or were members of the group. These sites offer translations of texts on various problems concerning the teachings and practices of this religious organization and reports by persons affected. Additionally, they quote biblical sayings and theological interpretations trying to defend their own understanding of Christian doctrine in contrast to that of Jehovah's Witnesses. Web sites from relatives' self-help groups, as for instance the Nihon Maikon Kenkyūkai[12] and Firippusu no Mirai,[13] are – according to their Web masters – useful tools for getting into contact with counselors or other people facing similar problems. Communication in both groups is mainly Net-based, although there are face-to-face meetings once or twice a year.[14]

Jehovah's Witnesses' members themselves are discouraged from accessing Internet sources of information and especially from using chat rooms and newsgroups:

> Extreme caution must be exercised when using a computer chat room ... Adults who are lonely or who are unhappily married should not turn to computer chat rooms for emotional support. There is danger in turning to strangers.[15]

According to a recent unpublished Japanese survey, the Internet played an important role as a source of information in the decision-making process of defecting members. The Internet offers members easier access to other voices and information, particularly to information critical of the organization. As a participant of the survey revealed: "The threshold is lower to browse the Internet with just a single click than to pick up an apostate book. Nevertheless, when typing the URL my finger trembled."[16]

In Japan, counseling for apostatized Witnesses is most often done by pastors of Baptist churches. Some Web sites of Baptist churches include pages on Jehovah's Witnesses containing critiques about their teachings and refutations of the Witnesses' interpretations of the Bible – as well as offering counseling for those who intend to defect. Nakazawa Keisuke, a counselor for Jehovah's Witnesses and a senior pastor of the Ōno Kirisuto Kyōkai, a Baptist church in Kanagawa Prefecture, confirmed that the Internet played an important role in the process of apostatizing, although counseling itself is never undertaken by e-mail or other forms of computer-mediated communication.[17]

In addition, there are several sites opposing and defending the Jehovah's Witnesses, which consist merely of links to bulletin board systems and mailing lists. Thus, private communication among adherents as well as with apostates appears not only to be conducted through Web sites, but also via bulletin board systems, mailing lists and guestbooks.

SEICHŌ NO IE

The Seichō no Ie, literally "House of Growth," is a new religion founded in 1930 by Taniguchi Masaharu (1893–1985). Taniguchi's forty-volume *Seimei no jissō* is the Seichō no Ie's central doctrinal text, but publications by the current president Taniguchi Seichō (born 1919) and other leaders are also highly treasured. The Seichō no Ie publications include the monthly *Seishimei* newspaper and four magazines aimed at men, women, young people and children respectively, totaling over one-and-a-half million copies per month.[18]

The Seichō no Ie's teaching is syncretic, containing elements of Shintō, Buddhist and Christian doctrine. Religious practice focuses on studying the Seichō no Ie's publications, practicing *shinsōkan*, a form of meditation, participating in regular sessions of local men's, women's or youth groups, taking part in spiritual training seminars and attending monthly or yearly lecture meetings. Faith, filial piety and especially gratitude are believed to result in benefits such as healing, financial improvement and harmony.[19]

The Seichō no Ie explicitly promotes the use of computer networks for both internal communication and external representation. Annual policy statements have for several years urged the use of computer networks resulting in a network of loosely related Web sites presenting and propagating the Seichō no Ie to the outside world as well as in a private forum used by leading members of staff to more effectively circulate news.[20]

The Seichō no Ie's official sites constitute a network whose structure mirrors that of the religion's offline organization. The Japanese Religious Juridical Persons Law (*shūkyō hōjinhō*) enables religious organizations to encompass any number of legally independent institutions, known as comprehended religious juridical persons (*hihōkatsu shūkyō hōjin*), under one umbrella organization which is then referred to as comprehensive religious juridical person (*hokatsu shūkyō hōjin*).[21] The Seichō no Ie's umbrella organization consists of the combined administrative and ceremonial headquarters of Tokyo and Nagasaki, the comprehended institutions are its fifty-nine regional headquarters and three further centers of nationwide importance. The Seichō no Ie also has numerous centers overseas, especially in Brazil.

The Seichō no Ie's official sites accordingly include those by the headquarters and by its various branches and centers linked together through one distinct index-site containing the official collection of links.[22] Although the centers in Tokyo and Nagasaki are legally united into one religious corporation, they nonetheless maintain separate sites, each reflecting its own main functions: the Tokyo head-quarter's site contains excerpts from the newspaper *Seishimei*, introductions to the Seichō no Ie's history and doctrine, and an extensive section of questions and answers.[23] The site of Nagasaki's Sōhonzan, on the other hand, features photos of its precincts, ceremonies and environmental activities as well as practical infor-mation on upcoming events.[24] To date only three regional and several overseas headquarters maintain their own Web sites, but their number is likely to increase considering the Seichō no Ie's positive attitude towards the Internet.

Another essential part of the Seichō no Ie's horizontal structure is sections for men, women, young people and managers. The headquarters – and Web masters – of these organizations are situated in the main headquarters in Tokyo, but activities are conducted nationwide. Official sites for these sections are technically part of the main index-site, but because of the sections' significance links to their sites figure prominently on the Tokyo headquarters' site. Compared with the magazines issued by these organizations, the Web sites offer relevant practical information, but considerably fewer and shorter reports, testimonials and articles.[25]

All of the official sites are produced independently and mostly by local members of staff without professional help or cooperation between Web masters of different branches. In some cases – notably the training centre in Tobitakyu and one of the US sites – members happened to include Internet specialists who offered their expertise. Staff at the Tokyo headquarters examine local sites before including them in the official collection of links, thereby acknowledging them as official Web sites.[26]

Because the official sites are created by various members and members of staff, layout and contents vary considerably and reflect the respective section's or center's individuality. The training center in Tobitakyu, for instance, only completed renovating its main building in spring 2001 and started its site explicitly to publicize the new facilities and invite visitors and participants.[27] The head of Seichō no Ie's International Spiritual Training Seminar (ISTS) in New York similarly explained his reasons for opening a site:

> We have opened it because we want to let the people in the world know about ISTS's fascination. In the near future we want to enrich the contents so that one can for example register online to participate [in seminars].[28]

The youth organization's site includes many photos of its local members in order to familiarize potential newcomers with those they will meet.[29] The Hōzō shrine for worship of ancestors and aborted or stillborn babies figures as prominently in Uji's real center as on its Web site.[30]

Apart from the official sites there are numerous other sites published by local branches or individual members. Even more than the official sites they reflect their authors' personalities and, unsurprisingly, differ considerably from one another in content, layout and usability. Vice-president Taniguchi Masanobu's (born 1951) site attests to his enthusiasm for computers by being updated several times a week.[31] There is also the site of the Kiga Kyūsai, an online-only organization founded by members of the managers' section who donate 20 yen each to the World Food Programme for each visitor to the site.[32]

Official and also private Seichō no Ie sites generally provide abundant information on dates, contents, addresses and prices of seminars and other activities as well as countless photos of members enjoying these activities. Most sites contain testimonials of benefits received through membership or short messages praising individual events. Surprisingly little information is presented on the founder Taniguchi Masaharu. The express purpose of the sites is to make the Seichō no Ie known to as many people as possible and to invite them to experience for themselves the benefits highlighted. The main target group, therefore, is those participating in Seichō no Ie activities for the very first time.

The second result of the Seichō no Ie's policy of promoting Internet use is the private forum established with Nifty, one of Japan's leading providers for Internet services, for the use of leading members of staff and heads of regional centers. It is text-based and contains reports of meetings, important pieces of news or noteworthy testimonials. It improved not so much communication, but the flow of information to overcome problems such as those arising out of cases of temporary absence.[33]

Internal communication has changed due to the ubiquitous use of e-mails, especially among leading members of staff in Japan and overseas and among members of the youth organization. Communication by e-mail in other areas, especially the women's organization or local branches, remains difficult, because addressees cannot necessarily be expected to have access to a computer. About 80 per cent of the youth organization's senior high school and university students own mobile telephones and are thus increasingly informed of upcoming events through e-mails directed to mobile telephones.[34]

Although the form of communication has changed through the use of the Internet, it does not seem to have intensified, as the increasing use of e-mails replaces a similar amount of telephone calls, facsimiles and letters. While e-mails may be conveniently employed to convey pieces of information, they only rarely – and mainly with the young – substitute for the telephone in encouraging or counseling members.[35]

ONLINE SELF-REPRESENTATION BY JEHOVAH'S WITNESSES AND SEICHŌ NO IE

Comparing the Web sites of the Seichō no Ie and the Jehovah's Witnesses reveals that sites by these religions differ on the scale between "official" and "private;" whereas sites by the Seichō no Ie are located continuously along the scale, sites by Jehovah's Witnesses and their antagonists cluster at either end with nothing in between. This is due to the fact that Jehovah's Witnesses are generally discouraged from using the Internet to avoid being influenced by potentially negative opinions. Consequently, regional headquarters do not publish their own Web sites. A second reason is the two religions' contrasting position towards religious truth: the Japanese Watch Tower Bible and Tract Society disseminates its exclusive religious truth through its single official Web site and argues that nothing else should be trusted when making conversions or speaking to the faithful. Private sites by members inevitably state that they are merely private sites. The Seichō no Ie, on the other hand, argues through its many testimonials that benefits gained through participation in its activities will convince anybody, independent of their religious background.

In analyzing the structure of Internet sites scholars distinguish between hierarchic and interlinked structures.[36] In contrast to interlinked structures, hierarchic structures closely resemble those of traditional texts and do not employ horizontal connections between parts of the site. Although there are differences between individual sites, most sites by the Seichō no Ie and Jehovah's Witnesses are hierarchic rather than interlinked. Some sites for and against Jehovah's Witnesses present very long texts with few adaptations to the needs of the Internet. These traditionally structured long texts indicate that the requirements of Internet publication are not always met, nor are the novel possibilities of Internet publishing fully explored.

Research has shown that it is possible to infer from a great number of reciprocal external links – links connecting to sites with a different base URL – that the organizations in question constitute a close-knit network.[37] External links are found in considerably greater numbers on sites of groups protective of the Jehovah's Witnesses than on sites opposing them. We may thus conjecture that more effort is made by members of the group protective of the Jehovah's Witnesses to establish online networks than by opponents. In the Seichō no Ie's case real-life efforts at community-building through joint events and magazines are mirrored by their interlinked Web sites. External links mainly lead to other Seichō no Ie sites, in particular those related to publications.

Internal communication within both organizations does not seem to have increased with the use of the Internet. Although computer networks in part have come to replace other means of communication between members of staff, this has not yet affected communication among members, as access to computers cannot be taken for granted and older forms of communication are well established.[38] An exception are the Jehovah's Witnesses' self-help groups and groups of apostates whose internal communication increased significantly due to the Internet.

Sites by and about both organizations contain numberless testimonials but their contents and aims differ considerably. In Jehovah's Witnesses' testimonials spouses of members defend the organization, denying that membership causes tensions within their families, whereas Seichō no Ie testimonials enthusiastically praise membership as being beneficial, as seen in this testimonial by a young mother:

> Since the beginning of this year I have become able to feel very joyful because I started participating in the Mothers' Classroom. I have learned to easily express feelings of gratitude. I want to keep on using many positive words.[39]

This difference may in part be explained through society's attitudes to these religions since public opinion is much more critical of the Jehovah's Witnesses than of the Seichō no Ie.

Both religious organizations place considerable importance on reading, selling and personally distributing their books and magazines. The Seichō no Ie's publications-related sites are accordingly very big and frequently linked to its other sites.[40] Surprisingly, the Watch Tower Bible and Tract Society's official site merely quotes several articles from its magazines without attempting to sell or even advertise them online.

Thus, despite obvious similarities – international operation, emphasis on written materials and proselytization – Seichō no Ie and the Watch Tower Bible and Tract Society present themselves very differently on the Internet. Whereas the Seichō no Ie encourages individualistic self-representations by its various centers, the Watch Tower Bible and Tract Society presents itself as one globally homogeneous organization.[41] This pattern is identical to other forms of self-representation as the Jehovah's Witnesses' print publications are identical the world over whereas the Seichō no Ie's differ considerably depending on the country and language of publication.

Whereas the Seichō no Ie actively employs the Internet as a means of advertising its benefits and attracting new members, the Jehovah's Witnesses, although actively engaged in missionizing, do not officially proselytize through the Internet. Their private sites use it almost exclusively to counter public misapprehensions and defend themselves against criticism.

CONCLUSION

To date we have examined as case studies several sites of new religions, Shintō shrines, Buddhist temples and Christian churches. Results of these studies have been published on the project's Web site.[42] Our case studies of established and new religions have shown that there is no one homogeneous way of self-representation by religious organizations on the Japanese Internet. We can, however, tentatively recognize some patterns:

1   Presentations on the Internet do not offer much radically new content or induce religious organizations to enter upon fields of activity radically different from earlier roles and activities. Rather, self-representation of religious movements on the Internet often seems to reflect their situations and practices in real-life: sites of Buddhist temples advertise funerals, sites of shrines promote weddings, and both advertise their premises as worth a sightseeing excursion.[43]

2   Although some organizations use the Internet as one step towards proselytization, most new religious movements recruit primarily through word-of-mouth using pre-existing social networks. Impersonal contacts as offered by the Internet hardly suffice to establish lasting relationships between religious organizations and potential members.[44] The Internet can, however, create awareness of a religious organization and thus sow the seeds for later conversion through personal contacts.

3   Some religious organizations attempt to strengthen existing networks between branches and age groups or gender groups of members by interlinking their various sites. Organizations such as the Seichō no Ie thereby encourage members to browse the sites of branches other than their own and simultaneously present themselves as one entity to Internet users outside of their community.

4   The Internet offers easy access to voices critical of a particular religion or organization and may thus stimulate members to defect. Possible reactions by the organizations include actively disseminating their own points of view on the Internet registering them prominently with search engines. This form of "countering negative publicity"[45] through online publications can be observed among new religions such as Falun Gong, Scientology and Aleph, all of which run several large and professionally designed Web sites.[46] The other possible reaction is to minimize the use of the Internet and discourage members from browsing it. Thus, despite the Internet's potential as a proselytizing medium, not all religious movements actively involved in missionizing in real-life make extensive use of the Internet.

5   Internal communication within established religious organizations, including new religions, does not appear to have increased notably due to the Internet since computer-mediated communication did not reduce the number of face-to-face meetings. It rather replaces other non-personal forms of communication such as letters, facsimiles and telephone calls. For groups and organizations not firmly established whose structures of internal communication have not yet fully developed, however, the Internet through its e-mails, bulletin board

systems and mailing lists can function as a substantial and powerful means of communication.[47]

6   Although statements about future developments can be but speculative, it may be expected that concomitant with the growing establishment of Internet technology Internet use may become a matter of everyday life. Seichō no Ie's proactive attitude toward the Internet will surely cause an increase of official and semi-private sites, whereas the Jehovah's Witnesses may either keep their retroactive attitude or, as their colleagues in Germany, France and Russia have done, establish regional official sites after all. It may also be expected that once a thorough foundation of online information has been laid, time and know-how will allow topics surpassing today's presentation of activities to be addressed and result in a fuller exploitation of the Internet's interactive possibilities.

ACKNOWLEDGMENT

The authors of this chapter would like to thank the German Research Council, the project's supervisor, Professor Klaus Antoni, and the other members of the project for their generous support.

NOTES

1   Cf. *New Religions' Index*, at: *http://www.uni-tuebingen.de/cyberreligion/nr/kataplan.htm*, (21 March 2002).

2   Recent monographs on new religions in Japan include P. Clarke, *Japanese New Religions in Global Perspective*, Richmond, Japan Library, 2000; N. Inoue (ed.), *Shin shūkyō jiten* (Dictionary of new religions), Tokyo, Kōbundō, 1990; N. Inoue, *Shin shūkyō no kaidoku* (Understanding new religions), Tokyo, Chikuma shoten, 1992; I. Reader, *Religion in Contemporary Japan*, London, Macmillan, 1991; I. Reader, *Religious Violence in Contemporary Japan: The Case of the Aum Shinrikyō*, Richmond, UK, Curzon Press, 2000; S. Shimazono, *Gendai kyūsai shūkyō ron* (Salvation religions in contemporary society), Tokyo, Seikyūsha, 1992.

3   Research for this work is based on a project on the "Self-representation and self-understanding of religious communities on the Japanese Internet: the WWW as a source for Japanese Studies." The project was funded by the German Research Council between December 1999 and December 2001 and conducted at Tübingen University's Japanese Department. The project's Web site is at: *http://www.uni-tuebingen.de/cyberreligion/* (21 March 2002).

4   References to "personal" or "online communication" relate to face-to-face or e-mail interviews conducted with those in charge of organizations' Internet presentation. Personal interviews for this paper were conducted in September 2001 in Japan.

5   Even in the 1920s many sermons of J.F. Rutherford, the organization's second leader, were recorded on gramophone and today back numbers of many of their publications are available on CD-ROM. See G.D. Chryssides, "New Religions and the Internet," *Diskus WebEdition*, vol. 4, no. 2. Online. Available HTTP: *http://www.uni-marburg.de/religionswissenschaft/journal/diskus/chryssides_3.html* (17 October 2001).

6   *Watchtower*, Official Web site of Jehovah's Witnesses, at: *http://www.watchtower.org/languages/japanese/index.html* (11 July 2001).

7   *Watchtower*, Official Web site of Jehovah's Witnesses.

8 Translated from *Ehoba no Shōnin Kisha Kurabu*, at: *http://www1.plala.or.jp/tompoppo/info01. html* (6 February 2002).

9 Translated from *Ehoba no Shōnin Kisha Kurabu*.

10 Online communication with a member of Jehovah's Witnesses, throughout October 2001.

11 Translated from *Hokkaidō no Ehoba no Shōnin no Nisei no Hōmupēji* at: *http://www6.tok2.com/ home/miyauchi2001/jw2hokkaido/* (6 February 2002).

12 *Ehoba no Shōnin no Otokotachi, at: http://www.jca.apc.org/~resqjw/* (15 July 2001).

13 *Ehoba no Shōnin no Kodomotachi,* at: *http://www.alles.or.jp/~philip/jw%20child. html* (15 July 2001).

14 Personal communication with members of the self-helf groups Nihon Maikon Kenkyūkai and Firippusu no Mirai, 22 September 2001 and 23 September 2001.

15 *Mezame yo!* 8 June 2000, p. 10; *Awake!* 8 June 2000. Online. Available HTTP: *http://www. watchtower.org/library/* (11 July 2001). Nonetheless, a member of staff of the Japanese Watch Tower Bible and Tract Society's public affairs section told the authors that all members of the organization are allowed to use the Internet without any restrictions. (Personal communication with two members of staff of the Japanese Watch Tower Bible and Tract Society's public affairs section, 27 September 2001).

16 Y. Inose, "Dakkai no jōken – dakkai kaunseringu to jihatsuteki dakkai" (Conditions for defecting: defection counseling and voluntary defection), unpublished paper presented at the annual meeting of The Japanese Association of the Study of Religion and Society, 2001.

17 Personal communication with Nakazawa Keisuke, 23 September 2001.

18 "Kōdokusha ga nenkan kyūzenninzō" (Subscribers increased by 9,000 within one year), *Seishimei*, September 2001, p. 1.

19 General information on the Seichō no Ie is based on several months of field research conducted by Birgit Staemmler in 1997.

20 Personal communication with members of staff at Seichō no Ie's Tokyo headquarters, 18 September 2001.

21 *Religious Juridical Persons and Administration of Religious Affairs.* Online. Available HTTP: *http://www.bunka.go.jp/English/11/XI-1.html* (24 October 2001).

22 *Seichō no Ie*, at: *http://www.sni-honbu.or.jp* (20 August 2001).

23 *Seichō no Ie*, at: *http://www.sni.or.jp* (20 August 2001).

24 *Seichō no Ie Sōhonzan, at: http://sou.sni-honbu.or.jp* (20 August 2001).

25 *Seichō no Ie*, at: *http://www.sni-honbu.or.jp/brha*; *http://www.sni-honbu.or.jp/whda*; *http://www. sni-honbu.or.jp/yyaa; http://www.sni-honbu.or.jp/sukk* (20 August 2001).

26 Personal communication, with members of staff at Seichō no Ie's Tokyo headquarters, 18 September 2001.

27 *Seichō no Ie Tobitakyū Rensei Dōjō*, at: *http://www.sni-tobitakyu.or.jp/shisetsu/shisetsu.html* (20 August 2001); and personal communication, September 2001.

28 *Seichō no Ie*, at: *http://www.sni-or.jp/honbu/html/sn010709.htm* (20 August 2001).

29 *Seichō no Ie*, at: *http://www.sni-honbu.or.jp/yyaa/KatudoKyoten.htm* (20 August 2001) and personal communication with members of the youth organization's staff at Seichō no Ie's Tokyo headquarters, 18 September 2001.

30 *Seichō no Ie Uji Bekkaku Honzan*, at: *http://uji.sni.or.jp/* (20 August 2001).

31 *Taniguchi Masanobu*, at: *http://homepage2.nifty.com/masanobu-taniguchi/* (10 October 2001).

32 *Kiga Kyusai*, at: *http://kiga.be.happy.net* (20 August 2001).

33 Personal communication with members of staff at Seichō no Ie's Tokyo headquarters, 18 September 2001.

34 Personal communication with members of the youth organization's staff at Seichō no Ie's Tokyo headquarters, 18 September 2001.

35 Personal communication with members of staff at Seichō no Ie's Tokyo headquarters, 18 September 2001.

36 P. Rössler, "Standardisierte Inhaltsanalysen im World Wide Web: Überlegungen zur Anwendung einer Methode am Beispiel einer Studie zu Online-Shopping-Angeboten," in K. Beck and

G. Vowe (eds), *Computernetze: ein Medium öffentlicher Kommunikation?*, Berlin, Wissenschaftsverlag Volker Spiess, 1997, p. 263.

37 See P. Rössler, "Standardisierte Inhaltsanalysen im World Wide Web."

38 See Shitara Minoru's statement about the Shinnyoen in N. Inoue (ed.), *Intānetto jidai no shūkyō* (Religion in the Internet age), Tokyo, Shinshokan, 2000, p. 67. Aleph, on the other hand, makes extensive use of computer networks for internal communication as it is still difficult for members to assemble in real life. (Personal communication with Aleph's public affairs spokesman, 27 September 2001).

39 *Seichō no Ie*, at: *http://www.sni-honbu.or.jp/whda/Hahaoya/hahaoya10.html* (20 August 2001).

40 *Nihon Kyōbunsha*, at: *http://www.kyobunsha.co.jp* (20 August 2001), and *Sekai Seiten Fukyū Kyōkai*, at: *http://www.ssfk.or.jp* (20 August 2001).

41 In March 2001 the Church of Jesus Christ of Latter-Day Saints, another Christian new religion that is controversial in Japan, issued a letter calling for the discontinuation of local unit Web sites until a policy was established to govern them. Local Churches all over the world were told not to create or sponsor Web sites. Online. Available HTTP: *http://lds.org/news/archiveday/ 0,5287,4688,FF.html* (24 October 2001).

42 See *http://www.uni-tuebingen.de/cyberreligion/public/public.htm* (24 October 2001).

43 P. Kienle and B. Staemmler, "Cyberreligion: Selbstdarstellungen japanischer Religionsgemein-schaften im Internet mit einer Analyse ausgewählter Beispiele," in G. Schucher (ed.), *Asien und das Internet*, Hamburg, Institut für Asienkunde, 2002, pp 183–93.

44 L. Dawson and J. Hennebry, "New religions and the Internet: recruiting in a new public space," *Journal of Contemporary Religion*, vol. 14, no.1, 1999, pp. 17–39, especially pp. 26–30. See also M. Castells, *The Information Age: Volume I, The Rise of the Network Society*, Malden, MA, Blackwell Publishers, 2000, p. 388–9.

45 See S. Horsfall, "How religious organizations use the Internet: a preliminary inquiry," in J.K. Hadden, and D.E. Cowan, (eds), *Religion on the Internet: Research Prospects and Promises*, New York, Elsevier Science, 2000, p. 174.

46 See lists of links in *Falun Dafa Information Center*, at: *http://faluninfo.net/* (13 March 2002); *Related Sites*, at: *http://related.scientology.org* (13 March 2002); *Arefu*, at: *http://www.aleph.to/ information/link.html* (13 March 2002).

47 Cf. M. Castells, *The Information Age: Volume I, The Rise of the Network Society*, p. 393.

Bibliography

Abbate, J., *Inventing the Internet*, Cambridge, MIT Press, 1999.

Adams, B., J. Duyvendak and A. Krouwel (eds), *The Global Emergence of Gay and Lesbian Politics: National Imprints of a Worldwide Movement*, Philadelphia, Temple University Press, 1999.

Akaishi, C., *"Shinguru Mazāzu Fōramu"* (Single mothers' forum), *Onna tachi no 21 Seiki*, no. 20, 1999, p. 28.

Akita, M., *Noise War: Noizu myūjikku to sono tenkai* (Noise war: Noise music and its outlook), Tokyo, Seikyūsha, 1992.

Allison, A., "Memoirs of the Orient," *Journal of Japanese Studies*, vol. 27, no. 2, 2001, pp. 381–98.

Allison, A., *Nightwork: Sexuality, Pleasure, and Corporate Masculinity in a Tokyo Hostess Club*, Chicago, University of Chicago Press, 1994.

Alt.books, *SEX no arukikata: Tōkyō fūzoku kanzen gaido* (How to find your way around sex: the complete guide to Tokyo's sex scene), Tokyo, Alt.books, 1998.

Anderson, B., *Imagined Communities: Reflections on the Origin and Spread of Nationalism*, London, Verso, 1983.

Ang, I., *On Not Speaking Chinese: Living between Asia and the West*, London, Routledge, 2001.

Appadurai, A., *Modernity at Large: Cultural Dimensions of Globalization*, Minneapolis, University of Minnesota Press, 1996.

Appadurai, A., "Disjuncture and difference in the global cultural economy," in M. Featherstone (ed.), *Global Culture: Nationalism, Globalization and Modernity*, London, Sage, 1990.

Appadurai, A., "The production of locality," in R. Fardon, (ed.), *Counterworks: Managing the Diversity of Knowledge*, London, Routledge, 1995, pp. 204–25.

Appiah, K.A. and H.L. Gates Jr (eds), *Identities*, Chicago, University of Chicago Press, 1995.

Ariadone (ed.), *Shikō no tame no intānetto* (The Internet for academic purposes), Tokyo, Chikuma Shobo, 1999.

Atkins, E. T., *Blue Nippon: Authenticating Jazz in Japan*, Durham, Duke University Press, 2001.

Axford, B., "The transformation of politics or anti-politics?" in B. Axford and R. Huggins (eds), *New Media and Politics*, London, Thousand Oaks, New Delhi, Sage Publications, 2001, p. 15.

Bachnik, J. and C. Quinn (eds), *Situated Meaning: Inside and Outside in Japanese Self, Society, and Language*, Princeton, Princeton University Press, 1994.

Baudrillard, J., *Simulacra and Simulation* (trans. S.F. Glaser), Ann Arbor, University of Michigan Press, 1994 [1981].

Baym, N., *Tune In, Log On: Soaps, Fandom and Online Community*, Thousand Oaks, Sage, 2000.

Bech, H., *When Men Meet: Homosexuality and Modernity*, Oxford, Polity Press, 1997.

Befu, H., *Hegemony of Homogeneity: An Anthropological Analysis of Nihonjinron*, Melbourne, Trans Pacific Press, 2001.

Bell, D. and B. Kennedy (eds), *The Cybercultures Reader*, London, Routledge, 2000.

Bell, D., "Cybersubcultures: introduction," in D. Bell and B. Kennedy (eds), *The Cybercultures Reader*, London, Routledge, 2000, pp. 205–9.

Bell, D., "Pleasure and danger: the paradoxical spaces of sexual citizenship," *Political Geography*, vol. 14, no. 2, 1995, pp. 139–53.

Bell, V., "Performativity and belonging: an introduction," in V. Bell (ed.), *Performativity and Belonging*, London, Sage Publications, 1999, pp. 1–10.

Bensfeld, J. and R. Lilienfeld, *Between Public and Private*, New York, The Free Press, 1979.

Berger, P.L., B. Berger and H. Kellner, *The Homeless Mind*, Harmondsworth, Penguin, 1974.

Berry, C. and F. Martin, "Queer 'n' Asian on – and off – the Net: the role of cyberspace in queer Taiwan and Korea," in D. Gauntlett (ed.), *Web.Studies*, London, Arnold, 2000, pp. 74–81.

Berry, C., F. Martin and A. Yue (eds), *Mobile Cultures: New Media in Queer Asia*, Durham, Duke University Press, 2003.

Bhabha, H., *The Location of Culture*, London, Routledge, 1994.

Birch, D., T. Schirato and S. Srivastava, *Asia: Cultural Politics in the Global Age*, Sydney, Allen and Unwin, 2001.

Birch, D., "An 'open' environment? Asian case studies in the regulation of public culture," *Continuum*, vol. 12, no. 3, 1998, pp. 335–348.

Botting, G., "IT revolution bastardizing the Japanese language," *Sunday Mainichi*, 19 November 2000, p. 7.

Bridges, B., *Japan and Korea in the 1990s*, Cambridge, Cambridge University Press, 1993.

Brod, H. and M. Kaufman (eds), *Theorizing Masculinities*, Thousand Oaks, CA, Sage Publications, 1994.

Buckley, S. (ed.), *Broken Silence: Voices of Japanese Feminism*, Berkeley, University of California Press, 1997.

Buckley, S. and V. Mackie, "Women in the new Japanese state," in G. McCormack and Y. Sugimoto (eds), *Democracy in Contemporary Japan*, Sydney, Hale & Iremonger, 1986, pp. 173–85.

Buruma, I., *A Japanese Mirror: Heroes and Villains of Japanese Culture*, London, Jonathan Cape, 1984.

Butler, J., "Imitation and gender insubordination," in L. Nicholson (ed.), *The Second Wave: A Reader in Feminist Theory*, London, Routledge, 1997.

Butler, J., *Gender Trouble: Feminism and the Subversion of Identity*, London, Routledge, 1990.

Carby, H., "In body and spirit: representing black women musicians," *Black Music Research Journal*, vol. 11, no. 2, 1991, pp. 177–92.

Carby, H., "It jus be's dat way sometimes," *Radical America*, vol. 20, 1988, pp. 9–24.

Caspary, C., "Das japanische Noise-Netzwerk. Eine Diskussion über die heutigen Möglichkeiten Alternativen zu Mainstreamkulturpraktiken zu schaffen" (The Japanese Noise Alliance), Vienna, 2000, unpublished MA thesis accepted by Vienna University (in German).

Castells, M., *The Information Age: Volume II, The Power of Identity*, second edition, Malden, MA, Blackwell Publishers 1997.

Castells, M., *The Information Age: Volume III, End of Millennium*, Malden, MA, Blackwell Publishers, 1998.

Castells, M., *The Information Age: Volume I, The Rise of the Network Society*, 2nd edn, Malden, MA, Blackwell Publishers, 2000.

Castells, M., *The Internet Galaxy*, Oxford, Oxford University Press, 2001.

Chalmers, S., "Lesbian (in)visibility and social policy in Japanese society," in V. Mackie (ed.), *Gender and Public Policy in Japan*, London, Routledge, 2002.

Chao, A., "Global metaphors and local strategies in the construction of Taiwan's lesbian identities," *Culture, Health and Sexuality*, vol. 2, no. 4, October–December 2000, pp. 377–90.

Chiang, M., "Coming out into the global system," in D. Eng and A. Hom (eds), *Q & A: Queer in Asian America*, Philadelphia, Temple University Press, 1998, pp. 374–95.

Chryssides, G.D., "New religions and the Internet," *Diskus WebEdition*, vol. 4, no. 2, 1996. Online. Available HTTP: *http://www.uni-marburg.de/religionswissenschaft/journal/diskus/chryssides_3.html* (17 October 2001).

Clarke, P., *Japanese New Religions in Global Perspective*, Richmond, Japan Library, 2000.

Collins, P., *Black Feminist Thought: Knowledge, Consciousness, and the Politics of Empowerment*, New York, Routledge, 2000.

Comaroff, J. and J. Comaroff, *Of Revelation and Revolution: Christianity, Colonialism, and Consciousness in South Africa*, Chicago, University of Chicago Press, 1991.

Condry, I., *The Social Production of Difference: Imitation and Authenticity in Japanese Rap Music*. Online. Available HTTP: *http://www.yale.edu/~condryi/rap/* (24 April 1999).

Connell, R., *The Men and the Boys*, St Leonards NSW, Allen & Unwin, 2000.

Connell, R., *Masculinities*, St Leonards, NSW, Allen & Unwin, 1995.

Cooper, M., *They Came to Japan: An Anthology of European Reports on Japan, 1543–1640*, Berkeley, University of California Press, 1965.

Corliss, M., "Wetland conservation efforts gain ground," *Japan Times*, 6 September 2001, p. 2.

Craig, T. (ed.), *Japan Pop! Inside the World of Japanese Popular Culture*, Armonk, NY, M.E. Sharpe, 2000.

Crystal, D., *Language and the Internet*, Cambridge, Cambridge University Press, 2001.

Currah, P., "Searching for immutability: homosexuality, race and rights discourse," in A. Wilson (ed.), *A Simple Matter of Justice? Theorizing Lesbian and Gay Politics,* London, Cassell, 1995, pp. 51–90.

Daily Yomiuri, "Poll: Japanese youth have more pagers, less respect," *Daily Yomiuri*, 11 April 1997, p. 3.

Daily Yomiuri, "Creature comfort," *Daily Yomiuri*, 25 March 1999, p. 16.

Daily Yomiuri, "Cell-phone users hang up on notions of time," *Daily Yomiuri*, 21 July 2001, p. 8.

Dale, P.N., *The Myth of Japanese Uniqueness*, London, Routledge, 1986.

Dasgupta, R., "Performing masculinities? The 'salaryman' at work and play," *Japanese Studies*, vol. 20, no. 2, 2000, pp.189–200.

Davis, A., *Blues Legacies and Black Feminism*, New York, Random House, 1998.

Dawson, L. and J. Hennebry, "New religions and the Internet: recruiting in a new public space," *Journal of Contemporary Religion* vol. 14, no. 1, 1999, pp. 17–39.

De Certeau, M., *The Practice of Everyday Life*, Berkeley, University of California Press, 1984.

Doi, T., "Amae: a key concept for understanding Japanese personality structure," in T. Lebra and W. Lebra (eds), *Japanese Culture and Behavior: Selected Readings – Revised Version*, Honolulu, University of Hawaii Press, 1986, pp. 121–9.

Donath, J., "Identity and deception in the virtual community," in M. Smith and P. Kollock (eds), *Communities in Cyberspace*, London, Routledge, 1999, pp. 29–59.

Ducke, I., *Status Power: Japan's Foreign Policy toward Korea*, New York, Routledge (forthcoming).

Durkheim, E., *The Division of Labor in Society* (trans. G. Simpson), New York, The Free Press, 1933.

Elias, N., *The History of Manners: The Civilizing Process*, vol. 1, New York, Pantheon Books, 1978 [1939].

Escobar, A., "Welcome to Cyberia: notes on the anthropology of cyberculture," in D. Bell and B. Kennedy (eds), *The Cybercultures Reader*, London, Routledge, 2000, pp. 56–76.

Ess, C. and F. Sudweeks, *Culture, Technology, Communication: Towards an Intercultural Global Village*, Albany, NY, SUNY, 2001.

Everard, J., *Virtual States: The Internet and the Boundaries of the Nation-State*, London, Routledge, 2000.

Fallows, J., "The Japanese are different from you and me," *The Atlantic*, vol. 258, September 1986, pp. 35–42.

Featherstone, M., "Localism, globalism, and cultural identity," in R. Wilson, and W. Dissanayake (eds), *Global/Local: Cultural Production and the Transnational Imaginary*, Durham, Duke University Press, 1996, pp. 46–77.

Feldman, E., *The Ritual of Rights in Japan: Law, Society, and Health Policy*, Cambridge, Cambridge University Press, 2000.

Fenster, M.A., *The Articulation of Difference and Identity in Alternative Popular Music Practice*, Ann Arbor, UMI Dissertation Service, 1992.

Fernback, J. and B. Thompson, "Virtual communities: abort, retry, failure?" Online. Available HTTP: *http://www.rheingold.com/texts/techpolitix/VCcivil.html*, 1995 (8 October 2001).

Fiske, J., *Understanding Popular Culture*, London and New York, Routledge, 1989.

Foreign Press Center Japan, *Facts and Figures of Japan*, Tokyo, 2000.

Foucault, M., *The History of Sexuality: Volume 1, An Introduction*, New York, Random House, 1978.

Foucault, M., *Discipline and Punish: The Birth of the Prison*, New York, Vintage Books, 1979.

Fox, R., *Recapturing Anthropology: Working in the Present*, Santa Fe, School of American Research Press, 1991.

Fraser, N., "Rethinking the public sphere: a contribution to the critique of actually existing democracy," in C. Calhoun (ed.), *Habermas and the Public Sphere*, Cambridge, MA, MIT Press, 1991, pp. 56–80.

Freeman, L.A., *Closing the Shop: Information Cartels and Japan's Mass Media*, Princeton, NJ, Princeton University Press, 2000.

Friedland, L., "Electronic democracy and the new citizenship," *Media, Culture and Society*, 18, 1996, pp. 185–212.

Fukami, A., "'Iron John' movement appears in Japan," *Japan Times Weekly International Edition*, April 27 – May 3 1992, p. 14.

Garnham, N., *Capitalism and Communication: Global Culture and the Economics of Information*, London and Newbury Park, CA, Sage, 1990.

Gauntlett, D (ed.), *Web Studies,* London, Arnold, 2000.

Ghosh, A., "Phone wars: episode3G," *Time*, 27 November 2000, pp. 43–8.

Giddens, A., *Modernity and Self-Identity: Self and Society in the Late Modern Age*, Stanford, CA, Stanford University Press, 1991.

Giddens, A., *The Consequences of Modernity*, Stanford, CA, Stanford University Press, 1990.

Goffman, E., *Behavior in Public Places*, New York, The Free Press, 1963.

Goffman, E., *Stigma: Notes on the Management of Spoiled Identity*, Engelwood Cliffs, NJ, Prentice-Hall, Inc., 1963.

Goffman, E., *The Presentation of the Self in Everyday Life*, Garden City, NJ, Doubleday Anchor Books, 1959.

Golden, A., *Memoirs of a Geisha*, London, Vintage, 1998.

Gottlieb, N., "Discriminatory language in Japan: Burakumin, the disabled and women," *Asian Studies Review* vol. 22, no. 2, 1998, pp. 157–73.

Gottlieb, N., *Word-Processing Technology in Japan: Kanji and the Keyboard*, Richmond, UK, Curzon Press, 2000.

Grossberg, L., *We Gotta Get Out of this Place. Popular Conservatism and Postmodern Culture*, New York, Routledge, 1992.

Gupta, A. and J. Ferguson (eds), *Anthropological Locations: Boundaries and Grounds of a Field Science*, Berkeley, University of California Press, 1997.

Habermas, J., *The Structural Transformation of the Public Sphere. An Inquiry into a Category of Bourgeois Society*, Cambridge, MA, MIT Press, 1991.

Hall, S. and P. du Gay (eds), *Questions of Cultural Identity*, London, Sage Publications, 1996.

Hamada, J., "The right to information: the core of the network society," in *Review of Media, Information and Society*, 4, 1999, pp. 69–78.

Hamasuna, M., "Adding personality to cellular phones," *Daily Yomiuri*, 3 April 1999, p. 8.

Hammond, P. (ed.), *Cultural Difference, Media Memories: Anglo-American Images of Japan*, London, Cassell, 1997.

Hanna, J., "Sequels living up to expectations," *Daily Yomiuri*, 21 January 1999, p. 14.

Hara, T., "Kisha kurabu mondai" (The press club problem) in K. Isshiki *et al.* (eds), *Shin masu komi gaku ga wakaru* (Understanding new mass media studies), Tokyo, Asahi Shinbunsha, 2001, pp. 22–5.

Haruo, A., S. Itō and Y. Murase (eds), *Nihon no otoko wa doko kara kite, doko e iku no ka?* (Where have Japanese men come from, where will they go?), Tokyo, Jūgatsusha, 2001.

Hashimoto, N., *Women and Men in Japan*, Tokyo, Zenkoku-kyōiku bunka kaikan, 2001.

Hawisher, G. and C. Selfe, *Global Literacies and the World-Wide Web*, London, Routledge, 2000.

Hebdige, D., *Subculture, the Meaning of Style*, London, Methuen, 1979.

Hicks, G., *The Comfort Women*, St Leonards, NSW, Allen & Unwin, 1995.

Hine, C., *Virtual Ethnography*, Thousand Oaks, CA, Sage Publications, 2000.

Holden, T.J.M., "Surveillance – Japan's sustaining principle," *Journal of Popular Culture,* vol. 28, no. 1, Summer 1994, pp. 193–208.

Holert, T. and M. Terkessidis (eds), *Mainstream der Minderheiten. Pop in der Konsumgesellschaft* (Mainstream of minorities. Pop in consumer society), Berlin, Edition ID-Archiv, 1996.

Holeton, R., *Composing Cyberspace: Identity, Community and Knowledge in the Electronic Age*, Boston, McGraw-Hill, 1998.

Hood, J. (ed.), *Men, Work, and Family*, Newbury Park, Sage Publications, 1993.

Hooks, B., *Black Looks: Race and Representation*, Boston, South End Press, 1992.

Horsfall, S., "How religious organizations use the Internet: a preliminary inquiry," in J.K. Hadden and D.E. Cowan (eds), *Religion on the Internet: Research Prospects and Promises,* New York, Elsevier Science, 2000.

Hoyt, E., *The New Japanese: A Complacent People in a Corrupt Society*, London, Robert Hale, 1991.

Huggins, R., "The transformation of the political audience," in B. Axford and R. Huggins (eds), *New Media and Politics*, London, Thousand Oaks, New Delhi, Sage Publications, 2001, p. 134.

Humphries, L., *Tearoom Trade: A Study of Homosexual Encounters in Public Places*, London, Gerald Duckworth & Co., 1970.

Ikegami, E., *The Taming of the Samurai: Honorific Individualism and the Making of Modern Japan*, Cambridge MA, Harvard University Press, 1995.

Inose, Y., "Dakkai no jōken – dakkai kaunseringu to jihatsuteki dakkai" (Conditions for defecting: defection counselling and voluntary defection), unpublished paper presented at the annual meeting of The Japanese Association of the Study of Religion and Society, 2001.

Inoue, H., *Josei ni yasashii Intānetto no hon* (A simple Internet book for women), Tokyo, CQ Shuppan, 1999.

Inoue, N. (ed.), *Shin shūkyō jiten* (Dictionary of new religions), Tokyo, Kōbundō, 1990.

Inoue, N., *Shin shūkyō no kaidō* (Understanding new religions), Tokyo, Chikuma shoten, 1992.

Inoue, N. (ed.), *Intānetto jidai no shūkyō* (Religion in the Internet age), Tokyo, Shinshokan, 2000.

Ishibashi, H., "17sai, rekishi toi netto kōgeki" (At 17, Net attack on history question), *Asahi Shimbun*, 2 April 2002, p. 34.

Ishii-Kuntz, M., "Japanese fathers: work demands and family roles," in J. Hood (ed.), *Men, Work, and Family*, Newbury Park, Sage Publications, 1993.

Itō, H., *Tsūshin kaisen no on to ofu–aru ongaku fan no baai* (On and off in data transmission networks). Unpublished manuscript. 1998. Online. Available HTTP: *http://member. nifty.ne.jo/haruki/works/bz/onoff.htm* (1 February 2002).

Itō, K., *Danseigaku nyūmon* (An introduction to men's studies), Tokyo, Sakuhinsha, 1996.

Itō, K., *"Otokorashisa" no yukue: danseibunka no bunkashakaigaku* (Tracing "machoness/masculinity:" the cultural sociology of male culture), Tokyo, Shinyosha, 1997.

Itō, K. and K. Muta (eds), *Jendā de manabu shakaigaku* (The study of sociology through gender), Kyoto, Sekai Shisōsha, 1998.

Itō, S. and R. Yanase, *Coming Out in Japan* (trans. F. Conlan), Melbourne, Trans Pacific Press, 2000.

Itō, S., *Dōseiai no kiso chishiki* (Basic information about homosexuality), Tokyo, Ayumi shuppan, 1996.

Itō, T. and N. Chisako, "Attachment to mobile phones reaching point of addiction," *Daily Yomiuri*, 8 July 2001, p. 14.

Itō Y., "The birth of Joho Shakai and Johoka concepts in Japan and their diffusion outside Japan," *Keiō Communication Review*, 13, pp. 3–12.

Itoi, K., "Rising daughters," *Newsweek*, 3 April 2000, CXXXV, no.14, pp. 40–5.

Iwamoto, N., "Nikkan kankei no mirai o kanjisaseru Shinjuku rittoru koria" (Little Korea in Shinjuku, where you feel the future of Japanese – South Korean relations), *Jitsugyō no Nihon*, May 1997, pp. 50–2.

Jackson, P., "Global queering in Thailand: peripheral genders and the limits of queer theory," paper presented at International Association for the Study of Society Culture and Sexuality (IASSCS) Conference on Sexual Diversity and Human Rights, Manchester Metropolitan University, 21–24 July 1999.

Japan Times, "Virtual Pets II," *Japan Times*, 25 March 1999, p. 11.

Japan Times, "Mobiles big among high schoolers: poll," *Japan Times*, 26 December 2000, p. 3.

Japan Times, "Subscribers to i-mode top 20 million," *Japan Times*, 6 March 2001, p. 12.

Japan Times, "One in four Tokyo elementary, junior high kids has a cellphone," *Japan Times*, 7 March 2001, p. 9.

Japan Times, "Most schools ignore disputed text," *Japan Times*, editorial, August 2001, p. 2.

Japan Times, "The Diet that set a precedent," *Japan Times*, editorial, 11 December 2001, p. 7

Jones, G., *Corregidora*, Boston, Beacon Press, 1986.

Jones, J., "The accusatory space," in M. Wallace (ed.), *Black Popular Culture*, Seattle, Bay Press, 1992.

Jones, M., "Noise," in A. Roman (ed.), *Japan Edge*, San Francisco, Cadence Books, 1999, pp. 75–100.

Jordan, T., "Language and libertarianism: the politics of cyberculture and the culture of cyberpolitics," *Sociological Review*, vol. 49, no. 1, 2000, pp. 1–17.

Jordan, W., *White over Black: American Attitudes toward the Negro, 1550–1812*, Chapel Hill, University of North Carolina Press, 1968.

Kageyama,Y., "Cellphone makers focus on fun," *Japan Times,* 5 March 2001, p. 17.

Kageyama, Y., "Internet-capable phone firms target US," *Japan Times*, 18 January 2001, p. 9.

Kang, R., "A comparison of the foreign policy making process in Japan and South Korea, in the case of the loan negotiations 1981–3," Newcastle University, PhD Dissertation, 1994.

Kang, W., "The engine for the next economic leap: the Internet in Korea," in S. Rao and B. Klopfenstein (eds), *Cyberpath to Development in Asia: Issues and Challenges*, Westport, CT and London, Praeger, 2002, pp. 111–36.

Kashiwazaki, C., "The politics of legal status," in S. Ryang (ed.), *Koreans in Japan: Critical Voices from the Margin*, London and New York, Routledge, 2000, p. 26.

Kato, T., "From a class party to a national party," *AMPO*, vol. 29, no. 2, March 2000, pp. 11–13.

Katz, E., "Mass communication research and the study of popular culture: an editorial note on a possible future for this journal," *Studies in Public Communication*, 2, 1959, pp.1–6.

Kaufman, M., "Men, feminism, and men's contradictory experiences of Power," in H. Brod and M. Kaufman (eds), *Theorizing Masculinities*, Thousand Oaks, CA, Sage Publications, 1994, pp. 142–63.

Kawabe, R., "Finding the right approach to helping children," *Daily Yomiuri*, 2 December 2000, p. 7.

Kawahira, T., "Mejia o katsuyō shita shimin undō no ugoki" (Civic movements increasingly make use of the media), *Shūkan Kinyōbi*, no. 389, 23 November 2001, pp. 22–3.

Kawaura Y., '*Intānetto ni okeru nettowāku no tokuchō*' (Particularities of networks on the Web), in Nihon Shinri Gakkai, *Dai 63 kai taikai happyō ronbun shū*, Nagoya, Nihon Shinri Gakkai, 1999.

Kazuya, *Gei seikatsu manyuaru* (Gay lifestyle manual), Tokyo, Data House, 1998.

Keck, M. and K. Sikkink, *Activists Beyond Borders: Advocacy Networks in International Politics*, Ithaca, NY, Cornell University Press, 1998.

Keizai Kikakuchō Kokumin Seikatsukyoku (EPA Citizen's Lifestyle Bureau), *Shinkokumin Seikatsu Shihyō*, (People's Life Indicators), Tokyo, 2000.

Kendall, L., "'Oh no! I'm a nerd!' Hegemonic masculinity on an online forum," *Gender and Society*, vol. 14, no. 2, 2000, pp. 256–74.

Kienle, P. and B. Staemmler (2002), "Cyberreligion: Selbstdarstellungen japanischer Religionsgemeinschaften im Internet mit einer Analyse ausgewählter Beispiele," in G. Schucher (ed.), *Asien und das Internet*, Hamburg, Institut für Asienkunde, 2002, pp. 183–93.

Kim, P., "Global civil society remakes history: 'the Women's International War Crimes Tribunal 2000'," *Positions: East Asia Cultures Critique*, vol. 9, no. 3, Winter 2001, pp. 611–17.

Kimmel, M., "Masculinity as homophobia: fear, shame, and silence in the construction of gender identity," in H. Brod, and M. Kaufman (eds), *Theorizing Masculinities*, Thousand Oaks, CA, Sage Publications, 1994, pp. 119–41.

Kinmonth, E.H., *The Self-Made Man in Meiji Japanese Thought: From Samurai to Salaryman*, Berkeley, University of California Press, 1981.

Kinoshita, A., "Onna tachi no idobata kaigi" (Women's chatter), *Shakai kyōiku* (Social education), no. 30, 1986, p. 32.

Kinsella, S., *Adult Manga: Culture and Power in Contemporary Japanese Society*, Richmond, UK, Curzon Press, 2000.

Kinsella, S., "Japanese subculture in the 1990s: *otaku* and the amateur *manga* movement," *Journal of Japanese Studies* vol. 24, no. 2, 1998, pp. 289–316.

Kinsella, S., "Cuties in Japan," in L. Skov and B. Moeran (eds), *Women, Media and Consumption in Japan*, Richmond, UK, Curzon Press, 1995, pp. 220–54.

Kluver, R., "Globalization, informatization, and intercultural communication," *American Communication Journal*, vol. 3, no. 3, 2000. Online. Available HTTP: *http://acjournal. org/holdings/vol3/Iss3/spec1/kluver.htm* (8 October 2001).

Kogawa, T., "New trends in Japanese popular culture," in G. McCormack and Y. Sugimoto (eds), *The Japanese Trajectory: Modernization and Beyond*, Cambridge, Cambridge University Press, 1988, pp. 54–66.

Kondo, D., *Crafting Selves: Power, Gender and Discourses of Identity in a Japanese Workplace*, Chicago, University of Chicago Press, 1990.

Koyama, T., "Widening Net," *The Nikkei Weekly*, 12 November 2001, p. 3.

Kumagai, H., "A dissection of intimacy: a study of 'bipolar posturing' in Japanese social interaction – 'amaeru' and amayakasu, indulgence and deference," *Culture, Medicine, and Psychiatry*, vol. 5, no. 3, 1981, pp. 249–72.

Kumisaka, S., "The current condition of minorities in Japan and challenges – the Buraku issue," *Buraku Liberation News*, 101, 1998.

Laidlaw, I., "Reflection on a research trip to Japan," *Buraku Liberation News*, 120, 2001.

Large, T., "Forget e-mail, get into twee mail," *Daily Yomiuri*, 25 February 1999, p. 7.

Larimer, T., "What makes DoCoMo go," *Time*, 27 November 2000, pp. 50–4.

Lax, S., "The Internet and democracy," in D. Gauntlett (ed.), *Web.Studies: Rewiring Media Studies for the Digital Age*, London, Arnold, pp. 159–69.

LeBlanc, R., *Bicycle Citizens: the Political World of the Japanese Housewife*, Berkeley, California University Press, 1999.

Lebra, T., *Japanese Patterns of Behavior*, Honolulu, University of Hawaii Press, 1976.

Lebra, T.S. and W.P. Lebra (eds), *Japanese Culture and Behavior: Selected Readings*, Honolulu, University of Hawaii Press, 1986 [1974].

Lee, O-Young, *The Compact Culture: the Japanese Tradition of "Smaller Is Better,"* Tokyo, Kodansha, 1982.

Lunsing, W., *Beyond Common Sense: Sexuality and Gender in Contemporary Japan*, London, Kegan Paul, 2001.

Lunsing, W., "Japan: finding its way?," in B.D. Adam, J. Duyvendak and A. Krouwel (eds), *The Global Emergence of Gay and Lesbian Politics: National Imprints of a Worldwide Movement*, Philadelphia, Temple University Press, 1999, pp. 293–325.

Lunsing, W., "Lesbian and gay movements: between hard and soft," in C. Derichs and A. Oziander (eds), *Soziale Bewegungen in Japan*, Hamburg, Mitteilungen der Gesellschaft für Natur und Volkerkunde, 1998, pp. 279–310.

Lunsing, W., "Gay boom in Japan? Changing views of homosexuality," *Thamyris*, no. 4, Autumn 1997, p. 284.

McChesney, R.W., *Rich Media, Poor Democracy: Communication Politics in Dubious Times*, Urbana, University of Illinois Press, 1999.

McClure, S., "Newest cellular phone fad rings a bell," *Daily Yomiuri*, 15 January 2001, p. 7.

McCombie, S.C., "AIDS in cultural, historical, and epidemiological context," in D. Feldman (ed.), *Culture and AIDS*, New York, Praeger, 1990, pp. 9–28.

Mackie, V., "The language of transnationality, globalisation and feminism," *International Feminist Journal of Politics*, vol. 3, no. 2, 2001, pp. 180–206.

Mackie, V., "Feminism and the media in Japan," *Japanese Studies Bulletin*, vol. 12, no. 2, 1992, pp. 23–31.

McLagan, M., "Computing for Tibet: virtual politics in the post-Cold War era," in G.E. Marcus (ed.), *Connected: Engagements with Media*, Chicago, University of Chicago Press, 1996, pp. 159–94.

McLelland, M., "The Newhalf Net: Japan's 'intermediate sex' online," *The International Journal of Sexuality and Gender Studies*, vol. 7, nos 2/3, April/July 2002, pp. 163–75.

McLelland, M., "Out on the global stage: authenticity, interpretation and orientalism in Japanese coming out narratives," *The Electronic Journal of Contemporary Japanese Studies*, 2001. Online. Available HTTP: *http://www.japanesestudies.org.uk/articles/McLelland.html* (4 April 2002).

McLelland, M., "Local meanings in global space: a case ctudy of women's 'boy love' Web sites in Japanese and English," *Mots Pluriels*, no. 19, Special Issue: The Net: New Apprentices and Old Masters, October 2001. Online. Available HTTP: *http://www.arts.uwa.edu.au/MotsPluriels/MP1901mcl.html* (16 May 2002).

McLelland, M., "Live life more selfishly: an online gay advice column in Japan," *Continuum*, vol. 15, no. 1, April 2001, pp. 103–16.

McLelland, M., *Male Homosexuality in Modern Japan: Cultural Myths and Social Realities,* Richmond, UK, Curzon Press, 2000.

McLelland, M., "Out and about on Japan's gay net," *Convergence: Journal of Research into New Media Technology*, vol. 6, no. 3, Autumn 2000, pp. 16–33.

McLelland, M., "No climax, no point, no meaning? Japanese women's boy-love sites on the Internet," *Journal of Communication Inquiry*, vol. 24, no. 3, July 2000, pp. 274–91.

McNeill, D., "Marching to war over history," *South China Morning Post*, (Focus section), 17 June 2001, p. 2.

McNeill, D., "An unwelcome visit from the Uyoku," *New Statesman*, 26 February 2001, pp. 32–3.

McVeigh, B.J. (n.d.), "Japan's esthetic of techno-cute: marrying the futuristic and the cuddly" (unpublished manuscript).

McVeigh, B.J., *Wearing Ideology: State, Schooling and Self-Presentation in Japan*. Oxford, Berg Publishers, 2000.

McVeigh, B.J., "How Hello Kitty commodifies the cute, cool, and camp: 'consumutopia' versus 'control' in Japan," *Journal of Material Culture*, vol. 5, no. 2, 2000, pp. 225–45.

McVeigh, B.J., "Commodifying affection, authority and gender in the everyday objects of Japan," *Journal of Material Culture*, vol. 1, no. 3, 1996, pp. 291–312.

Mainichi Shimbun, "Boshi kansen no osore ga genjitsu ni" (The hard facts on the dangers of mother-to-child infection: babies with high transmission rates), *Mainichi Shimbun*, 17 February 1987, p. 23.

Manalansan IV, M., "In the Shadows of Stonewall: Examining Gay Transnational Politics and the Diaspora Dilemma," *GLQ: A Journal of Lesbian and Gay Studies*, vol. 2, 1995, pp. 425–438.

Manzenreiter, W., "Japan's Digital Unite: Grundlagen und Grenzen des M-Commerce in internationalen Vergleich" (Japan's digital unite: conditions and limitations of mobile commerce in international comparison), in C. Erten and R. Pirker (eds), *Wirtschafts-macht Süd-Ost-Asien: Länderspezifische Erfolgsfaktoren für wirtschaftliches Handeln*, Vienna, Wirtschaftsverlag, 2002, pp. 187–204.

Marcus, G. (ed.), *Connected: Engagements with Media*, Chicago, University of Chicago Press, 1996.

Martinez, D.P., *The Worlds of Japanese Popular Culture: Gender, Shifting Boundaries and Global Cultures*, Cambridge, Cambridge University Press, 1998.

Mathews, G., *Global Culture/Individual Identity: Searching for Home in the Cultural Supermarket,* London, Routledge, 2000.

Matsuda, M., "Kētai ni yoru denshi mēru kyūzō to sono eikyō" (Rapid increase of mobile emails and its impact), *Nihongogaku* 17, October 2000. Online. Available HTTP: *http://www3.justnet.ne.jp/~misam/nihongogaku.html* (4 April 2002).

Matsui, Y., "Intānetto o Josei no te ni" (Let women take over the Internet), *Onna tachi no 21 Seiki*, no. 20, 1999, p. 7.

Matsumoto, Y., "Netto de tsunagaru onna no undō" (Women's movements linked up by the Net), *Nihon Joseigaku Kenkyūkai Nyūsu* (Japan Women's Studies Research Group newsletter), no. 217, 2001, pp. 6–10.

Matsuura, S., "Josei tachi ga intānetto de hajimete iru koto" (Women are beginning to use the Internet), *Onna tachi no 21 Seiki*, no. 20, 1999, p. 20.

Men's Center Japan (ed.), *"Otokorashisa" kara "jibunrashisa" e* (From "macho-ness/ masculinity" to "being yourself"), Kamogawa Booklet no. 95, Kyoto, Kamogawa Shuppan, 1996.

Men's Center Japan (ed.), *Otokotachi no "watashi" sagashi* (Men's search for "self"), Kamogawa Booklet no. 104, Kyoto, Kamogawa Shuppan, 1997.

Mihira, S., "Kanzennaru 'danseigaku' de otoko ni nare!" (Becoming a man through perfecting "men's studies"), *Bart*, vol. 7, no. 4, 14 July 1997, pp. 10–20.

Miller, D. and D. Slater, *The Internet: An Ethnographic Approach*, Oxford and New York, Berg, 2001.

Miller, S., "The (temporary?) queering of Japanese TV," *Journal of Homosexuality*, vol. 39, nos 3/4, 2000, pp. 83–109.

Ministry of Public Management, Home Affairs, Posts and Telecommunications, *Information and Communication in Japan, the 2001 White Paper*, Tokyo, Gyōsei, 2001.

Mitra, A., "Virtual Commonality: Looking for India on the Internet," in D. Bell and B. Kennedy (eds), *The Cybercultures Reader*, London, Routledge, 2000, pp. 676–94.

Miyata, K., "*Netto shakai no miraizu: nettowāku komyuniti kara mita seikatsu sekai no henyō*" (Future plans of the network society. Change of everyday life, seen from virtual communities), in *Shin Chōsa Geppō* 19, 1999, pp. 54–7.

Mouer, R. and Y. Sugimoto, *Images of Japanese Society: A Study of the Social Construction of Reality*, London and New York, Kegan Paul International, 1986.

Mohanty, C., "Under Western eyes: feminist scholarship and colonial discourses," *Boundary 2*, vol. 12, no. 3/vol. 13, no. 1, Spring/Fall 1984, pp. 333–58.

Monomi no Tō Seisho Sasshi Kyōkai (ed.), "Intānetto-poruno donna gai ga arimasu ka," (How harmful is Internet porn?), *Mezame yo!* vol. 81, no. 11, 8 June 2000, pp. 3–10.

Moog, S. and J. Sluyter-Beltrao, "The transformation of political communication," in B. Axford and R. Huggins (eds), *New Media and Politics*, London, Thousand Oaks, New Delhi, Sage Publications, 2001, p. 55.

Morgan, C., "My encounter with the Burakumin," *Buraku Liberation News*, 118, 2001.

Morgan, J., "'Some could suckle over their shoulders:' male travelers, female bodies, and the gendering of racial ideology, 1500–1770," *The William and Mary Quarterly*, vol. 54, no. 1, 1997, pp. 167–92.

Morikawa, S., "The significance of Afrocentricity for non-Africans: examination of the relationship between African Americans and the Japanese," *Journal of Black Studies*, vol. 31, no. 4, 2001, pp. 423–36.

Morioka, M., *Ishiki tsūshin: doriimu nabigeita no tanjō* (Conscious communication), Tokyo, Chikuma Shobō, 1993.

Morris-Suzuki, T., *Beyond Computopia. Information, Automation and Democracy in Japan*, London, Kegan Paul, 1988.

Morse, M., *Virtualities: Television, Media Art, and Cyberculture*, Bloomington, IN, Indiana University Press, 1998.

Moya, P. and M. Hames-Garcia (eds), *Reclaiming Identity: Realist Theory and the Predicament of Postmodernism*, Berkeley, University of California Press, 2000.

Murphy, P., "Veteran pirate brings leftism to Net night owls," *Asahi Shinbun*, 15–16 December 2001, p. 29.

Nagamine, Y., "Isolation fears lead to phone addiction," *Daily Yomiuri*, 13 July 2001, p. 6.

Nakarmi, L., "The power of the NGOs," *Asiaweek*, vol. 26, no. 511, February 2000.

Nees, G., "Letter to President," *New York Times*, Section A, 9 October 2001, p. 23.

Nishigaki, T., *Sei naru bācharu riaritii–jōhō shisutemu shakai ron* (Holy virtual reality: discourses on the information system society), Tokyo, Iwanami Shoten, 1995.

Nishiyama, C., "Josei seisaku to jendā" (Women's policies and gender), *Kanagawa Josei Jyānaru*, vol. 13, 1995, pp. 30–44.

N.N., "Zatsuon-gunron. Noizu ni torikumu ātistotachi no manifesto" (The group model of Noise: the Noise artists' manifesto), *Myūjikku Magajin*, no. 10, 1995, pp. 90–7.

Norris, P., *A Virtuous Circle: Political Communications in Postindustrial Societies*, Cambridge, Cambridge University Press, 2000.

Ogura, T., "Nihon ni okeru saibasupēsu no kisei" (The regulation of cyberspace in Japan), in T. Ogura and Y. Kuihara, *Shimin undō no tame no intānetto*, Tokyo, Shakai Hyōronsha, 1996, pp. 120–57.

Ogura, T., *Kanshi shakai to puraibashi* (Privacy and the surveillance society), Tokyo, Impacto shupansha, 2001.

Onojima, D., *Nyū senseishonzu. Nihon no arutanatibu rokku 1978–1998* (New sensations. Alternative rock in Japan 1978–1998), Tokyo, Myūjikku Magajin, 1998.

Ortmanns-Suzuki, A., "Japan und Südkorea: die Schulbuchaffäre" (Japan and South Korea: the textbook affair), *Japanstudien*, 1989, pp. 135–82.

Patton, C., *Inventing AIDS*, New York, Routledge, 1990.

Pharr, S., *Losing Face: Status Politics in Japan*, Berkeley and Los Angeles, CA, California University Press, 1990.

Plato, *The Republic of Plato*, Allan Bloom (ed.), New York, Basic Books, 1968.

Reader, I., *Religion in Contemporary Japan*, London, Macmillan, 1991.

Reader, I., *Religious Violence in Contemporary Japan: The Case of the Aum Shinrikyo*, Richmond, UK, Curzon Press, 2000.

Reid, E., "Hierarchy and power: social control in cyberspace," in M. Smith and P. Kollock (eds), *Communities in Cyberspace*, London, Routledge, 1999, pp. 107–33.

Rheingold, H., *The Virtual Community: Finding Connection in a Computerized World*, London, Secker & Warburg, 1994.

Rheingold, H., *The Virtual Community: Homesteading on the Electronic Frontier*, New York, HarperCollins, 1994.

Rimmer, P.J. and T. Morris-Suzuki, "The Japanese Internet: visionaries and virtual democracy," *Environment and Planning*, no. 31, 1999, pp. 1189–206.

Robins, K., "Cyberspace and the world we live in," in D. Bell and B. Kennedy (eds), *The Cybercultures Reader*, London, Routledge, 2000, pp. 77–95.

Rose, F., "Pocket monster," *Wired*, vol. 9. no. 9, September 2001, pp. 128–35.

Ross, A., "Hacking away at cyberculture," in D. Bell and B. Kennedy (eds), *The Cybercultures Reader*, London, Routledge, 2000, pp. 264–5.

Rössler, P., "Standardisierte Inhaltsanalysen im World Wide Web: Überlegungen zur Anwendung einer Methode am Beispiel einer Studie zu Online-Shopping-Angeboten," in K. Beck. and G. Vowe (eds), *Computernetze: ein Medium öffentlicher Kommunikation?*, Berlin, Wissenschaftsverlag Volker Spiess, 1997.

Rössler, P. and W. Eichhorn, "WebCanal – ein Instrument zur Beschreibung von Angeboten im World Wide Web," in B. Batinic, A. Werner, L. Gräf and W. Bandilla (eds), *Online Research: Methoden, Anwendungen und Ergebnisse,* Göttingen, Hogrefe, 1999.

Russell, J., "Race and reflexivity: the black other in contemporary Japanese mass culture," in J. Treat (ed.), *Contemporary Japan and Popular Culture*, Honolulu, University of Hawaii Press, 1996, pp. 17–40.

Russell, J., "Narratives of denial: racial chauvinism and the black other in Japan," *Japan Quarterly*, vol. 38, no. 4, 1991, pp. 416–28.

Ryang, S. (ed.), *Koreans in Japan: Critical Voices from the Margin*, London and New York, Routledge, 2000.

Sasaki-Uemura, W., "Competing publics: citizen groups, mass media and the state in the 1960s," *Positions: East Asia Cultures Critique* vol. 10, no. 1, 2002, pp. 79–110.

Sasaki-Uemura, W., *Organizing the Spontaneous: Citizen protest in Postwar Japan*, Honolulu, University of Hawaii Press, 2001.

Sassen, S., *Globalization and its Discontents*, New York, The New Press, 1998.

Sawa, K., "Mobile phones silence chatty students," *Mainichi Daily News*, 22 October 2000, p. 12.

Schilling, M., "Into the heartland with Tora-san," in T.J. Craig (ed.), *Japan Pop! Inside the World of Japanese Popular Culture*, Armonk, NY, M. E. Sharpe, 2000, pp. 245–55.

Screech, T., review of John Treat's *Great Mirrors Shattered: Homosexuality, Orientalism and Japan*, *The Journal of Asian Studies*, vol. 59, no. 3, August 2000, pp. 759–62.

Seishimei, "Kōdokusha ga nenkan kyūzenninzō" (Sudden increase in annual subscribers), *Seishimei*, September 2001, p. 1.

Shaw, D., "Gay men and computer communication: a discourse of sex and identity in cyberspace," in S. Jones (ed.), *Virtual Culture: Identity and Communication in Cybersociety*, London, Sage, 1998, pp. 133–45.

Schodt, F., *Manga! Manga! The World of Japanese Comics*, Tokyo, Kodansha International, 1986.

Shibui, T., *Anonimasu: netto o tokumei de tadayou,* (Anonymous: Adrift on the Net with a Pseudonym), Tokyo, Jōhō Sentā Shuppankyoku, 2001.

Shibuya, T., "'Feminisuto danseikenkyū' no shiten to kōsō: nihon no danseigaku oyobi danseikenkyū hihan o chūshin ni" (A view and vision of feminist studies on men and masculinities), *Shakaigaku Hyōron* (Japan Sociological Review), vol. 51, no. 4, 2001, pp. 447–63.

Shilling, C., *The Body and Social Theory*, London, Sage Publications, 1994.

Shimazono, S., *Gendai kyūsai shūkyō ron*, Tokyo, Seikyūsha, 1992.

Shiokura, Y., *Hikikomori* (Confining oneself indoors), Tokyo, Birēji Sentā Shuppan Kyoku, 2000.

Simmel, G., *The Sociology of Georg Simmel* (ed. and trans. K.H. Wolff), New York, The Free Press, 1950.

Sinha, I., *The Cybergypsies*, New York, Scribner, 2000.

Slevin, J., *The Internet and Society*, Oxford, Blackwell, 2000.

Smith, M., "Invisible crowds in cyberspace: mapping the social structure of the usenet," in M. Smith and P. Kollock (eds), *Communities in Cyberspace*, London, Routledge, 1999, pp. 195–219.

Smith, M. and P. Kollock, "Introduction," in M. Smith and P. Kollock (eds), *Communities in Cyberspace*, London, Routledge, 1999, pp. 3–28.

Smith, M. and P. Kollock, *Communities in Cyberspace*, London, Routledge, 1999.

Sonoda, K., *Health and Illness in Changing Japanese Society*, Tokyo, University of Tokyo Press, 1988.

Standish, I., *Myth and Masculinity in Japanese Cinema: Towards a Political Reading of the "Tragic Hero,"* Richmond, UK, Curzon Press, 2000.

Stocker, T., "The future at your fingertip," Online. Available HTTP: *http://www.tkai.com/ press/001004independent.htm* (20 September 2001).

Strom, S., "Japan suddenly feels widening social gap," *International Herald Tribune*, 5–6 January 2001, p.1

Sugimoto, Y., *An Introduction to Japanese Society*, New York, Cambridge University Press, 1997.

Summerhawk, B., C. McMahill and D. McDonald (eds), *Queer Japan: Personal Stories of Japanese Lesbians, Gays, Bisexuals and Transsexuals*, Norwich VT, New Victoria Press, 1998.

Surratt, C., *Netlife: Internet Citizens and their Communities*, Commack, NY, Nova Science Publishers Inc., 1998.

Tachibana, T., *Intānetto wa gurōbaru burein* (The Internet is the global brain), Tokyo, Kōdansha, 1997.

Taga, F., "Dansei/Danseigakukenkyū no sho chōryū" (Cross-currents within men and masculinities/men's studies), Paper presented at 5th Japan Gender Association Conference, Kyoto, Japan, 22 September 2001.

Takahara, K., "Art of communication or just cute? Nail salons ringing up cell phone profits," *Japan Times*, 18 September 1999, p. 3.

Takao, Y., "Welfare state retrenchment – the case of Japan," *International Public Policy*, vol. 19, no. 3, pp. 265–92.

Takatori, M., *Nihonteki-shiho no genkei* (Original types of Japanese thinking), Tokyo, Kōdansha, 1975.

Takeshita, K., "Shakai kagaku no tenkan to kindai seiō bunmei. Dejitaru shakai to gendai Ajia" (The transformation of social science and Western civilization. Digital society and contemporary Asia), *Keizai Ronshū*, vol. 50, no. 3, December 2000, pp. 35–62.

Tanaka, Y., *Japan's Comfort Women: Sexual Slavery and Prostitution during World War II and the US Occupation*, London, Routledge, 2002.

Taylor, M., M. Kent and W. White, "How activist organizations are using the Internet to build relationships," *Public Relations Review*, vol. 27, no. 3, 2001, p. 266.

Tepper, M., "Usenet communities and the cultural politics of information," in D. Porter (ed.), *Internet Culture*, New York, Routledge, 1997, pp. 39–54.

Terranova, T., "Post-human unbounded. Artificial evolution and high-tech subcultures," in D. Bell and B. Kennedy (eds.) *The Cybercultures Reader*, London, Routledge, 2000, pp. 268–79.

Time International, "Japan's shame: lawmakers as finally pushing legislation to help end the country's dubious distinction as the world's main source of child pornography," *Time International*, 19 April 1999, p. 34.

Time International, "Dating Game: Looking for Prince Charming? In Japan, Check your Cell Phone," *Time International*, vol. 157, issue 22, 4 June 2001, p. 88.

Trend, D., (ed.), *Reading Digital Culture*, Malden, MA, Blackwell Publishers, 2001.

Tsang, D., "Notes on queer 'n' Asian virtual sex," *Amerasia Journal*, vol. 20, no. 1, 1994, pp. 117–128.

Tsukurukai, *The Restoration of a National History*, Tokyo (pamphlet), 1998.

Turner, B.S., *Religion and Social Theory,* London, Heinemann Educational Books, 1983.

Umeda, T., M. Kihara, S. Hashimoto, S. Ishikawa, M. Kamakura and T. Shimamoto, "Nihon no iseikan seiteki sesshoku ni yoru eizu no tokucho-eizu sābeiransu ni yoru eikoku oyobi beikoku to no hikaku kenkyū" (Characteristics of AIDS contracted through heterosexual sexual contact in Japan: a comparative study with America and the United Kingdom by AIDS surveillance), *Nihon Koeishi* (Japanese Journal of Public Health), vol. 48, no. 3, 2001, pp. 200–7.

Umezawa, T., *Sarariiman no jikakuzō* (Salaryman self-images), Tokyo, Minerva Shobō, 1997.

Uno, S., "What are women's centers in Japan?," *Dawn: Newsletter of the Dawn Center*, 1997, pp. 6–8.

Valentine, J., "On the borderlines: the significance of marginality in Japanese Society," in E. Ben-Ari, B. Moeran and J. Valentine (eds), *Unwrapping Japan: Society and Culture in Anthropological Perspective*, Honolulu, University of Hawaii Press, 1990.

Vandenberg, A. (ed.), *Citizenship and Democracy in a Global Era*, London, Macmillan, 2000.

Van Wolferen, K., *The Enigma of Japanese Power: People and Politics in a Stateless Nation*, Tokyo, Tuttle, 1993.

Vervoorn, A., *ReOrient*, Melbourne, Oxford University Press, 1998.

Vidal, J., "Another coalition stands up to be counted," *The Guardian*, 19 November 2001, p. 13.

Wakefield, N., "New media, new methodologies: studying the Web," in D. Gauntlett (ed.), *Web.Studies: Rewiring Media Studies for the Digital Age*, London, Arnold, 2000, pp. 31–41.

Watanabe, K., "Kyanpasu sekusharu harasumento zenkoku nettowāku" (National campus sexual harrassment network), *Women's Asia 21*, no. 20, 1999, p. 29.

Watney, S., *Policing Desire: Pornography, AIDS, and the Media*, Minneapolis, University of Minnesota Press, 1989 [1987].

Watney, S., "The spectacle of AIDS," in D. Crimp (ed.), *AIDS: Cultural Analysis/Cultural Activism*, Cambridge, MIT Press, 1988, pp. 71–86.

Wilson, M., "Community in the abstract: a political and ethical dilemma?" in D. Bell and B. Kennedy (eds), *The Cybercultures Reader*, London, Routledge, 2000, pp. 644–57.

Wilson, R. and W. Dissanayake (eds), *Global/Local: Cultural Production and the Transnational Imaginary*, Durham: Duke University Press, 1996.

Woodland, R., "Queer spaces, modem boys and pagan statues: gay/lesbian identity and the construction of cyberspace," in D. Bell and B. Kennedy (eds), *The Cybercultures Reader*, London, Routledge, 2000, pp. 416–31.

Yajima, M., (ed.), *Danseidōseiaisha no raifuhisutorii* (Male homosexuals' life histories), Tokyo, Gakubunsha, 1997.

Yamamoto Seiichi, interviewed by Matsuyama Shinya, in D. Onojima, *Nyū senseishonzu. Nihon no arutanatibu rokku 1978–1998* (New sensations. Alternative rock in Japan 1978–1989), Tokyo, Myūjikku Magajin, 1998, p. 93.

Yamazaki, K., "Onnatachi ga tsukuru ōtanatibu media," (Alternative media made by women), *Onna tachi no 21 Seiki*, no. 10, 1997, p. 7.

Yoshida, J., *Intānetto kūkan no shakaigaku. Jōhō nettowāku to kōkyōken* (Sociology of Internet space. Information networks and the public sphere), Kyoto, Sekai Shisō Sha, 2000.

Yuzawa, N., (2001) "Masumedeia ni miru danseizō: 'otokorashisa' wa media ga tsukuru" (Masculine representations in the media: "macho-ness" is media constructed), in H. Asai, S. Itō and Y. Murase (eds), *Nihon no otoko wa doko kara kite, doko e iku no ka?* (Where have Japanese men come from, where will they go?) Tokyo, Jūgatsusha, 2001, pp. 150–69.

Index